BIOTECHNOLOGY AND GENETICS IN FISHERIES AND AQUACULTURE

BIOTECHNOLOGY & GENETICS IN FISHERIES AND AQUACULTURE

Bradley Brooks
Susie Carter
Debi Coffa

Kruger Brentt
Publishers
2023

Kruger Brentt Publishers UK. LTD.
Company Number 9728962

Regd. Office: 68 St Margarets Road, Edgware, Middlesex HA8 9UU

ISBN: 9781787150621

Disclaimer:

Every possible effort has been made to ensure that the information contained in this book is accurate at the time of going to press, and the publisher and author cannot accept responsibility for any errors or omissions, however caused. No responsibility for loss or damage occasioned to any person acting, or refraining from action, as a result of the material in this publication can be accepted by the editor, the publisher or the author. The Publisher is not associated with any product or vendor mentioned in the book. The contents of this work are intended to further general scientific research, understanding and discussion only. Readers should consult with a specialist where appropriate.

Every effort has been made to trace the owners of copyright material used in this book, if any. The author and the publisher will be grateful for any omission brought to their notice for acknowledgement in the future editions of the book.

For information on all our publications visit our website at http://krugerbrentt.com/

PREFACE

The role of aquaculture in increasing fish production is well recognized today. The contribution of aquaculture to global food production and security has been widely acknowledged since the start of the 21st century. Global aquaculture production has increased continuously over the last five decades, and particularly in India, aquaculture has become one of the fastest growing and most efficient agri-sector. Biotechnology can currently be considered of importance in aquaculture. The increase in the production of aquatic organisms over the last two decades through the use of biotechnology indicates that in a few generations' biotechnology may overtake conventional techniques, at least for the commercially more valuable species. At present, the most commonly used methods in fish biotechnology are the use of synthetic hormones in fish breeding, production of mono sex, uniparental and polyploid individuals which can be used to produce triploid, tetraploid, haploid, gynogenetic and androgenetic fish, molecular biology and transgenesis, biotechnology in aquaculture nutrition and health management, gene banking are chromosome manipulation and hormonal treatments.

Since the early 1980s, research in aquaculture and fisheries genetic biotechnology has steadily grown, and now research in this area is extremely active. The main vision of aquaculture biotechnology is to achieve improvements of aquaculture stock, preservation of genetic resources, disease diagnosis, and control of microbial/microalgal genetic engineering. In broad terms, biotechnology can be defined as any technological application that uses biological systems, living organisms or derivatives thereof, to make or modify products or processes for specific use. Biotechnology has the potential to enhance reproduction and the early developmental success of culture organism.

The present book contains eleven chapters covering all related disciplines. These chapters include *History of genetic biotechnology in aquaculture and fisheries, Genetic variation- an introduction, How can genetic variation be measured?, Polyploidy, Gynogenesis, androgenesis and nuclear transplantation, Sex reversal and breeding, Population genetics and interactions of hatchery and wild fish, Gene mapping, Quantitative trait locus mapping and marker-assisted selection, Gene expression, Isolation and cloning, Gene-transfer technology, Combining genetic enhancement programmes.* Related terminology is given at the end for ready reference. This book deals with some of the emerging topics on advanced concepts, which would an invaluable asset for the scientists, students and teachers working in the field of fishery and aquaculture. This book is also a valuable reference source for anyone interested in knowing more about fish genetics and biotechnology.

We are grateful to all those persons as well as various books, manuals, periodicals, magazines, journals etc. that helped in the preparation of this book. In spite of great care taken by me and publisher team, some errors may have escaped from our notice. We shall appreciate if such errors are brought to our notice. There will be always being some scope for further improvement. Further queries, constructive suggestions and criticisms for the improvement of the book are always welcome and shall be thankfully acknowledged.

Bradley Brooks
Susie Carter
Debi Coffa

CONTENTS

Preface v

1. **History of Genetic Biotechnology in Aquaculture and Fisheries** **1**
   - Introduction 1
   - Biotechnology and its various stages of development 3
   - Modern biotechnology 10
   - History of genome research 16
   - Traditional genetic biotechnologies for aquaculture 20
2. **Genetic Variation: An Introduction** **25**
   - Genetic Variation Causes 25
   - Deoxyribose nucleic acid 27
   - Ribose nucleic acid 31
   - Genetic code 37
   - Chromosomes 47
   - How does sexual reproduction produce variation? 49
   - Inheritance of mitochondrial and chloroplast DNA 55
   - Mitochondrial and chloroplast DNA 55
3. **How can Genetic Variation be Measured?** **59**
   - Introduction 59
   - The Basic Genetic Concepts and Theories 60
   - Genetic diversity in fishes 66
   - DNA fragment size variation Molecular markers 68
   - New Developing Markers in Fisheries and Aquaculture 74
   - Genetic Variation And Population Structure Study In Natural Populations 76

- Genetic Research Methods 79
- DNA marker methods 80
- Protein Conservation and Variation 84
- Allozymes 88
- Immunological identification of proteins 89
- Phenotypic Variance 90
- Phenotypic Variance: A Summary 94

4. Polyploidy 95

- Introduction 95
- Mechanisms of Polyploidy 95
- Polyploidy Induction in fish 96
- Polyploid Induction in Shellfish 98
- Induction of triploidy 100
- Triploid Fish Performance 101
- The Evolutionary Consequences of Polyploidy 104
- Production and performance of triploid oysters 112

5. Gynogenesis, Androgenesis and Nuclear Transplantation 119

- Microsatellite DNA analysis of gynogenetic families 120
- Androgenesis 124
- Performance of Gynogens and Androgens 126
- Gynogenesis in shrimps 130
- Gynogenesis in fishes 131
- Genetic material produced by gynogenesis 132
- Nuclear Transplantation 133

6. Sex Reversal and Breeding 141

- Introduction 141
- The Use of Sex Reversal to Understand and Control Fish Sex Determination or Differentiation 143
- Fish Sensitivity to Sex Reversal Treatments 145
- Hormonal treatment 147
- Sex-reversal in adult fish 148
- The importance of sex reversal 150
- Coexistence of Genetic and Environmental Influences 151
- Mechanisms of Sex Reversal 155
- New Actors of Sex Determination and Differentiation 156

- Revisiting Old Players of Sex Differentiation 158
- Brain Implication in the Sexual Phenotype 160
- Epigenetics: A Link between the Environment, Gene Expression, and Sex Reversal 162
- Consequences of Sex Reversal in Wild Populations 167
- Hormonal induction of sex reversal 170
- Genetics of Sex Determination 171

7. Population Genetics and Interactions of Hatchery and Wild Fish 175

- Introduction 175
- Genetic variation, population structure and biodiversity 175
- Effects of Geography and Environment on Population Variation 177
- Overview of the Impact of Climate Change on Fisheries 181
- Inter-specific Hybridization 186

8. Gene Mapping, Quantitative Trait Locus Mapping and Marker-assisted Selection 195

- Introduction 195
- Technique and approach used 197
- Scope and scale of application 197
- Accessibility 198
- Outcome and benefits 198
- Map construction 198
- Quantitative Trait Locus Mapping 204
- Marker-assisted Selection in Fish 208

9. Gene Expression, Isolation and Cloning 213

- Introduction 213
- Study of gene expression by using various techniques 215
- Gene expression in aquaculture management 216
- Expressed Sequence Tags 220
- Development of EST resources of aquaculture species 220
- Osmoregulation 221
- Epigenetic and its Mechanism 223
- Transposable Elements 226
- Ribosomes 230
- Protein expression proteomics. 232
- Separation and Isolation of Proteins 233

10. Gene-transfer Technology **239**

- Introduction 239
- History 240
- Different Gene Transfection Techniques 241
- Methods in the production of transgenic fish 242
- Gene Transfer Technology in Fish 245
- Gene transfer 247
- Gene Transfer Technology for Development of Transgenic Fishes 248
- Performance of Transgenic Fish 252
- Development of Transgenic Fishes 253
- Gene Flow 257
- Transgenic Fish Invasive Species 258
- Risk Mitigation 258
- Food Safety Issues 258

11. Combining Genetic Enhancement Programmes **259**

- Genetic diversity and the need for genetic resource management 259
- Sex Reversal and Triploidy 263
- Genetic Engineering and Crossbreeding 267
- Hybridisation vs crossbreeding (a fuzzy distinction) 272
- Gynogenesis, Selection and Hybridization 275
- The genetic improvement programs in aquaculture 283

Glossary **291**

Bibliography **299**

Index **313**

1

HISTORY OF GENETIC BIOTECHNOLOGY IN AQUACULTURE AND FISHERIES

INTRODUCTION

Biotechnology provides powerful tools for the sustainable development of aquaculture, fisheries, as well as the food industry. Increased public demand for seafood and decreasing natural marine habitats have encouraged scientists to study ways that biotechnology can increase the production of marine food products, and making aquaculture as a growing field of animal research. Biotechnology allows scientists to identify and combine traits in fish and shellfish to increase productivity and improve quality. Scientists are investigating genes that will increase production of natural fish growth factors as well as the natural defense compounds marine organisms use to fight microbial infections. Modern biotechnology is already making important contributions and poses significant challenges to aquaculture and fisheries development. It perceives that modern biotechnologies should be used as adjuncts to and not as substitutes for conventional technologies in solving problems, and that their application should be need-driven rather than technology-driven.

The use of modern biotechnology to enhance production of aquatic species holds great potential not only to meet demand but also to improve aquaculture. Genetic modification and biotechnology also holds tremendous potential to improve the quality and quantity of fish reared in aquaculture. There is a growing demand for aquaculture; biotechnology can help to meet this demand. As with all biotech-enhanced foods, aquaculture will be strictly regulated before approved for market. Biotech aquaculture also offers environmental benefits. When appropriately integrated with other technologies for the production of food, agricultural products and services, biotechnology can be of significant assistance in meeting the needs of an expanding and increasingly urbanized population in the next millennium. Successful development and application of biotechnology are possible only when a broad research and knowledge base in the biology, variation, breeding, agronomy, physiology, pathology, biochemistry and genetics of the manipulated organism exists.

Benefits offered by the new technologies cannot be fulfilled without a continued commitment to basic research. Biotechnological programmes must be fully integrated into a research background and cannot be taken out of context if they are to succeed.

Indian fisheries and aquaculture is an important sector of food production, providing nutritional security to the food basket, contributing to the agricultural exports and engaging about fourteen million people in different activities. With diverse resources ranging from deep seas to lakes in the mountains and more than 10% of the global biodiversity in terms of fish and shellfish species, the country has shown continuous and sustained increments in fish production since independence. Constituting about 4.4% of the global fish production, the sector contributes to 1.1% of the GDP and 4.7% of the agricultural GDP. The total fish production of 6.57 million metric tonnes presently has nearly 55% contribution from the inland sector and nearly the same from culture fisheries. Fish and fish products have presently emerged as the largest group in agricultural exports of India. (Marine products export review-MPEDA.,April 2006-March2007).The potential area of biotechnology in aquaculture include the use of synthetic hormones in induced breeding, transgenic fish, gene banking, uniparental and polyploidy population and health management.

What do stone-washed jeans and home pregnancy tests have in common with bread, cheese and wine? They are all examples of manipulating living organisms to create products or new processes. They are all products of biotechnology.

Humans have used biotechnology since the dawn of civilization. Egyptians used yeasts to bake leavened bread, the Chinese developed fermentation techniques for brewing and cheese making, and the Aztecs used Spirulina algae to make cakes. Today, when most people think of biotechnology, they probably think of recombinant DNA. Although much of modern biotechnology does deal with manipulating DNA, classical biotechnology began long before we even knew about genes or chromosomes. What began as recipes for production of food now includes technology to enhance everything from farming to pharmaceuticals.

What we think of as modern biotechnology began around the end of the nineteenth century. By then, Mendel's work on genetics was completed and institutes for investigating fermentation along with other microbial processes had been founded by Koch, Pasteur, and Lister.

At the beginning of the twentieth century, industry and agriculture started to incorporate biotechnology. The growing automobile industry, for example, used fermentation processes to produce acetone and paint solvents. By mid-century, major advances in genetics dominated biotechnology research. The discovery that DNA carries the genetic code and the structure of the "double helix" were described by Watson and Crick. Soon, new techniques were developed to allow manipulation of DNA. Genetically engineered plants, microbes, animals, and products like insulin, highlight the future of biotechnology.

The importance of the new genetics culminated in The Human Genome Project, an international project instituted by the Department of Energy and the National Institutes of Health to "map the human genome." The goal of this project is to identify the structure of the entire human genome, including its three billion base pairs and approximately 22,000 genes. The hope is that this knowledge will help scientists identify, prevent and treat many of the illnesses resulting from genetic malfunction.

Discussion of the ethical implications of recombinant experiments began several decades ago. James Watson was the elected as the first director of the Office of Human Genome Research at the NIH. With great insight, he suggested a portion of the budget of the human genome project be set aside to study the ethical, legal and social implications of the project. As genetic discoveries have progressed, the importance of regulating how the knowledge will be used has become of primary importance. Philosophers and ethicists now work side by side with research scientists and lawyers to help determine not only what we can do, but also what we should do with genetic knowledge.

Today, biotechnology is being used in countless areas including agriculture, clincial applications, bioremediation and forensics, where DNA sequencing is a common practice. Industry and medicine alike use the techniques of PCR, immunoassays and recombinant DNA. Genetic manipulation has been the primary reason that biology is now seen as the science of the future and biotechnology as one of the leading industries.

BIOTECHNOLOGY AND ITS VARIOUS STAGES OF DEVELOPMENT

It is well known that the technical application of biological material is considered to be biotechnology. To understand how biotechnology works it is important to think about the starting point or material for biotechnology processes. Generally, biotechnology utilizes living material or biological products to generate new products for use in various medical, agricultural, pharmaceutical and environmental applications. The ultimate goal of biotechnology is to benefit humanity by, for example, the production of resistant crops, vegetables, recombinant proteins, higher milk-producing animals, etc.

Different developmental stages have taken place in biotechnology to meet the various needs of humans at the time. Its development was principally based on observations, and the application of these observations to practical scenarios. Owing to the evolution of new technologies and a better understanding of various principles of life science, the complexity of biotechnology has increased. Table 1 gives a breakdown of historical events in biotechnology. The development of biotechnology can be divided into broad stages or categories, including:

- Ancient biotechnology (8000-4000 BC): Early history as related to food and shelter; includes domestication of animals.
- Classical biotechnology (2000 BC; 1800-1900 AD): Built on ancient biotechnology; fermentation promotes food production and medicine.

- 1900-1953: Genetics.
- 1953-1976: DNA research, science explodes.
- Modern biotechnology (1977): Manipulates genetic information in organisms; genetic engineering; various technologies enable us to improve crop yield and food quality in agriculture and to produce a broader array of products in industries.

Table 1. Historical events in biotechnology

Pre-1800 (Early applications and speculation)

6000 BC: Yeast was utilized to prepare beer (Sumerians and Babylonians).

4000 BC: In Egypt, a process was discovered to prepare leavened bread by means of yeast.

420 BC: Greek philosopher Socrates (470-399 BC) hypothesized on the similar characteristics between parents and their offspring.

320 BC: Greek philosopher Aristotle (384-322 BC) theorized that all inheritance originates from the father.

1000 AD: Hindus recognized that some illnesses may 'run in the family'. At the same time, the theory of abiogenesis, or spontaneous generation based on the idea that organisms arise from non-living matter, developed. According to this theory maggots could develop from horse hair.

1630: William Harvey explained that plants and animals are similar in their reproduction, i.e. they reproduce sexually.

1660-1675: Marcello Malpighi (1628-1694) investigated blood circulation in capillaries using a microscope and found that the brain is connected to the spinal cord by bundles of fibers which form the nervous system.

1673: Antonie van Leeuwenhoek (1632-1723) was the first researcher to explain micro-organisms such as protozoa and bacteria, and also identify that these micro-organisms play an active role in fermentation.

1701: Giacomo Pylarini found that the deliberate administration of smallpox could prevent its occurrence later in life, especially in children. Later, this procedure was termed 'vaccination' and a process that uses cowpox instead of smallpox was established as the most reliable treatment.

1800-1900(Significant advances in basic understanding)

1809: Nicolas Appert invent a technique using heat to can and sterilize food.

1827: In the field of heredity, there had long been a hunt for the so-called mammalian egg. It had proved elusive, however, in 1827 the first report of canine eggs offered a basic clue to major breakthroughs in reproduction, at first in lower animals

1850: Ignaz Semmelweis utilized epidemiological examinations to suggest the theory that puerperal fever could be transmitted from mother to mother by physicians. He also suggested that all physicians should wash their hands after investigating each patient. For this suggestion he was criticized by medical professionals and ultimately lost his employment.

1856: Carl Ludwig discovered a procedure for keeping animal organs alive under *in vitro* conditions. This was done by supplying blood to them. In contrast to the concepts of Justus von Liebig, Pasteur (1822-1895) suggested that microbes are responsible for fermentation.

1859: Charles Darwin (1809-1882) speculated that animal populations adapt their forms to eventually best utilize the surroundings, a process he described as 'natural selection'. During his stay in the Galapagos Islands, he saw how the finches' beaks on each island were adapted for the environment, especially regarding food sources.

1863: Pasteur discovered the method of pasteurization. In this method he heated wine enough to inactivate microbes (that would otherwise convert the 'vin' to 'vin aigre' or 'sour wine') and realized that this procedure did not affect the flavor of the wine.

Heinrich Anton de Bary established that a fungus was responsible for potato blight. A major challenge for researchers during this period was to differentiate whether a microbe was responsible for this or whether it was the outcome of a disease.

1865: Mendel (1822-1884) suggested the laws of heredity to the National Scientific Society (Brunn, Austria). Mendel anticipated that imperceptible core units of information were responsible for noticeable characteristics. He called these 'factors', which were later called genes (units that were inherited by one generation from its parents). The research done by Mendel was overlooked and not acknowledged due to Darwin's more sensational publication five years earlier, until 1900 when Hugo de Vries, Erich von Tschermak and Carl Correns supported Mendel's mechanism of heredity.

1868: Casimir Joseph Davaine cured plants suffering from bacterial infection by a novel heat treatment. While working in a hospital, Johannes Friedrich Miescher separated nuclein (a compound made of nucleic acid) from pus cells. These pus cells were derived from waste bandages.

1870: Walther Flemming discovered mitosis.

1871: During the period 1873-76 interest in DNA research began. DNA was initially derived from the sperm of trout (found in the river Rhine). During this period Koch investigated anthrax and explored certain techniques to identify, culture and stain micro-organisms. He also took images of them which were later supported by Gram, Cohn and Weigart.

1880: While working on fowl cholera, Louis Pasteur explored weakened (attenuated) strains of micro-organisms that might not be virulent but could nevertheless potentially prevent healthy individuals against severe forms of a similar disease.

1881: Koch explained techniques for harvesting bacterial colonies on potato slices, gelatin and agar medium. For the isolation of pure culture and for distinguishing the nutrients needed for genetic mutations, the agar technique was one of the most common methods. Thomas D Brock considered this breakthrough as the single most important discovery in the development of microbiology.

During the same period Pasteur explored the application of the attenuation process in the production of vaccines against certain bacterial pathogens, e.g. fowl cholera and anthrax; this was an early stage in immunology which led to the exploration of areas such as preventive medicine.

1884: Koch established his 'claims' for assessing whether a microbe or another agent is responsible for disease.

During the same period Pasteur established a rabies vaccine.

Gram described the differential staining technique for cellular peptidoglycan-containing bacteria now known as Gram staining.

Mendel passed away after 41 years of predominantly investigating the heredity 'factors' of pea plants. He did not receive any technical support during his lifetime, but said before his death, 'My time will come'

1900-1953 (Genetics:converging on DNA)

1900: Mendel's work finally took on importance

Mendel's work had given birth to genetic science. It was revived again by three researchers, de Vries, von Tschermak and Correns, who were working on the application of original work done by Mendel.

1902: Human genetics is born

Sutton found that chromosomes (paired) contain certain elements which are transferred from one generation to another. During this transfer, traits are transported through carriers called chromosomes. He also advised that Mendel's 'factors' are sited on chromosomes.

1905: X and Y chromosomes related to gender

Edmund Beecher Wilson and Nettie Stevens shared the same idea of separating X and Y chromosomes for the determination of sex. They also demonstrated that a single Y chromosome determines maleness, while two copies of the X chromosome decide femaleness.

1905-1908: William Bateson and R C Punnett, along with other researchers, found that several genes alter or modify the action of other genes.

1906: Paul Erlich also investigated atoxyl compounds and discovered the important features of Salvarsan (the first chemotherapeutic agent).

1907: Thomas Hunt Morgan started his investigation into fruit flies that would reveal that chromosomes have a defined role in heredity; additionally, he discovered mutation theory. This resulted in an understanding of the basic concepts and mechanisms of heredity.

1909: Mendel's laws to animals: Wilhelm Johannsen used the word 'gene' to mean the carrier/transporter of heredity. He also coined the terms 'genotype' and 'phenotype'; the genotype is the genetic composition/establishment of an organism, whereas the phenotype describes the actual organism or its morphological characteristics, resulting from a blend of the genotype and a range of external/environmental factors.

1910: Basis of modern genetics: Morgan also demonstrated that carriers of genetic information, called asor genes, are present on chromosomes, creating the basis for modern genetics. This work later assisted him in utilizing Drosophila fruit flies to examine heredity.

1911: During the same period Morgan established the separation of certain inherited features that are generally linked to the separation/breaking of chromosomes during the process of cell division. He also investigated the mapping of the genetic sites present on the chromosomes of the fruit fly.

1912: Crystallography era: William Lawrence Bragg discovered the application of x-rays in the determination of the molecular structure of crystalline substances.

1918: Herbert M Evans stated (mistakenly) that human genetic material is made up of 48 chromosomes.

1924: Eugenics in the United States

Several US diplomats, encouraged by the eugenics movement, accepted the US Immigration Act (1924), limiting the admission of illiterate refugees from Southern and Eastern Europe on the basis of their alleged genetic inferiority.

1926: Morgan published The Theory of the Gene.' This was based on Mendelian genetics (breeding investigations and optical microscopy).

Hermann Joseph Muller discovered that x-rays are responsible for genetic mutations in fruit flies taking place 1500 times faster than under normal conditions. This innovation offered researchers and scientists a procedure to induce mutations. Later, various mutagens were explored to understand the complexity behind different genotypes.

1928: Frederick Griffiths observed the 'transforming principle' in which a rough type of bacterium is transformed to a smooth type when a mysterious 'transforming element' from the smooth type is present. After 16 years, Oswald Theodore Avery discovered that 'transforming element' to be DNA.

Alexander Fleming studied an old culture of bacteria infected with fungal growth and found that it did not show any bacterial growth in a radius surrounding a piece of mold (fungi) in a petri dish. This breakthrough gave birth to the antibiotics era or penicillin age, and penicillin was accessible to patients 15 years later for therapeutic use.

1938: Proteins and DNA were studied by means of x-rays. This was the dawn of a new age of crystallography in which large molecular weight complex proteins can be studied by x-rays.

The term 'molecular biology' was coined.

1941: One gene, one enzyme: George Wells Beadle and Edward L Tatum examined Neurospora crassa, a mold that usually invades and grows on bread, and proposed 'one gene, one enzyme' theory: each gene encodes for or is translated into an enzyme to accomplish tasks within an organism.

1943: The Rockefeller Foundation (New York) collaborated with the Mexican government to start the Mexican Agricultural Program. This was the first step toward plant breeding at a global level.

1943-1953: Cortisone (17?,21-dihydroxypregn-4-ene-3,11,20-trione), a pregnane (21-carbon) steroid hormone, was first produced in great amounts. Cortisone is considered as the first biotech product.

1944: Selman Abraham Waksman (a Ukrainian-American researcher) explored streptomycin, an active antibiotic against TB.

1945: The United Nations Food and Agriculture Organization was established in Quebec, Canada, with the objective of encouraging agricultural practices.

1945-1950: For the first time, animal cell cultures were harvested in laboratories, giving birth to the field of animal tissue culture.

1947: Barbara McClintock first demonstrated 'transposable elements' known as 'jumping genes' with the capability to move (or jump) from one site on the genome to another site. Scientific society did not welcome the implications of her discovery at the time.

1950: Erwin Chargaff discovered that the same levels of adenine and thymine are present in DNA, as are the same levels of guanine and cytosine. These associations were later named 'Chargaff's rules'. Later, Chargaff's rules functioned as an important principle for James Watson and Francis Crick in measuring different models for the structure of DNA.

DNA research, science explodes (1953-1976)

1953-1976: Expanding the boundaries of DNA research: The discovery of the structure of DNA finally resulted in an explosion of research into molecular biology and genetics, providing the resources for biotechnology development.

1953: The journal Nature published Watson and Crick's article based on unfolding the double-helix structure of DNA.

1953: Based on his technical exposure George Otto Gey developed the HeLa human cell line. Cells taken from cancer patient Henrietta Lacks (who died in 1951) became the first immortal human cells and were cultured to develop a polio vaccine.

1957: Central dogma of DNA-how DNA makes a protein: Crick and Gamov studied 'central dogma', demonstrating how DNA functions to construct protein.

1959: François Jacob and Jacques Lucien Monod documented the veracity of gene-based regulation. They explained gene mapping with mappable control functions sited on the chromosome in the DNA sequence which they later named the 'repressor' and 'operon'.

1962: Watson and Crick were awarded the Nobel Prize in Physiology or Medicine with Maurice Wilkins. Disappointingly, Rosalind Franklin, who actually contributed to the discovery of the double-helical structure of DNA, died before this date, and Nobel Prize conventions do not permit a prize to be awarded posthumously.

1966: Genetic code cracked: The genetic code was explored by several researchers. Marshall Warren Nirenberg, J Heinrich Matthaei and S Ochoa reported that a genetic sequence of three nucleotide bases (called codons) decides each of 20 amino acids.

1967: Arthur Kornberg reported a study using single-stranded natural viral DNA to assemble 5300 nucleotide building blocks, and at the same time his Stanford group synthesized viral DNA.

1970: Oncogenes: Virologists Peter H Duesberg and Peter K Vogt identified the first oncogene in a virus. This gene can be utilized to study various human cancers

1972: First recombinant DNA molecule: Paul Berg, a biochemist, utilized a restriction enzyme to cut DNA into fragments. He employed a ligase enzyme to join two DNA strands concurrently to form a hybrid circular molecule. This was the first recombinant DNA (rDNA) molecule synthesized.

1972: NIH guidelines for rDNA: Berg and other researchers at the National Institutes of Health (NIH) worked hard to establish guidelines to sanction the strategy for DNA splicing. Their concerns resulted in the Asilomar Conference (1975).

1973: Ames test: Bruce Nathan Ames, a biochemist at UC Berkeley, developed an investigation to distinguish chemicals that damage DNA. Later, the Ames test became extensively used to identify cancer-causing substances.

1975: rDNA moritorium: A global meeting was held in Asilomar, California, with the objective of approving guidelines regulating rDNA experimentation. All the scientists involved discussed the development of 'safe' bacteria and plasmids.

1976: More about oncogenes

J Michael Bishop and Harold Varmus at the University of California, San Francisco (UCSF) established that cancer-causing genes called oncogenes become visible on animal chromosomes, and modifications in their structure or expression can result in metastatic growth.

1976: Release of NIH guidelines

The NIH released the first set of guidelines for rDNA experimentation. Later, these guidelines restricted several types of trials.

1977-present (modern biotechnology) The dawn of biotech

With the advent of genetic engineering it was possible to produce human protein in bacteria for the first time. Biotech-based organizations started focusing more on the applications of genetic engineering. In 1978, Herbert W Boyer at UCSF synthesized synthetic human insulin by introducing the insulin gene into the bacterium Escherichia coli. This breakthrough opened the gateway for further developments in DNA sequencing and cloning techniques.

1977: Genentech Inc. was the first organization to achieve the synthesis of a human protein (somatostatin) in a bacterium. Somatostatin is a human growth hormone (hGH)-releasing inhibitory factor. A synthetic, recombinant gene was for the first time employed to clone a protein. Several researchers believed that this was the beginning of the age of modern biotechnology.

1978: Recombinant insulin: Genentech Inc. announced that its laboratory had achieved the synthesis of human insulin using rDNA technology.

1980: Patents allowed: The US Supreme Court granted that genetically modified living organisms could be patented. According to a Supreme Court decision (1980) the Exxon oil company was allowed to patent an oil-eating micro-organism.

Kary Mullis and other researchers at UC Berkeley, California, established a tool for multiplying DNA sequences *in vitro* using the polymerase chain reaction (PCR).

1982: Site-directed mutagenesis: Genentech Inc. signed an agreement from the US Food and Drug Administration (FDA) to further market genetically engineered human insulin. In 1982 the

FDA allowed the first genetically engineered drug in the form of human insulin produced by bacteria.

Michael Smith at the University of British Columbia, Vancouver, established a procedure for producing precise amino acid changes anywhere in a protein.

1983: Site-directed mutagenesis: Eli Lilly obtained a license to make and sell insulin.

1985: During this period genetic fingerprinting stepped into the court room.

Cal Bio produced a gene by a cloning method that encodes human lung surfactant protein, an important step toward reducing premature birth complications.

For the first time, genetically modified plants that resistant to insects, viruses and bacteria were examined.

The NIH published guidelines for performing experiments in gene therapy on humans.

1986: Chiron Corp. obtained FDA approval for the production of the first recombinant vaccine for hepatitis.

A genetically modified crop (the tobacco plant) was allowed by the Environmental Protection Agency (EPA).

1987: Calgene Inc. obtained a patent for the tomato polygalacturonase DNA sequence, which was later used to synthesize an antisense RNA sequence that can further extend the shelf life of fruit.

1988: Harvard molecular geneticists Philip Leder and Timothy A Stewart were granted the first patent based on a genetically modified animal (a mouse that is highly susceptible to breast cancer).

1990: UCSF and Stanford University achieved their 100th rDNA patent license. At the end of the 1991 financial year, both organizations had received $40 million from the patent.

1990: Patents and money: The first gene-based treatment was performed on a four-year-old girl suffering from an immunological disorder known as adenosine deaminase deficiency (ADA) deficiency. Gene therapy emerged, however ethical concerns surrounding gene therapy were highly debated.

Commencement of the Human Genome Project, with the global objective to plot all of the genes in the human body. The expected cost was $13 billion.

Michael Crichton's novel Jurassic Park was released, in which bioengineered dinosaurs wander in a paleontological theme park; the project goes wrong, with deadly outcomes.

1992: The US Army started taking blood and tissue samples from all new employees as part of a 'genetic dog-tag'. This course of action was intended for better identification of soldiers killed in battle.

1993: Researcher Kary Mullis won the Nobel Prize in Chemistry for inventing the tool of PCR.

1996: A groundbreakingly efficient diagnostic biosensor test allowed for the first time the instant detection of the toxic strain of E. coli (strain 0157:H7), the bacteria responsible for several food poisoning outbreaks. The possibility of its use against anthrax and other bioterrorism agents was also assessed.

The discovery of a gene linked to Parkinson's disease offered researchers a significant new chance for the determination of the cause of, and potential treatments for, the incapacitating neurological disorder.

Reports showed that there were public concerns about research into the human genome and gene therapy, with a combination of fear and mistrust.

1997: Researchers at the Roslin Institute in Scotland announced that they had cloned a sheep called Dolly from the cell of an adult ewe. Dolly was the first mammal cloned by a technique called nuclear transfer technology. Nuclear transfer allows the introduction of complete genetic material from one cell into another enucleated unfertilized egg cell.

1998: A group of researchers succeeded in culturing embryonic stem cells.A number of researchers at Japan's Kinki University cloned eight identical calves by means of cells taken from a single adult cow.A rough draft of the human genome map was created, presenting the sites of more than 30 000 genes.

1999: A fatal neurological disease called bovine spongiform encephalopathy (BSE), also known as mad cow disease, that spread from cattle to humans, was diagnosed by a new medical diagnostic examination that facilitated the quick detection of BSE/Creutzfeldt-Jakob disease (CJD).

MODERN BIOTECHNOLOGY

A major obstacle to scientific discoveries was the Second World War. After the war, some essential discoveries were explored. These discoveries form the basis for modern biotechnology and have brought this field to its current status. Some of the prominent events of the modern age of biotechnology are highlighted in Table 2.

Table 2. Historical and current events that form the basis of modern biotechnology.

Year	Discoveries
The 1950s	
1952:	George Otto Gey created a continuous cell line taken from a human cervical carcinoma. This cell line, known as HeLa, is still used in therapeutic research.
1953:	Watson and Crick explored DNA as a genetic material, and discovered its structure, called the double-helix.
1954:	Joseph Murray carried out the first kidney transplant between identical twins.
1957:	Scientists revealed that sickle-cell anemia occurs due to an alteration in a single amino acid in hemoglobin cells.
1958:	Arthur Kornberg created DNA in a test tube for the first time. The first mechanical protein sequencer, the Moore-Stein amino acid analyzer, is developed.
The 1960s	
1960:	A French researcher discovered messenger RNA (mRNA).
1961:	François Jacob and Jacques Monad demonstrated the concept of Operon.
1962:	Osamu Shimomura explored the green fluorescent protein in the jellyfish Aequorea victoria. He afterward developed it into a technique for examining formerly invisible cellular processes.
Year	Discoveries
1963:	Samuel Katz and John F Enders developed the first vaccine for measles.
1963:	Autonomous groups in the USA, Germany and China produced insulin, a pancreatic hormone.
1964:	The existence of reverse transcriptase was predicted.
1967-71:	Maurice Hilleman made the first American vaccine for mumps.
	The first vaccine for rubella was developed.
	Rubella was combined with the measles and mumps vaccines to yield the measles/mumps/rubella (MMR) vaccine.
1968:	How the arrangement of nucleotides in nucleic acids regulates the cell's synthesis of proteins was discovered.

The 1970s

1970: Restriction enzymes were discovered.

1972: DNA ligases, which join DNA fragments together, were used for the first time.

The DNA composition of humans was discovered to be 99% similar to that of chimpanzees and gorillas.

The purified enzyme reverse transcriptase was first employed to prepare complementary DNA from purified messenger RNA in a test tube.

1973: Stanley Cohen and Herbert Boyer used bacterial genes to perform the first successful rDNA experiment.

Sir Edwin Mellor Southern developed a blotting technique for DNA called the Southern blot.

1974: The NIH formed a Recombinant DNA Advisory Committee to supervise recombinant genetic research.

The first vaccine for chicken pox was developed in Japan.

1975: Colony hybridization and Southern blotting were explored for identifying specific DNA sequences.

The first monoclonal antibodies were prepared.

César Milstein, Georges Jean Franz Kohler and Niels Kaj Jerne explored the monoclonal antibody technique by fusing immortal tumor cells with antibody-producing B-lymphocyte cells to generate hybridomas that constantly produce identical antibodies.

1975: The theory of cytoplasmic hybridization was proposed, and the first ever monoclonal antibodies were synthesized.

1976: The NIH published the first guidelines for rDNA research.

Molecular hybridization was employed for the prenatal diagnosis of alpha thalassemia. Yeast genes were expressed in *E. coli* bacteria.

1977: Procedures were developed to swiftly sequence long sections of DNA. Genetically engineered bacteria were employed to manufacture the human growth protein somatostatin, marking the first time a synthetic recombinant gene was employed to clone a protein. Several believed this to be the arrival of the 'age of biotechnology'.

R Austrian *et al.* at the University of Pennsylvania developed the first vaccine for pneumonia.

Year Discoverie

1978: Boyer synthesized the human insulin gene (i.e. a synthetic version of it) and inserted it into the bacterium E. coli, allowing the bacterium to produce human insulin.

Louise Brown, the first test-tube baby, was born in the UK. The first vaccine for meningococcal meningitis was developed.

The 1980s

1980:According to US Supreme Court, genetically altered life forms could be patented, creating vast possibilities for commercially exploiting genetic engineering.

The first patent of this nature was awarded to the Exxon oil company to patent an oil-eating micro-organism, which would afterward be employed in the 1989 cleanup of the Exxon oil spill at Prince William Sound, Alaska. S Cohen and D H Boyer received a US patent for gene cloning.

	The first automatic gene machine was developed in California.
	Launch of Amgen, which would grow to become the world's largest biotechnology medicines company.
1981:	Baruch Blumberg and Irving Millman developed the first vaccine for hepatitis B.
	Researchers in Switzerland cloned mice.
	The first transgenic animals were produced by transforming genes from other animals into mice.
1982:	The FDA supported the first recombinant protein.
1983:	Luc Montagnier of the Pasteur Institute in Paris isolated the AIDS virus.
	Kary Mullis discovered the polymerase chain reaction (PCR), a technique for multiplying DNA sequences.
	PCR was identified as the most innovative molecular biology technique. The FDA sanctioned a monoclonal antibody-based diagnostic analysis to identify Chlamydia trachomatis.
	The first artificial chromosome was produced and the first genetic markers for specific inherited diseases were discovered.
1984:	The DNA fingerprinting technique was discovered. When a restrictive enzyme is applied to DNA from various individuals, the ensuing sets of fragments sometimes vary noticeably from one person to the next. Such differences in DNA are called restriction fragment length polymorphisms and are particularly helpful in genetic investigations.
	The first genetically engineered vaccine was discovered for hepatitis B. The whole genome of the human immunodeficiency virus (HIV) virus was cloned and sequenced.
1985:	Genetic fingerprinting stepped into the courtroom.
	Genentech became the first biotechnology organization to launch its own biopharmaceutical product.
	Genetically engineered plants resistant to viruses, insects and bacteria were field-tested for the first time.
	Cloning of the gene that encodes human lung surfactant protein was achieved. This was a major step toward reducing premature birth problems.
	The NIH sanctioned guidelines for executing trials of gene therapy on humans.
Year	Discoveries
1986:	Peter G Schultz from UC Berkeley explained how to conjugate antibodies and enzymes (abzymes) to create therapeutics.
	The automated DNA sequencer was discovered in California. The FDA sanctioned the first monoclonal antibody treatment to fight kidney transplant rejection.
	The FDA sanctioned the first biotech-derived interferon drugs to treat cancer. Drugs to treat Kaposi's sarcoma, a complication of AIDS, were discovered.
	The FDA sanctioned the first genetically engineered human vaccine to avert hepatitis B.
1987:	The FDA sanctioned a genetically engineered tissue plasminogen activator to treat heart attacks.
	Maynard Olson and colleagues at Washington University discovered yeast artificial

chromosomes, which are expression vectors for large proteins.

Reverse transcription and the PCR were linked to augment messenger RNA sequences. DNA microarray technology, the use of a set of various DNAs in arrays for expression outline, was first explained.

The collection of DNA was used to recognize genes whose expression is altered by interferon.

The FDA sanctioned a diagnostic serum tumor marker test for ovarian cancer.

1988: Congress financed the Human Genome Project, a huge international attempt to map and sequence the human genetic code as well as the genomes of other species.

The first contract between two organizations with parallel patents for cross-licensing of biotech products occurred and became the example.

1989: The FDA sanctioned Amgen's first biologically derived human therapeutic.

Oil-eating bacteria were employed to clear up the Exxon Valdez oil spill.

A gene responsible for cystic fibrosis was explored.

The 1990s

1990: The first nationally sanctioned gene therapy treatment was executed effectively on a four-year-old girl suffering from an immune disorder called adenosine deaminase deficiency.

The Human Genome Project was launched.

The FDA approved the first hepatitis C antibody test, which helped to guarantee the purity of blood bank products.

The FDA sanctioned a bioengineered form of the protein interferon gamma to treat chronic granulomatous disease.

The FDA sanctioned a modified enzyme for enzyme replacement therapy to treat severe combined immunodeficiency disease. It was the first successful application of enzyme replacement therapy for an inherited disease.

1992: The US Army accumulated blood and tissue tests from all new recruits as part of a genetic dog-tag plan intended to achieve better recognition of soldiers killed in combat.

The FDA sanctioned the first genetically engineered blood-clotting factor-a recombinant protein used to treat hemophilia A.

Year Discoveries

The FDA sanctioned a recombinant protein to treat renal cell cancer. American and British researchers revealed a technique for analyzing embryos *in vitro* for genetic abnormalities, e.g. cystic fibrosis and hemophilia.

1993: The FDA approved a recombinant protein to treat multiple sclerosis-marking the first new multiple sclerosis treatment in 20 years.

A global research team, led by Daniel Cohen from the Center for the Study of Human Polymorphisms, Paris, created a rough map of all 23 pairs of human chromosomes.

Two minor trade organizations merged together to form the Biotechnology Industry Organization, an international biotechnology support group.

1994: The FDA sanctioned a recombinant protein to deal with growth hormone (GH) deficiency.

Mary-Claire King at UC Berkeley explored the first breast cancer gene, BRCA1. The FDA sanctioned a modified enzyme to deal with Gaucher's disease.

A number of genes, human and otherwise, were identified and their functions explained. These comprised: Ob, a gene inclining to obesity; BCR, a breast cancer receptiveness gene; BCL-2, a gene linked to apoptosis (programmed cell death); Hedgehog genes (named because of their shape), which synthesize proteins that direct cell differentiation in complex organisms; and Vpr, a gene regulating the reproduction of the HIV virus.

Genetic linkage studies recognized the role of genes in a variety of disorders, including bipolar disorder, cerulean cataracts, melanoma, dyslexia, prostate cancer, thyroid cancer, hearing loss, sudden infant death syndrome and dwarfism.

The FDA sanctioned a genetically engineered description of human DNase, which breaks down protein accretion in the lungs of cystic fibrosis patients. This corresponded to the first new therapeutic drug for treating cystic fibrosis.

1995: The first baboon-to-human bone marrow transplant was executed on an AIDS patient.

The first vaccine for hepatitis A was explored.

The NIH, the US Army and the Centers for Disease Control and Prevention were considerably involved in the growth and clinical testing of the vaccine.

Researchers at the Institute for Genomic Research completed the first full gene sequence of a living organism for the bacterium Haemophilus influenzae.

A European study group determined a genetic defect that turned out to be the most frequent cause of deafness.

1996: Researchers at the Department of Biochemistry at Stanford University and Affymetrix developed the gene chip, a small glass or silica microchip that contains thousands of individual genes that can be examined simultaneously. This symbolized a scientific breakthrough in gene expression and DNA sequencing technology.

Research groups sequenced the complete genome of a complex organism, Saccharomyces cerevisiae, otherwise known as baker's yeast. The accomplishment symbolizes the entire sequencing of the largest genome to date. A novel, economic diagnostic biosensor test was developed to hasten the detection of a toxic strain of E. coli, the bacteria responsible for several food-poisoning outbreaks.

Year Discoveries

1997: The first human artificial chromosome was discovered.

A mixture of natural and synthetic DNA was used to synthesize a genetic cassette that could possibly be adapted and employed in gene therapy.

The FDA sanctioned a recombinant follicle stimulating hormone to deal with infertility.

The FDA permitted the first bloodless HIV-antibody analysis, which used cells from patients' gums. Researchers at the Institute for Genomic Research sequenced the entire genome of the Lyme disease pathogen, Borrelia burgdorferi, along with the genome for the organism associated with stomach ulcers, Helicobacter pylori.

Researchers at the University of Wisconsin-Madison sequenced the E. coli genome.

The FDA permitted the first therapeutic antibody to treat cancer in the USA. It was employed for patients with non-Hodgkin's lymphoma.

1998:	Human skin was created in the laboratory for the first time.
	Two research groups cultured embryonic stem cells.
	Embryonic stem cells were employed to regenerate tissue and produce disorders mimicking diseases.
	Researchers at the Sanger Institute in the UK and at the Washington University School of Medicine in St Louis, USA, sequenced the first whole animal genome for the Caenorhabditis elegans worm.
	A rough draft of the human genome map was created, displaying the sites of more than 30 000 genes.
	The first vaccine for Lyme disease was discovered.
	The FDA sanctioned a novel monoclonal antibody to treat Crohn's disease.
	A monoclonal antibody therapy employed against breast cancer had positive outcomes, indicating a new age of management based on the molecular targeting of tumor cells.
	Support for using the HER2 inhibitor for the management of breast cancer in patients who had tested positive for the HER2 mutation brought personalized medicine to oncology.
The 2000s	
2000	Har Gobind Khorana synthesized DNA in a test tube.
	Kary Mullis added value to Har Gobind Khorana's findings by amplifying DNA in a test tube, to create a thousand times more than the original amount of DNA.
	Sir Ian Wilmut cloned an adult sheep and called the cloned sheep 'Dolly'.
	Craig Venter sequenced the human genome; the first publicly accessible genomes would later be those of James Watson and Ventor.
	Researchers at Celera Genomics and the Human Genome Project successfully completed a rough draft of the human genome.
2001	The journals Science and Nature reported the human genome sequence, making it feasible for researchers all over the world to start investigating innovative treatments for diseases that have genetic origins, e.g. heart disease, cancer, Parkinson's and Alzheimer's.
2002	An era of very rapid shotgun sequencing of major genomes was completed. Included were the mouse, chimpanzee, dog and hundreds of other species.
Year	Discoveries
2003	Celera and the NIH successfully finished the sequencing of the human genome.
2004	The FDA supported the first monoclonal antibody (i.e. antiangiogenic, inhibiting blood vessel formation or angiogenesis) for cancer therapy.
	The FDA approved a DNA microarray analysis system, which helped in selecting medications for different conditions. This was a significant step toward modified medicine.
2006	The FDA sanctioned a recombinant vaccine against human papillomavirus, which causes genital warts and cervical cancer.
	Researchers established the three-dimensional (3D) structure of HIV, which causes AIDS.
2007	Researchers discovered how to use human skin cells to produce embryonic stem cells.

2008	Venter replicated a bacterium's genetic structure completely from laboratory chemicals, taking a step nearer to generating the world's first living artificial organism.
2008	Japanese chemists developed the first DNA molecule made nearly entirely of artificial parts. The finding could be used in areas of gene therapy.
2009	Former US President Barack Obama signed an administrative order releasing national funding for broader research on embryonic stem cells.
	Scientists identified three new genes connected with Alzheimer's disease, paving the way for possible new diagnostics and therapeutics.
	Geron commenced the first FDA approved clinical trial by means of embryonic stem cells.
2010	The FDA sanctioned a modified prostate cancer medicine that improves a patient's immune cells to distinguish and attack cancer cells.
	The FDA sanctioned an osteoporosis treatment that was one of the first medicines based on genomic investigations.
	Craig Venter showed that a synthetic genome could duplicate alone.
2011	A trachea developed from stem cells was transplanted into a human recipient. Progress in 3D printing technology resulted in 'skin-printing'.
	The FDA sanctioned the first cord blood therapy to be employed in hematopoietic stem cell transplantation protocols in patients with disorders influencing the hematopoietic system.
2012	The FDA issued draft regulations for biosimilar drugs.

In the modern era, researchers had almost all the basic tools available to them for their applications. With these tools the majority of basic concepts were elucidated, which fast-forwarded the path to important scientific discoveries. These studies and discoveries have unlimited implications and applications.

Conclusively, biotechnology has brought humanity to this level of comfort; looking ahead, the next question is, where will it take us? Biotechnology has both beneficial and destructive potential. It is we who should now decide how to use this technology to help humanity rather than to destroy it.

HISTORY OF GENOME RESEARCH

The discussion of sequencing the entire human genome started approximately in the mid-1980s. Such a task was daunting at the time because of the limitations of sequencing technologies. The most efficient automated DNA sequencer was able to run 96 samples at a time, and approximately 500-800 base pairs of nucleotides could be read per sample. The sequencing reactions themselves were not fully automated and were expensive. With this technology, it was estimated that the project could take at least 15 years to complete.

With many planning activities involving government and scientific communities with various workshops, the project was officially launched in 1990 with a single goal of sequencing and assembling the entire genome of a single human individual containing three billion base pairs of DNA. The project started in the United States, but became an international collaborative project involving the participation of universities and research centres in China, France, Germany, Japan, the United Kingdom of Great Britain and Northern Ireland, and the United States of America.

It should be noted that a parallel project was conducted by a private corporation, the Celera Corporation, which formally launched in 1998. It is widely believed that this parallel run of the project from the private sector significantly increased competition, allowing speeding up of the project as well as enhancement of the sequence quality.

Development of genomics as a new branch of science

Into the mid-phase of the 13-year human genome project, it was becoming evident that it would not be possible to put the human genome together with just DNA sequences generated with sequencing technologies. This was because the vast majority of the three billion base pairs of human DNA are repetitive in nature. These repetitive elements of the genome are different, but highly similar sequences. Computer software would not lend the capability to assemble all such sequences correctly into their proper place.

Therefore, the demands on the development of additional technologies and research to meet this challenge were tremendously high. A new branch of science, genomics, was born. The scope of genomics is to apply the techniques of genetics and molecular biology for the understanding of genome structure, organization, expression and functions. Starting from its beginnings, genomics was meant to be a branch of science that uses principles and methods of multidisciplines, including those of biochemistry, molecular biology, genetics, statistics and informatics, to name a few. It is clear, as detailed below, that genomics is now the centre of life sciences.

Just as the "molecularization" of biology in the past 40 years, now many scientific disciplines have been or are being "genomicized". This is largely because genomics has led to major changes on how science is conducted. Rather than on the traditional process of proposing a hypothesis based on existing knowledge, testing the hypothesis experimentally, and accepting or rejecting the hypothesis as the route of scientific research, genomics and its related large data sets have allowed the change from hypothesis-driven research to datadriven research. All the data can be suddenly created, and it is the researchers' role to figure out the simple facts from the complex data.

As progress of the human genome project was being made, it became obvious that sequences themselves can be assembled by extension of the overlapping DNA sequences into longer contiguous sequences (contigs). However, the size of the contigs is limited by the presence of the repetitive elements. Thus, the human genome sequences could be assembled into many - hundreds of thousands - small contigs. Although the sequences within the contigs were known, the relationship or the positions of the contigs in the genome could not be determined by sequence assembly.

This demanded the development of methods to place and order the contigs onto the chromosomes and eventually to the genome. This was the start of the development of genome mapping technologies, including genetic linkage mapping, physical mapping, radiation hybrid mapping, and optical mapping. Within the area of physical mapping, several technologies were developed, with the most popular being radiation hybrid mapping and large-insert libraries such as BAC (bacterial artificial chromosome)-based fingerprinting for mapping the physical genomic DNA. Similar to the situation of the human genome project, genome projects of aquaculture species require similar technological support to assemble billions of base pairs of DNA together correctly.

Aquaculture genomics, a historical review

Aquaculture is a relatively new but increasingly growing sector of agriculture. It is very important not only for economic interests, but is also of social and cultural significance. In many Asian countries, serving seafood at the dinner table is seen as highly prestigious. Seafood accounts for 20 percent of animal protein sources for the world population. The seafood industry involves a total of 144 million tonnes annual production with 44 million fishers and fish farmers and 2.1 million vessels, contributing US$166 billion to the world economy and over US$25 billion of international trade annually.

Currently, aquaculture accounts for over 40 percent of seafood consumed in the world, and China is the only country where aquaculture produces more than 50 percent of consumed seafood.

Aquaculture genomics officially started in the 1990s, although related genome research was conducted in the 1980s. The start of aquaculture genomics was marked by the first Aquaculture Genomics Workshop, held in 1997 in Dartmouth, Massachusetts, United States of America. This workshop targeted a group of six species for genome research in the United States: salmonids, catfish, tilapia, striped bass, oysters and shrimps.

As any other agricultural sector, sustained production requires research on basic biology, including growth, nutrition, reproduction, physiology, and genetics and genomics. One interesting observation is that all these research fields are being unified through the use of genomic technologies or, that is, they are being "genomicized".

The availability of a draft whole genome sequence significantly enhances genome research and applications of genome-based technologies to improve agricultural production and quality. As such, whole genome sequences are available or nearly available for major livestock species, including cattle, swine, chicken and horse. Whole genome sequencing used to be nearly "utopian" because of its involved huge costs. Take the human genome as an example; it took the major genome centres over a decade to complete the draft genome sequencing of the first human individual, from 1986 to 2000.

The direct cost involved in the human genome sequencing was at the level of hundreds of millions of dollars. Along with the sequencing effort, related human genome research had cost much more. Human genome related research probably spent in excess of US$6 billion-US$8 billion. Thereafter, the sequencing of the cattle genome cost over US$60 million, much less expensive than the human genome sequencing, but yet is still at a cost level unthinkable for many aquaculture species. It was because of such huge costs whole genome sequencing used to be regarded as a milestone of a lifetime achievement.

Aquaculture genomics, a historical review

Aquaculture is a relatively new but increasingly growing sector of agriculture. It is very important not only for economic interests, but is also of social and cultural significance. In many Asian countries, serving seafood at the dinner table is seen as highly prestigious. Seafood accounts for 20 percent of animal protein sources for the world population. The seafood industry involves a total of 144 million tonnes annual production with 44 million fishers and fish farmers and 2.1 million vessels, contributing US$166 billion to the world economy and over US$25 billion of international trade annually. Currently, aquaculture accounts for over 40 percent of seafood consumed in the world, and China is the only country where aquaculture produces more than 50 percent of consumed seafood.

Aquaculture genomics officially started in the 1990s, although related genome research was conducted in the 1980s. The start of aquaculture genomics was marked by the first Aquaculture Genomics Workshop, held in 1997 in Dartmouth, Massachusetts, United States of America. This workshop targeted a group of six species for genome research in the United States: salmonids, catfish, tilapia, striped bass, oysters and shrimps.

As any other agricultural sector, sustained production requires research on basic biology, including growth, nutrition, reproduction, physiology, and genetics and genomics. One interesting observation is that all these research fields are being unified through the use of genomic technologies or, that is, they are being "genomicized".

The availability of a draft whole genome sequence significantly enhances genome research and applications of genome-based technologies to improve agricultural production and quality. As such, whole genome sequences are available or nearly available for major livestock species, including cattle, swine, chicken and horse. Whole genome sequencing used to be nearly "utopian" because of its involved huge costs. Take the human genome as an example; it took the major genome centres over a decade to complete the draft genome sequencing of the first human individual, from 1986 to 2000.

The direct cost involved in the human genome sequencing was at the level of hundreds of millions of dollars. Along with the sequencing effort, related human genome research had cost much more. Human genome related research probably spent in excess of US$6 billion-US$8 billion. Thereafter, the sequencing of the cattle genome cost over US$60 million, much less expensive than the human genome sequencing, but yet is still at a cost level unthinkable for many aquaculture species. It was because of such huge costs whole genome sequencing used to be regarded as a milestone of a lifetime achievement.

Genomes of at least two dozens of aquaculture species have been sequenced, or are now being sequenced. Of the six aquatic species groups included in the United States Animal Genome NRSP-8 Program, a whole genome has been sequenced for tilapia, rainbow trout, Atlantic salmon, catfish, striped bass, and oysters and shrimps. While it is extremely exciting for aquaculture geneticists to have the whole genome sequenced for many of the important aquaculture species, it is pivotally important to address some of the most important issues related to whole genome sequencing.

These include the issues related to effective assembly of the whole genome sequences and the issues related to the effective uses of the whole genome sequences. The usefulness of the whole genome sequence relies on the efficient sequence assembly and adequate sequence annotation, which in turn depends on the availability of a number of genome resources, including transcriptome sequencing and assembly and characterization of the non-coding part of the genome.

TRADITIONAL GENETIC BIOTECHNOLOGIES FOR AQUACULTURE

Selective breeding

The goal of selective breeding is to improve production and performance traits in farmed animals via the selection of individuals for desirable phenotypic characteristics. For aquaculture species, these traits may include faster growth, feed conversion efficiency, disease resistance, processing yields, low oxygen tolerance, stress tolerance, robustness, morphology and sexual maturation, among other traits of interest.

Aquaculture breeding programmes have a short history. Selective breeding programmes for strain enhancement began in the late 1960s with the development of Atlantic salmon breeding programmes in Norway. Domestication of many aquaculture species is still a recent event. Therefore, a number of techniques are effective, including strain selection, cross breeding, hybridization, and within-strain selection.

However, the main goal of genetic improvement programmes for production traits in modern breeding schemes is within-strain selection. This is typically achieved within a well-managed, commercial programme of family and pedigree tracking combined with extensive trait measurements on selection candidates or their relatives. Selective breeding programmes exist for various aquaculture species, such as Atlantic salmon, rainbow trout, tilapia, common carp, grass carp, silver carp, crucian carp, rohu carp, yellowtail, sea bream, channel catfish, European seabass, turbot, Asian seabass, Pacific and eastern oyster, shrimps, scallops and pearl oysters, among 60 some species.

Selective breeding has proven to be very effective in enhancing the traits of agricultural plants and animals, including aquaculture species, through the use of genetics principles. For instance, the genetic gain has been greater than 12 percent per generation for growth rate and for disease resistance when challenge tests are applied. The main reasons for the large genetic gains observed for aquatic species are their relatively high fertility and the natural existence of broad genetic variation for economically important traits, both of which allow a very high selection intensity to be applied. However, the genetic improvement of species grown in aquaculture is far behind that achieved for animals and plants. Less than 10 percent of aquaculture production is based on genetically improved stocks.

Polyploidy

Polyploidy is lethal in mammals and birds, but has led to the development of many productive plant varieties such as domesticated wheat. Triploid fish are viable and are usually sterile, while tetraploid fish are usually viable and fertile.

Through chromosome manipulations, aquaculture organisms can be obtained with various levels of chromosome sets. A regular diploid organism has two sets of chromosomes. The polyploid state refers to individuals with extra sets of chromosomes.

Triploid fish have three sets of chromosomes, and tetraploid fish have four sets of chromosomes. In addition to full sets of chromosomal compositions, aneuploids have at least a diploid set of chromosomes with one or more additional chromosomes, but not a full complement to the set.

Triploids are organisms with three sets of homologous chromosomes. They are found spontaneously in both wild and cultured populations and can be induced in many commercial species of fish and shellfish. Triploidy is induced by allowing normal fertilization and then forcing retention of the second polar body. The second polar body is retained by applying temperature (hot shocks or cold shocks), hydrostatic pressure, anaesthetics or chemical shocks shortly after fertilization. Additionally, in some cases such as in rainbow trout, triploidy can be induced by applying high pH and high calcium to either the sperm or eggs. Many reviews and books are available covering the technical aspects of polyploidy, and therefore this paper will not cover this extensively

The performance of triploid fish varies. Triploidy can affect growth, feed conversion efficiency, disease resistance and other traits. For growth, triploid fish can grow faster, at the similar rate, or slower. However, even for those that grow faster, this advantage is not obvious until sexual maturity. It is apparent in sterile fish that energy otherwise exspended on development of gonads and gametes is converted into growth.

In addition to direct induction of triploid fish and shellfish such as oysters, the production of tetraploid fish has been a way for mass production of triploid fish. First-generation tetraploids can be produced by hydrostatic pressure treatment before the first cleavage and raised until the adult stage. With rainbow trout, survival and growth of tetraploids is severely depressed when compared with the diploid control.

However, when the tetraploid male fish were mated with normal diploid female fish to produce triploid fish, the progenies were consistently normal with high survival rates.

The progenies were found to be almost all triploids by karyology, which failed to detect a significant rate of aneuploidies. Therefore, the use of tetraploid as a means of producing triploid fish is a viable option, although the fertilizing ability of tetraploid males was always low, 0 to 97 percent of the control, with an average of 40 percent compared with the diploid controls.

Gynogenesis

Gynogenesis is a form of all-female inheritance. In fish species, ultraviolet (UV) irradiation has been used to inactivate the sperms, and such UV-inactivated sperms are used to trigger gynogenetic development without contributing the paternal genome to the progeny. Practically, sperms from a closely related, but different species, are used to reduce the possibility of real fertilization in case the sperms are not completely inactivated.

The developing embryo is initially haploid, but the diploidy is recovered by inhibiting either the second polar body extrusion or the first cleavage. The gynogen produced by inhibiting the second polar body extrusion is called meiotic gynogen, while the gynogen produced by inhibiting the first cleavage is called mitotic gynogen. Mitotic gynogens are completely homozygous, while meiotic gynogens are partially heterozygous. The gene-centramere recombination makes the distal part of the chromosome from the centromere more heterozygous. Meitic gynogens are easier to produce than the mitotic gynogens because the survival rate of mitotic gynogens is very low.

One of the practical goals of gynogenesis is the production of genetically identical populations, i.e. the clonal lines. The clonal lines may be important for breeding programmes, and they can certainly be important for genome research because they provide a homozygous template for genome sequencing, such as the doubled haploid of channel catfish that was used as the sequencing template for whole genome sequencing. Although clonal lines have been produced with aquaculture species such as ayu, amago salmon and hirame, their large-scale aquaculture has not been realistic. The major purpose of gynogen production has been for research.

Androgenesis

Androgenesis refers to all-paternal inheritance. Androgens can be produced by irradiating eggs and then doubling the paternal genome. Androgens are more difficult to produce than gynogens, presumably because of the extremely low survival rate of irradiated eggs. Diploidy can be recovered only by blockage of first cell cleavage.

Like gynogenesis, androgenesis can be used to produce clonal populations or monosex populations for the purpose of breeding programmes or to elucidate sex-determining mechanisms. If the male is heterogametic, XX and YY androgens will result in equal proportion. If the male is the homogametic, then the androgens will be 100 percent ZZ and all male. YY individuals are viable. As a matter of fact, the production of YY fish through androgenesis followed by regular mating with a normal XX female is a major way of producing all male populations in fish. In many species, males grow faster, and use of monosex populations is of interest to aquaculturists.

Gene transfer

Gene transfer is a process of transferring one or a few foreign gene(s) into an organism. However, the foreign gene can be from other organisms or from the organism itself. The concept is relatively straightforward: if the functions of a gene are well known, then the gene can be transferred into the organism to deliver the functions. For instance, the growth hormone gene was well studied and known of its functions to be promoting growth, then the transfer of the growth hormone into an organism should promote growth. This simple concept was demonstrated when transgenic mice with the growth hormone gene grew much larger, up to 2.5 times larger than the nontransgenic controls.

The first successful gene transfer in fish was demonstrated in 1985 when Zhu, He and Chen transferred the human growth hormone gene in goldfish. Since then, transgenic fish have been produced with various aquaculture species, including rainbow trout, channel catfish, Nile tilapia and northern pike.

A number of techniques were developed for transferring the genes of interest into fish, including microinjection and electroporation. Either method works well. However, transgenic technologies suffer from several major lines of shortcomings:

i. the doses of gene transfer cannot be controlled. Most often, multiple copies, often in the form of tandem head to tail arrays are integrated into the genome;
ii. the integration sites are random, and such sites can be within a functional gene; and
iii. the pleiotropic effect of genes cannot be controlled. The major concern of transgenic fish was the use of foreign DNA, including the promoters and the regulatory DNA sequences such as poly A signals.

Researchers introduced the concept of "all-fish vectors" that contain all sequences from fish for transfer into fish.

Transgenic fish have generally performed the way scientists expected them to. For instance, significantly enhanced growth rates were observed with transgenic fish with growth hormone genes. This has been demonstrated in goldfish, channel catfish, northern pikes, Atlantic salmon, rainbow trout, tilapia and carp, among many other species, although failure to observe enhanced growth was also reported. The best example of growth enhancement was demonstrated with transgenic salmon.

In addition to the growth trait, gene transfer was used to improve several other traits, including cold tolerance and disease resistance. For instance, transgenic expression of viral coat protein genes, or antisense of viral early phase genes, has been explored in shrimp. Attempts for improving disease resistance were made by expression transgenes of lytic peptides, and organisms containing these genes exhibited enhanced disease resistance. Antimicrobial peptide genes were also used to enhance disease resistance. For instance, transgenic fish expressing cecropins were found to be more resistant against bacterial diseases.

In addition to the purpose of enhancing performance traits of aquaculture species, fish have been considered for the production of pharmaceuticals as biological factories.

Fish have potential advantages as bioreactors compared with mammals. These advantages include a short generation interval, low cost of maintenance of the animals, easy maintenance, large numbers of individuals, high-density culture, and mammalian viruses and prions that are not found in fish.

Public acceptance of transgenic fish has been relatively low because of two lines of concerns: (i) food safety concerns; and (ii) ecological safety concerns. The question on whether it is safe to consume transgenic fish has been one major question of consumers.

As aquaculture species have aquatic living environments, tracking of transgenic aquatic animals is more difficult, and therefore the concerns over ecological safety have been serious.

One commonality shared by all the traditional genetic technologies is that researchers are operating in a black box without the knowledge of the genome, how the genome works, and how the genetic changes made will affect the expression and functions of the genome.

Modern genetic technologies depend on the knowledge of the genome, genome maps, genome composition, organization, expression and function. Therefore, in the next sections, a great level of detail will be provided to cover how genomes are being studied, biotechnologies that were developed for the understanding of the genomes, transcriptome, and gene networks and functions. Using genome information, genome-based technologies have been developed and are continuously being developed for genetic gains in aquaculture species.

2

GENETIC VARIATION: AN INTRODUCTION

Genetic variation is important to the processes of natural selection and biological evolution. The genetic variations that arise in a population happen by chance, but the process of natural selection does not. Natural selection is the result of the interactions between genetic variations in a population and the environment. The environment determines which genetic variations are more favorable or better suited for survival. As organisms with these environmentally selected genes survive and reproduce, more favorable traits are passed on to the population as a whole.

GENETIC VARIATION CAUSES

Genetic variation occurs mainly through DNA mutation, gene flow (movement of genes from one population to another), and sexual reproduction. Due to the fact that environments are unstable, populations that are genetically variable will be able to adapt to changing situations better than those that do not contain genetic variation.

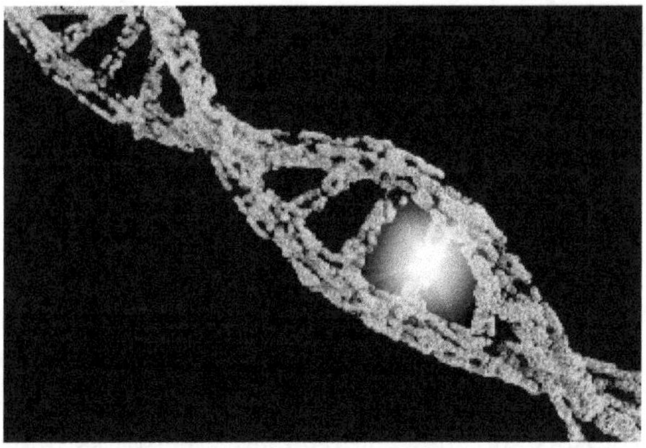

Computer graphic illustration depicting a point mutation. A point mutation is a genetic mutation where a single nucleotide base is changed.

- **DNA Mutation:** A mutation is a change in the DNA sequence. These variations in gene sequences can sometimes be advantageous to an organism. Most mutations that result in genetic variation produce traits that confer neither an advantage or disadvantage. Mutations lead to genetic variation by altering genes and alleles in a population. They may impact an individual gene or an entire chromosome. Although mutations change an organism's genotype (genetic makeup), they may not necessarily change an organism's phenotype.

- **Gene Flow:** Also called gene migration, gene flow introduces new genes into a population as organisms migrate into a new environment. New gene combinations are made possible by the availability of new alleles in the gene pool. Gene frequencies may also be altered by the emigration of organisms out of a population. The immigration of new organisms into a population may help organisms better adapt to changing environmental conditions. The migration of organisms out of a population could result in a lack of genetic diversity.
- **Sexual Reproduction:** Sexual reproduction promotes genetic variation by producing different gene combinations. Meiosis is the process by which sex cells or gametes are created. Genetic variation occurs as alleles in gametes are separated and randomly united upon fertilization. The genetic recombination of genes also occurs during crossing over or the swapping of gene segments in homologous chromosomes during meiosis.

Genomic information provides powerful tools to enhance physiological research, the results of which may be used for optimization of husbandry practices, feeding and feed formulations, breeding technologies, or non-genetic selection or screening (e.g., epigenetics, proteomics, and metabolomics). Whole genome sequences, in various states of assembly, are now available for many aquaculture species, enabling the identification of genomic variations such as insertions/deletions, single nucleotide polymorphisms (SNPs), copy number variations, and differentially methylated regions. However, this information is only useful when used to predict phenotypes that have a positive impact on production or product quality.

For this reason, genetic mapping, quantitative trait loci (QTL) analysis, genome-wide association studies (GWAS), expression profiling, and bioinformatic analysis can be used to identify genotypic variants associated with particular phenotypic traits, which can then be exploited in breeding programs.

For some aquaculture species we have reached the point where genome-based technologies such as marker-assisted and whole genome selection can be applied to enhance aquaculture traits and research is beginning to shift toward understanding functional polymorphisms and the gene regulatory networks underlying commercially important traits. A more complete understanding of the gene networks underlying growth, reproduction, and disease resistance will provide the knowledge-base for developing more robust and productive genetic stocks for the aquaculture industry.

The degree to which genome enabled technologies and genomic information have been or can be applied in genetic improvement programs varies across aquaculture species. Private sector investment in research and development for the implementation of new technologies is dependent on unique industry structure (e.g., overall size of the industry, size of individual companies) and the level of vertical integration. In addition, the approach used for germplasm improvement and status of existing breeding programs dictates whether and which genome enabled technologies are suitable for a given industry.

Industries with centralized breeding, such as rainbow trout and salmon, have greater potential to benefit from new technologies compared to industries where breeding activities are widely distributed. Finally, the current demand for species-specific genomic tools (such as high through-put genotyping assays) among the diverse aquaculture industry sectors is low, rendering them commercially unaffordable. This forces some industries interested in genetic improvement to rely on the public sector for resources that enable application of state-of-the-art genomic technologies.

Here we review the development of genomic tools and application of genome enabled technologies for the genetic improvement of aquaculture species. Specifically, we review the status of genome mapping and sequencing, identify gaps in our current knowledge, and highlight the need to implement new technologies in aquaculture. We then propose a set of priorities for future research in aquaculture genomics, genetics, and breeding.

DEOXYRIBOSE NUCLEIC ACID

DNA is made up of molecules called nucleotides. Each nucleotide contains three components: a phosphate group, which is one phosphorus atom bonded to four oxygen atoms; a sugar molecule; and a nitrogen base. The four types of nitrogen bases are adenine (A), thymine (T), guanine (G) and cytosine (C), and together, these serve as the "letters" that make up the genetic code of our DNA.

Nucleotides are attached together to form two long strands that spiral to create a structure called a double helix. If you think of the double-helix structure as a ladder, the phosphate and sugar molecules would be the sides, while the base pairs would be the rungs. The bases on one strand pair with the bases on another strand: Adenine pairs with thymine (A-T), and guanine pairs with cytosine (G-C).

Deoxyribonucleic acid (DNA) vaccination is based on the administration of the gene encoding the vaccine antigen, rather than the antigen itself. Subsequent expression of the antigen by cells in the vaccinated hosts triggers the host immune system. Among the many experimental DNA vaccines tested in various animal species as well as in humans, the vaccines against rhabdovirus diseases in fish have given some of the most promising results. A single intramuscular (IM) injection of microgram amounts of DNA induces rapid and long-lasting protection in farmed salmonids against economically important viruses such as infectious haematopoietic necrosis virus (IHNV) and viral haemorrhagic septicaemia virus (VHSV).

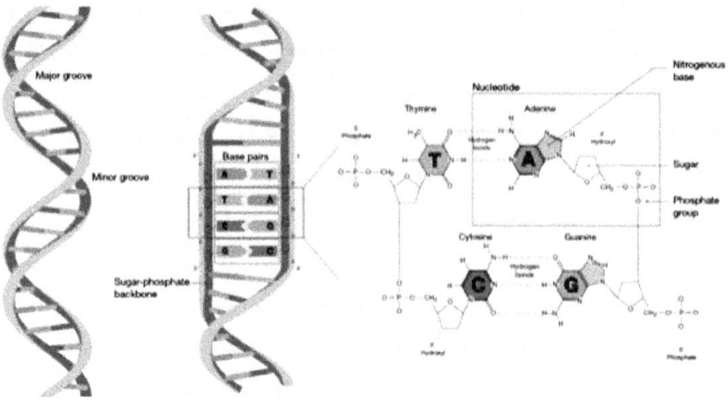

Fig. 2. Deoxyribonucleic acid (DNA)

DNA vaccines against other types of fish pathogens, however, have so far had limited success. The most efficient delivery route at present is IM injection, and suitable delivery strategies for mass vaccination of small fish have yet to be developed. In terms of safety, no adverse effects in the vaccinated fish have been observed to date.

As DNA vaccination is a relatively new technology, various theoretical and long-term safety issues related to the environment and the consumer remain to be fully addressed, although inherently the risks should not be any greater than with the commercial fish vaccines that are currently used. Present classification systems lack clarity in distinguishing DNA-vaccinated animals from genetically modified organisms (GMOs), which could raise issues in terms of licensing and public acceptance of the technology.

The potential benefits of DNA vaccines for farmed fish include improved animal welfare, reduced environmental impacts of aquaculture activities, increased food quality and quantity, and more sustainable production. Testing under commercial production conditions has recently been initiated in Canada and Denmark.

Similar to the way that letters in the alphabet can be arranged to form words, the order of nitrogen bases in a DNA sequence forms genes, which, in the language of the cell, tell cells how to make proteins. The shorthand for this process is that genes "encode" proteins.

But DNA is not the direct template for protein production. To make a protein, the cell makes a copy of the gene, using not DNA but ribonucleic acid, or RNA. RNA shares a similar structure to DNA, except it contains only one strand, rather than two — so it looks like just one half of a ladder. In addition, while RNA has three of the four nitrogen bases in common with DNA, it uses a base called uracil rather than thymine to pair with adenine.

As a cell prepares to build a new protein, its DNA unzips to expose one strand of the gene with the instructions to build said protein. Then, an enzyme zooms in and constructs a new RNA molecule whose sequence mirrors that of the unzipped gene. This RNA copy, called messenger RNA (mRNA), tells the cell's protein-making machinery which amino acids to string together into a protein, according to "Biochemistry(opens in new tab)".

DNA molecules are long - so long, in fact, that they can't fit into cells without the right packaging. To fit inside cells, DNA is coiled tightly to form structures called chromosomes. Each chromosome contains a single DNA molecule, wrapped tightly around spool-like proteins called histones, which provide chromosomes their structure. Humans have 23 pairs of chromosomes, which are found inside each cell's nucleus, the control center of the cell.

Most chromosomes look like microscopic Xs; that said, humans and most other mammals carry a pair of sex chromosomes that can be either X or Y-shaped, according to the National Human Genome Research Institute. In general, females carry two X sex chromosomes in each body cell and males carry one X and one Y. But there is some natural variation in the number of sex chromosomes people carry — sometimes, there may be extra sex chromosomes, or one might be missing, so other patterns, such as X, XXX, XXY and XXYY can also occur.

DNA was first observed by Swiss biochemist Friedrich Miescher in 1869, according to a paper Developmental Biology. Miescher used biochemical methods to isolate DNA - which he then called nuclein - from white blood cells and sperm, and determined that it was very different from protein. (The term "nucleic acid" derives from "nuclein.")

In 1952, chemist Rosalind Franklin, who was working in the lab of biophysicist Maurice Wilkins, used X-ray diffraction - a way of determining the structure of a molecule by the way X-rays bounce off it - to learn that DNA had a helical structure. Franklin documented this structure in what became known as Photo 51.

In 1953, Wilkins showed the photo to biologists James Watson and Francis Crick - without Franklin's knowledge. Armed with the information that DNA was a double helix and previous reports that the bases adenine and thymine occurred in equal amounts within DNA, as did guanine and cytosine, Watson and Crick published a landmark 1953 paper in the journal Nature. In that research, they proposed the iconic double-helix model of DNA as we now know it, with sugar-phosphate sides and rungs made up of A-T and G-C base pairs. They also suggested that, based on their proposed structure, DNA could be copied — and, therefore, passed on.

Watson, Crick and Wilkins were awarded the Nobel Prize in medicine in 1962 "for their discoveries concerning the molecular structure of nucleic acids and its significance for information transfer in living material." Franklin was not included in the award, even though her work was integral to the research.

NA sequencing involves technology that allows researchers to determine the order of bases in a section of DNA. The technology can be used to determine the order of bases in genes, chromosomes or an entire genome. In 2000, researchers completed a "working draft" sequence of the human genome, according to the National Human Genome Research Institute, and finished the project in 2003.

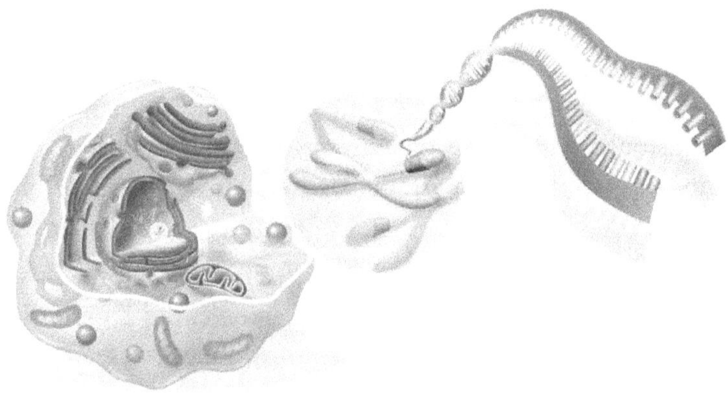

Fig. 3. This illustration shows how DNA is packaged inside chromosomes, which then fit inside a cell's nucleus.

DNA sequencing and genotyping

Researchers followed the protocol to prepare DNA for RAD sequencing. DNA libraries were standardized to a concentration of 10 nM and pooled such that 96 individuals were sequenced per lane of an Illumina HiSeq 4000 at the University of Oregon Genomics Core Facility. The resulting sequences were single-end and had a length of 100 bp.

Researchers used Stacks v. 1.46 to analyse sequencing data and genotype samples. Loci were filtered for missing data, minor allele frequency and potential paralogues. As a final quality control measure and in the absence of a genome assembly for Pacific herring, we aligned loci to the chromosome-level assembly of the Atlantic herring genome using Bowtie2 v. 2.2.6 (–sensitive alignment) and only retained loci that aligned with a mapping quality greater than or equal to 20.

Individual herring were assessed for missing data and removed if they had more than 10% missing genotypes. We screened individuals for intraspecific DNA contamination by calculating individual multilocus heterozygosity (H_I) and comparing it to a set of clean reference samples. Individual samples were removed from the dataset if they had H_I values higher than clean reference samples ($H_I > 0.32$).

Current efforts to develop vaccines, particularly for aquacultured species, have turned largely to biotechnology because it provides the means to inexpensively produce sufficient quantities of the immunoprotective antigen. These efforts have resulted in several prototype vaccines for fish and the publication of a large number of articles on the subject.

However, there are only a few recombinant DNA-based vaccines for aquaculture in the licensing pipeline. Continued funding of research on recombinant DNA vaccines comes from the recognition by industry and government funding agencies that this research can lead to an increased understanding of the mechanisms in protective immunity. This is especially important for fish and shellfish species since our knowledge of the immune mechanisms in these animals is pitifully meagre.

This presentation discusses the relative merits of the different recombinant DNA technologies that have been used to produce viral vaccines for fish and the promising approaches that are under consideration to increase the efficacy of these vaccines. There are many approaches to antigen production by recombinant DNA techniques including:

i. the preparation of purified antigenic proteins produced from the cloned viral genes in a variety of vector/host expression systems,
ii. chemical synthesis or the use of fusion vectors to produce peptides corresponding to known epitopes,
iii. defined attenuations, i.e. specific genetic alterations, of live virus vaccines,
iv. the use of live bacterial or viral vectors to deliver resistance genes or viral antigens,
v. anti-idiotype antibodies, and
vi. DNA vaccines where purified plasmid DNA expressing the pathogen gene under a eucaryotic promoter is injected.

All of these technologies have been used more or less successfully in the development of vaccines for aquacultural species. However, the requirements for safety, effectiveness, ease of application and low cost/dose restrict their commercial development for aquaculture. The ideal viral vaccine for aquaculture must be effective in preventing death, be inexpensive to produce and license, provide immunity of long duration, and be easily administered.

In addition, these vaccines must not only provide protection against the lethal effects of virus infection but prevent the formation of virus persistence. This is especially true for infectious haematopoietic necrosis virus (IHNV) which has been shown to persist in survivors in the presence of high antibody levels. Since resolution of virus persistence is thought to be correlate with cell-mediated immunity, vaccines designed to augment the cell-mediated immunity must be developed for fish.

Approaches that are being considered include the use of cytokines in combination with subunit vaccines and the use of specific MHC-I inducer adjuvants with the vaccine. The "tailoring" of vaccine immunogenicity using different combinations of antigen and adjuvant will be presented.

RIBOSE NUCLEIC ACID

Introduction

Nucleic acids, and DNA in particular, are key macromolecules for the continuity of life. DNA bears the hereditary information that's passed on from parents to children, providing instructions for how (and when) to make the many proteins needed to build and maintain functioning cells, tissues, and organisms.

How DNA carries this information, and how it is put into action by cells and organisms, is complex, fascinating, and fairly mind-blowing, and we'll explore it in more detail in the section on molecular biology. Here, we'll just take a quick look at nucleic acids from the macromolecule perspective.

Roles of DNA and RNA in cells

Nucleic acids, macromolecules made out of units called nucleotides, come in two naturally occurring varieties: deoxyribonucleic acid (DNA) and ribonucleic acid (RNA). DNA is the genetic material found in living organisms, all the way from single-celled bacteria to multicellular mammals like you and me. Some viruses use RNA, not DNA, as their genetic material, but aren't technically considered to be alive (since they cannot reproduce without help from a host).

DNA in cells

In eukaryotes, such as plants and animals, DNA is found in the **nucleus**, a specialized, membrane-bound vault in the cell, as well as in certain other types of organelles (such as mitochondria and the chloroplasts of plants). In prokaryotes, such as bacteria, the DNA is not enclosed in a membranous envelope, although it's located in a specialized cell region called the nucleoid.

In eukaryotes, DNA is typically broken up into a number of very long, linear pieces called **chromosomes**, while in prokaryotes such as bacteria, chromosomes are much smaller and often circular (ring-shaped). A chromosome may contain tens of thousands of **genes**, each providing instructions on how to make a particular product needed by the cell.

From DNA to RNA to proteins

Many genes encode protein products, meaning that they specify the sequence of amino acids used to build a particular protein. Before this information can be used for protein synthesis, however, an RNA copy (transcript) of the gene must first be made. This type of RNA is called a messenger RNA (mRNA), as it serves as a messenger between DNA and the ribosomes, molecular machines that read mRNA sequences and use them to build proteins. This progression from DNA to RNA to protein is called the "central dogma" of molecular biology.

Importantly, not all genes encode protein products. For instance, some genes specify ribosomal RNAs (rRNAs), which serve as structural components of ribosomes, or transfer RNAs (tRNAs), cloverleaf-shaped RNA molecules that bring amino acids to the ribosome for protein synthesis. Still other RNA molecules, such as tiny microRNAs (miRNAs), act as regulators of other genes, and new types of non-protein-coding RNAs are being discovered all the time.

Nucleotides

DNA and RNA are polymers (in the case of DNA, often very long polymers), and are made up of monomers known as nucleotides. When these monomers combine, the resulting chain is called a polynucleotide (poly- = "many").

Each nucleotide is made up of three parts: a nitrogen-containing ring structure called a nitrogenous base, a five-carbon sugar, and at least one phosphate group. The sugar molecule has a central position in the nucleotide, with the base attached to one of its carbons and the phosphate group (or groups) attached to another.

Nitrogenous bases

The nitrogenous bases of nucleotides are organic (carbon-based) molecules made up of nitrogen-containing ring structures.

Each nucleotide in DNA contains one of four possible nitrogenous bases: adenine (A), guanine (G) cytosine (C), and thymine (T). Adenine and guanine are **purines**, meaning that their structures contain two fused carbon-nitrogen rings.

Cytosine and thymine, in contrast, are pyrimidines and have a single carbon-nitrogen ring. RNA nucleotides may also bear adenine, guanine and cytosine bases, but instead of thymine they have another pyrimidine base called uracil (U). As shown in the Fig. 4, each base has a unique structure, with its own set of functional groups attached to the ring structure.

In molecular biology shorthand, the nitrogenous bases are often just referred to by their one-letter symbols, A, T, G, C, and U. DNA contains A, T, G, and C, while RNA contains A, U, G, and C (that is, U is swapped in for T).

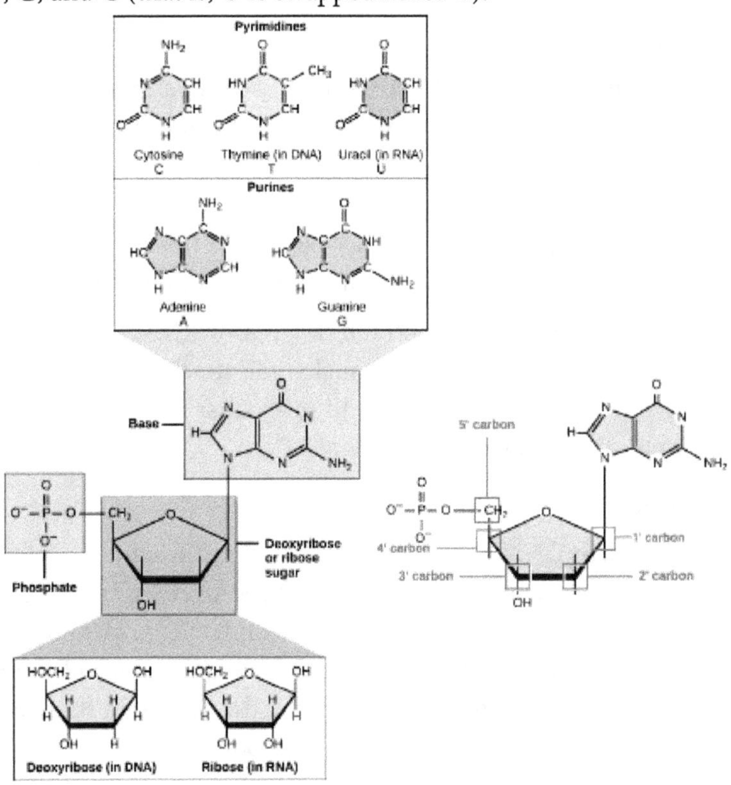

Fig. 4. Nitrogenous bases

Sugars

In addition to having slightly different sets of bases, DNA and RNA nucleotides also have slightly different sugars. The five-carbon sugar in DNA is called **deoxyribose**, while in RNA, the sugar is ribose. These two are very similar in structure, with just one difference: the second carbon of ribose bears a hydroxyl group, while the equivalent carbon of deoxyribose has a hydrogen instead. The carbon atoms of a nucleotide's sugar molecule are numbered as 12, 22, 32, 42, and 52 (12 is read as "one prime"), as shown in the figure above. In a nucleotide, the sugar occupies a central position, with the base attached to its 12 carbon and the phosphate group (or groups) attached to its 52 carbon.

Phosphate

Nucleotides may have a single phosphate group, or a chain of up to three phosphate groups, attached to the 5' carbon of the sugar. Some chemistry sources use the term "nucleotide" only for the single-phosphate case, but in molecular biology, the broader definition is generally accepted.

In a cell, a nucleotide about to be added to the end of a polynucleotide chain will bear a series of three phosphate groups. When the nucleotide joins the growing DNA or RNA chain, it loses two phosphate groups. So, in a chain of DNA or RNA, each nucleotide has just one phosphate group.

Polynucleotide chains

A consequence of the structure of nucleotides is that a polynucleotide chain has directionality – that is, it has two ends that are different from each other. At the 5' end, or beginning, of the chain, the 5' phosphate group of the first nucleotide in the chain sticks out. At the other end, called the **3' end, the 3'** hydroxyl of the last nucleotide added to the chain is exposed. DNA sequences are usually written in the 5' to 3' direction, meaning that the nucleotide at the 5' end comes first and the nucleotide at the 3' end comes last.

As new nucleotides are added to a strand of DNA or RNA, the strand grows at its 3' end, with the 5' phosphate of an incoming nucleotide attaching to the hydroxyl group at the 3' end of the chain. This makes a chain with each sugar joined to its neighbors by a set of bonds called a phosphodiester linkage.

Properties of DNA

Deoxyribonucleic acid, or DNA, chains are typically found in a double helix, a structure in which two matching (complementary) chains are stuck together, as shown in the diagram at left. The sugars and phosphates lie on the outside of the helix, forming the backbone of the DNA; this portion of the molecule is sometimes called the sugar-phosphat**e** backbone. The nitrogenous bases extend into the interior, like the steps of a staircase, in pairs; the bases of a pair are bound to each other by hydrogen bonds.

The two strands of the helix run in opposite directions, meaning that the 52 end of one strand is paired up with the 32 end of its matching strand. (This is referred to as antiparallel orientation and is important for the copying of DNA.)

So, can any two bases decide to get together and form a pair in the double helix? The answer is a definite no. Because of the sizes and functional groups of the bases, base pairing is highly specific: A can only pair with T, and G can only pair with C, as shown below. This means that the two strands of a DNA double helix have a very predictable relationship to each other.

For instance, if you know that the sequence of one strand is 5'-AATTGGCC-3', the complementary strand must have the sequence 3'-TTAACCGG-5'. This allows each base to match up with its partner (Fig. 5).

When two DNA sequences match in this way, such that they can stick to each other in an antiparallel fashion and form a helix, they are said to be complementary (Fig. 6).

Properties of RNA

Ribonucleic acid (RNA), unlike DNA, is usually single-stranded. A nucleotide in an RNA chain will contain ribose (the five-carbon sugar), one of the four nitrogenous bases (A, U, G, or C), and a phosphate group. Here, we'll take a look at four major types of RNA: messenger RNA (mRNA), ribosomal RNA (rRNA), transfer RNA (tRNA), and regulatory RNAs.

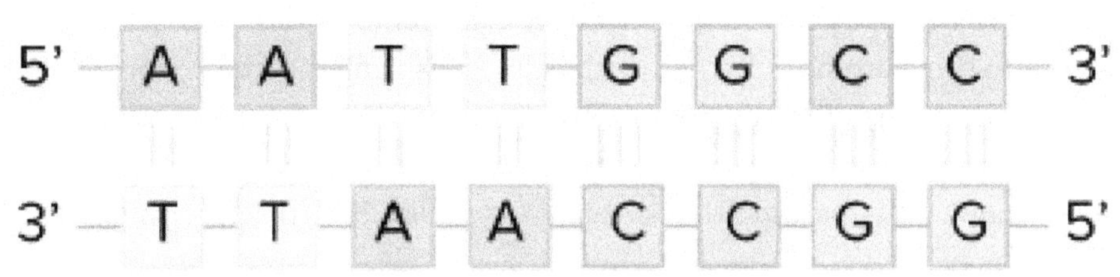

Fig. 5

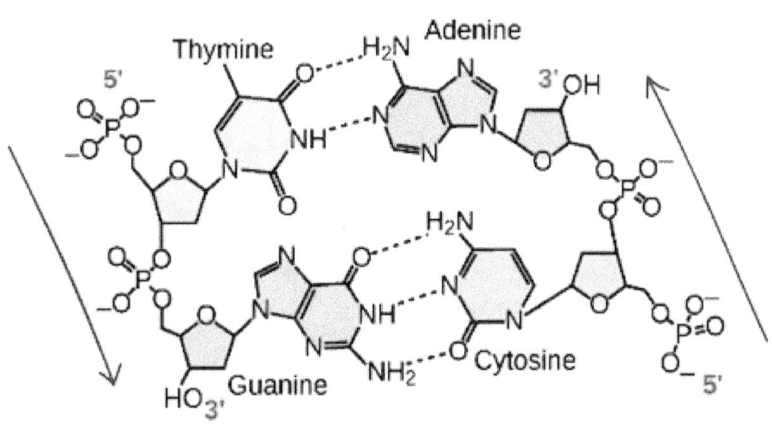

Fig. 6

Messenger RNA (mRNA)

Messenger RNA (mRNA) is an intermediate between a protein-coding gene and its protein product. If a cell needs to make a particular protein, the gene encoding the protein will be turned "on," meaning an RNA-polymerizing enzyme will come and make an RNA copy, or transcript, of the gene's DNA sequence. The transcript carries the same information as the DNA sequence of its gene. However, in the RNA molecule, the base T is replaced with U. For instance, if a DNA coding strand has the sequence 5'-AATTGCGC-3', the sequence of the corresponding RNA will be 5'-AAUUGCGC-3'.

Once an mRNA has been produced, it will associate with a ribosome, a molecular machine that specializes in assembling proteins out of amino acids. The ribosome uses the information in the mRNA to make a protein of a specific sequence, "reading out" the mRNA's nucleotides in groups of three (called **codons**) and adding a particular amino acid for each codon.

Ribosomal RNA (rRNA) and transfer RNA (tRNA)

Ribosomal RNA (rRNA) is a major component of ribosomes, where it helps mRNA bind in the right spot so its sequence information can be read out. Some rRNAs also act as enzymes, meaning that they help accelerate (catalyze) chemical reactions – in this case, the formation of bonds that link amino acids to form a protein. RNAs that act as enzymes are known as **ribozymes**.

Transfer RNAs (tRNAs) are also involved in protein synthesis, but their job is to act as carriers – to bring amino acids to the ribosome, ensuring that the amino acid added to the chain is the one specified by the mRNA. Transfer RNAs consist of a single strand of RNA, but this strand has complementary segments that stick together to make double-stranded regions. This base-pairing creates a complex 3D structure important to the function of the molecule.

Regulatory RNA (miRNAs and siRNAs)

Some types of non-coding RNAs (RNAs that do not encode proteins) help regulate the expression of other genes. Such RNAs may be called regulatory RNAs.

For example, microRNAs (miRNAs) and small interfering RNAs siRNAs are small regulatory RNA molecules about 22 nucleotides long. They bind to specific mRNA molecules (with partly or fully complementary sequences) and reduce their stability or interfere with their translation, providing a way for the cell to decrease or fine-tune levels of these mRNAs.

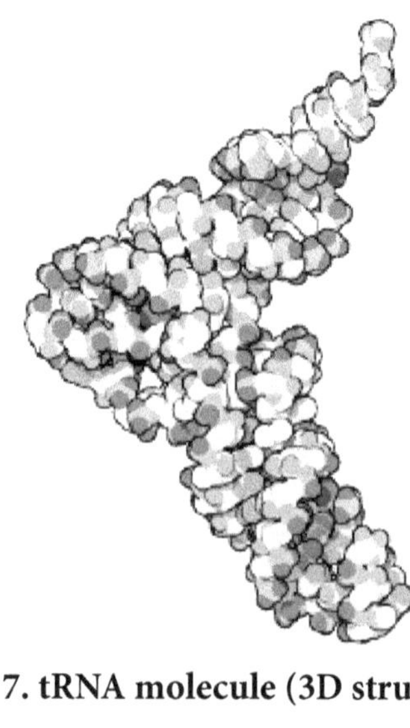

Fig. 7. tRNA molecule (3D structure)

These are just some examples out of many types of noncoding and regulatory RNAs. Scientists are still discovering new varieties of noncoding RNA.

GENETIC CODE

The genetic code can be defined as the set of certain rules using which the living cells translate the information encoded within genetic material (DNA or mRNA sequences). The ***ribosomes*** are responsible to accomplish the process of translation. They link the amino acids in an mRNA-specified (messenger RNA) order using tRNA (transfer RNA) molecules to carry amino acids and to read the mRNA three nucleotides at a time.

the sequence of nucleotides in deoxyribonucleic acid (DNA) and ribonucleic acid (RNA) that determines the amino acid sequence of proteins. Though the linear sequence of nucleotides in DNA contains the information for protein sequences, proteins are not made directly from DNA. Instead, a messenger RNA (mRNA) molecule is synthesized from the DNA and directs the formation of the protein.

RNA is composed of four nucleotides: adenine (A), guanine (G), cytosine (C), and uracil (U). Three adjacent nucleotides constitute a unit known as the codon, which codes for an amino acid. For example, the sequence AUG is a codon that specifies the amino acid methionine. There are 64 possible codons, three of which do not code for amino acids but indicate the end of a protein. The remaining 61 codons specify the 20 amino acids that make up proteins.

The AUG codon, in addition to coding for methionine, is found at the beginning of every mRNA and indicates the start of a protein. Methionine and tryptophan are the only two amino acids that are coded for by just a single codon (AUG and UGG, respectively). The other 18 amino acids are coded for by two to six codons. Because most of the 20 amino acids are coded for by more than one codon, the code is called degenerate.

The genetic code, once thought to be identical in all forms of life, has been found to diverge slightly in certain organisms and in the mitochondria of some eukaryotes. Nevertheless, these differences are rare, and the genetic code is identical in almost all species, with the same codons specifying the same amino acids. The deciphering of the genetic code was accomplished by American biochemists Marshall W. Nirenberg, Robert W. Holley, and Har Gobind Khorana in the early 1960s.

Genetic Code Table

The complete set of relationships among ***amino acids*** and ***codons*** is said to be a genetic code which is often summarized in a table (Fig. 8).

It can be seen that many amino acids are shown in the table by more than one codon. For example, there are six ways to write leucine in mRNA language.

Note: *A **codon** is a sequence of three nucleotides which together form a unit of genetic code in a DNA or RNA molecule.*

A key point of the genetic code is its universal nature. This indicates that virtually all species with minor exceptions use the genetic code for protein synthesis.

In other words, genetic code is defined as the nucleotide sequence of the base on DNA which is translated into a sequence of amino acids of the protein to be synthesized.

Properties of Genetic Code

- Triplet code
- Non-ambiguous and Universal
- Degenerate code
- Nonoverlapping code

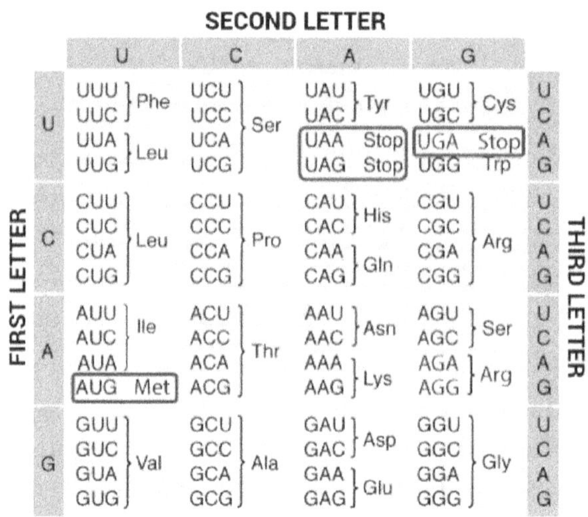

Fig. 8 Genetic Code Table

- Commaless
- Start and Stop Codons
- Polarity

These properties of genetic code are explained below.

Triplet code

A codon or a code word is defined as a group of bases that specify an amino acid. There is strong evidence, which proves that a sequence of three nucleotides codes for an amino acid in the protein, i.e., the code is a ***triplet***.

The four bases of nucleotide i.e, (A, G, C, and U) are used to produce three-base codons. The 64 codons involve sense codons (that specify amino acids). Hence, there are 64 codons for 20 amino acids since every codon for one amino acid means that there exist more than code for the same amino acid.

Commaless code

No room for punctuation in between which indicates that every codon is adjacent to the previous one without any nucleotides between them.

Nonoverlapping code

The code is read sequentially in a group of three and a nucleotide which becomes a part of triplet never becomes part of the next triplet.

For example

5'-UCU-3' codes for Serine

5'-AUG-3' codes for methionine

Polarity

Each triplet is read from 5' '! 3' direction and the beginning base is 5' followed by the base in the middle then the last base which is 3'. This implies that the codons have a fixed polarity and if the codon is read in the reverse direction, the base sequence of the codon would reverse and would specify two different proteins.

Degenerate code

Every amino acid except tryptophan (UGG) and methionine (AUG) is coded by various codons, i.e, a few codons are synonyms and this aspect is known as the degeneracy of genetic code. For instance, UGA codes for tryptophan in yeast mitochondria.

Start and Stop Codons

Generally, *AUG codon* is the initiating or start codon. The polypeptide chain starts either with eukaryotes (methionine) or prokaryotes (N- formylmethionine).

On the other hand, *UAG, UAA* and *UGA* are called as termination codons or stop codons. These are not read by any tRNA molecules and they never code for any amino acids.

Non-ambiguous and Universal

The genetic code is non-ambiguous which means a specific codon will only code for a particular amino acid. Also, the same genetic code is seen valid for all the organisms i.e. they are universal.

Exceptions to the Code

The genetic code is universal since similar codons are assigned to identical amino acids along with similar START and STOP signals in the majority of genes in microorganisms and plants. However, a few exceptions have been discovered and most of these include assigning one or two of the STOP codons to an amino acid.

Apart from this, both the codons ***GUG*** and ***AUG*** may code for methionine as a starting codon, although GUG is meant for valine. This breaks the property of non-ambiguousness. Thus, it can be said that few codes often differs from the universal code or non-ambiguous code.

Protein structure

The effects of variants on protein structure can vary dramatically depending on the type of protein and the extent of variation.

Take, for instance, a large, multi-domain protein with multiple variants. One variant could involve the deletion of a large region of protein sequence, corresponding to an individual domain. If the core function of these domains are independent, the function of the remaining domains may not be affected at all. Another variant of this protein, however, may involve the changing of only a single amino acid in the protein sequence.

Despite the relative insignificance of this change to the overall protein sequence, if this residue is key to the function of the protein (for instance at the active site of an enzyme) then this could completely eradicate protein function, or modify it in some way, for instance to change the specificity of an enzyme.

Proteins are the end products of the decoding process that starts with the information in cellular DNA. As workhorses of the cell, proteins compose structural and motor elements in the cell, and they serve as the catalysts for virtually every biochemical reaction that occurs in living things. This incredible array of functions derives from a startlingly simple code that specifies a hugely diverse set of structures.

In fact, each gene in cellular DNA contains the code for a unique protein structure. Not only are these proteins assembled with different amino acid sequences, but they also are held together by different bonds and folded into a variety of three-dimensional structures. The folded shape, or conformation, depends directly on the linear amino acid sequence of the protein.

What Are Proteins Made Of?

The building blocks of proteins are amino acids, which are small organic molecules that consist of an alpha (central) carbon atom linked to an amino group, a carboxyl group, a hydrogen atom, and a variable component called a side chain. Within a protein, multiple amino acids are linked together by peptide bonds, thereby forming a long chain. Peptide bonds are formed by a biochemical reaction that extracts a water molecule as it joins the amino group of one amino acid to the carboxyl group of a neighboring amino acid. The linear sequence of amino acids within a protein is considered the primary structure of the protein.

Proteins are built from a set of only twenty amino acids, each of which has a unique side chain. The side chains of amino acids have different chemistries. The largest group of amino acids have nonpolar side chains. Several other amino acids have side chains with positive or negative charges, while others have polar but uncharged side chains. The chemistry of amino acid side chains is critical to protein structure because these side chains can bond with one another to hold a length of protein in a certain shape or conformation. Charged amino acid side chains can form ionic bonds, and polar amino acids are capable of forming hydrogen bonds. Hydrophobic side chains interact with each other via weak van der Waals interactions. The vast majority of bonds formed by these side chains are noncovalent. In fact, cysteines are the only amino acids capable of forming covalent bonds, which they do with their particular side chains. Because of side chain interactions, the sequence and location of amino acids in a particular protein guides where the bends and folds occur in that protein (Fig. 9).

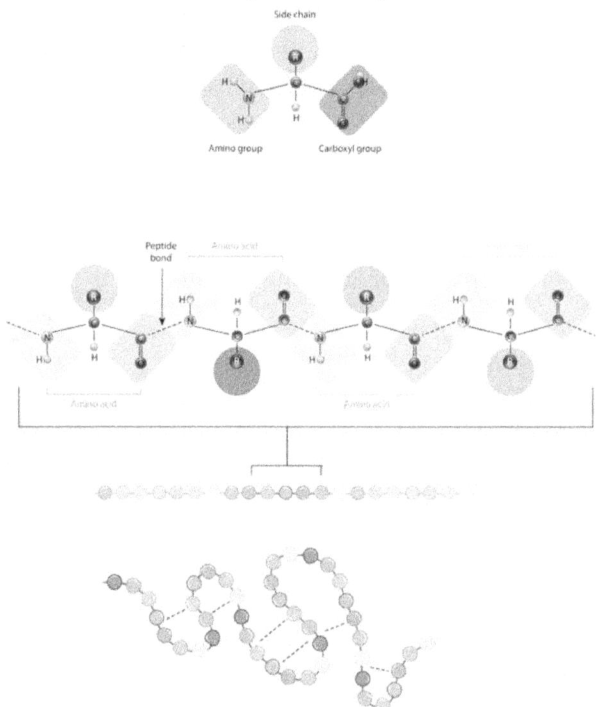

Fig. 9: The relationship between amino acid side chains and protein conformation

The primary structure of a protein — its amino acid sequence — drives the folding and intramolecular bonding of the linear amino acid chain, which ultimately determines the protein's unique three-dimensional shape.

Hydrogen bonding between amino groups and carboxyl groups in neighboring regions of the protein chain sometimes causes certain patterns of folding to occur. Known as alpha helices and beta sheets, these stable folding patterns make up the secondary structure of a protein. Most proteins contain multiple helices and sheets, in addition to other less common patterns (Fig. 10).

The ensemble of formations and folds in a single linear chain of amino acids - sometimes called a polypeptide - constitutes the tertiary structure of a protein. Finally, the quaternary structure of a protein refers to those macromolecules with multiple polypeptide chains or subunits.

The final shape adopted by a newly synthesized protein is typically the most energetically favorable one. As proteins fold, they test a variety of conformations before reaching their final form, which is unique and compact. Folded proteins are stabilized by thousands of noncovalent bonds between amino acids. In addition, chemical forces between a protein and its immediate environment contribute to protein shape and stability.

For example, the proteins that are dissolved in the cell cytoplasm have hydrophilic (water-loving) chemical groups on their surfaces, whereas their hydrophobic (water-averse) elements tend to be tucked inside. In contrast, the proteins that are inserted into the cell membranes display some hydrophobic chemical groups on their surface, specifically in those regions where the protein surface is exposed to membrane lipids. It is important to note, however, that fully folded proteins are not frozen into shape. Rather, the atoms within these proteins remain capable of making small movements.

Even though proteins are considered macromolecules, they are too small to visualize, even with a microscope. So, scientists must use indirect methods to figure out what they look like and how they are folded. The most common method used to study protein structures is X-ray crystallography. With this method, solid crystals of purified protein are placed in an X-ray beam, and the pattern of deflected X rays is used to predict the positions of the thousands of atoms within the protein crystal.

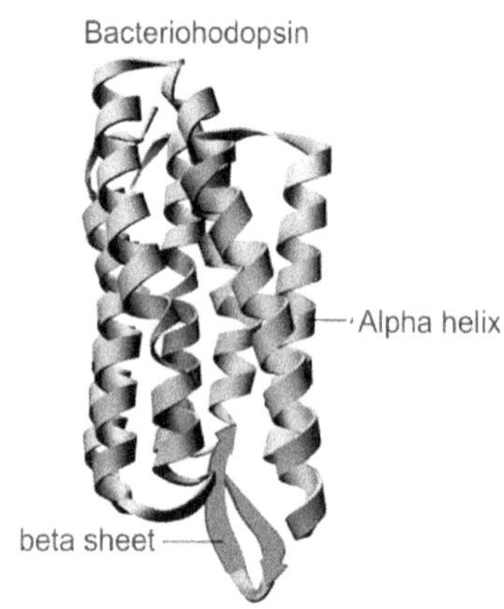

Fig. 10: The structure of the protein bacteriorhodopsin

How Do Proteins Arrive at Their Final Shapes?

In theory, once their constituent amino acids are strung together, proteins attain their final shapes without any energy input. In reality, however, the cytoplasm is a crowded place, filled with many other macromolecules capable of interacting with a partially folded protein. Inappropriate associations with nearby proteins can interfere with proper folding and cause large aggregates of proteins to form in cells. Cells therefore rely on so-called chaperone proteins to prevent these inappropriate associations with unintended folding partners.

Chaperone proteins surround a protein during the folding process, sequestering the protein until folding is complete. For example, in bacteria, multiple molecules of the chaperone GroEL form a hollow chamber around proteins that are in the process of folding. Molecules of a second chaperone, GroES, then form a lid over the chamber. Eukaryotes use different families of chaperone proteins, although they function in similar ways.

Chaperone proteins are abundant in cells. These chaperones use energy from ATP to bind and release polypeptides as they go through the folding process. Chaperones also assist in the refolding of proteins in cells. Folded proteins are actually fragile structures, which can easily denature, or unfold. Although many thousands of bonds hold proteins together, most of the bonds are noncovalent and fairly weak.

Even under normal circumstances, a portion of all cellular proteins are unfolded. Increasing body temperature by only a few degrees can significantly increase the rate of unfolding. When this happens, repairing existing proteins using chaperones is much more efficient than synthesizing new ones. Interestingly, cells synthesize additional chaperone proteins in response to "heat shock."

What Are Protein Families?

All proteins bind to other molecules in order to complete their tasks, and the precise function of a protein depends on the way its exposed surfaces interact with those molecules. Proteins with related shapes tend to interact with certain molecules in similar ways, and these proteins are therefore considered a **protein family**. The proteins within a particular family tend to perform similar functions within the cell.

Proteins from the same family also often have long stretches of similar amino acid sequences within their primary structure. These stretches have been conserved through evolution and are vital to the catalytic function of the protein.

For example, cell receptor proteins contain different amino acid sequences at their binding sites, which receive chemical signals from outside the cell, but they are more similar in amino acid sequences that interact with common intracellular signaling proteins. Protein families may have many members, and they likely evolved from ancient gene duplications. These duplications led to modifications of protein functions and expanded the functional repertoire of organisms over time.

Classification of Proteins

Based on the molecular shape, proteins can be classified into two types.

Fibrous Proteins:

When the polypeptide chains run ***parallel*** and are held together by hydrogen and disulfide bonds, then the fiber-like structure is formed. Such proteins are generally insoluble in water. These are water-insoluble proteins.

Example – keratin (present in hair, wool, and silk) and myosin (present in muscles), etc.

Globular Proteins

This structure results when the chains of polypeptides ***coil around*** to give a spherical shape. These are usually soluble in water.

Example – Insulin and albumins are common examples of globular proteins.

Levels of Protein Structure

Primary Structure of Protein

- The Primary structure of proteins is the exact ordering of amino acids forming their chains.
- The exact sequence of the proteins is very important as it determines the final fold and therefore the function of the protein.
- The number of polypeptide chains together form proteins. These chains have amino acids arranged in a particular sequence which is characteristic of the specific protein. Any change in the sequence changes the entire protein.

Figure 11 represents the primary protein structure (an amino acid chain). As you might expect, the amino acid sequence within the polypeptide chain is crucial for the protein's proper functioning. This sequence is encrypted in the DNA genetic code. If mutation is present in the DNA and the amino acid sequence is changed, the protein function may be affected.

Fig. 11. Primary Structure of Protein

The protein 's primary structure is the amino acid sequence in its polypeptide chain. If proteins were popcorn stringers designed to decorate a Christmas tree, a protein 's primary structure is the sequence in which various shapes and varieties of popped maize are strung together.

Covalent, peptide bonds which connect the amino acids together maintain the primary structure of a protein.

All documented genetic disorders, such as cystic fibrosis, sickle cell anemia, albinism, etc., are caused by mutations resulting in alterations in the primary protein structures, which in turn lead to alterations in the secondary , tertiary and probably quarterly structure.

Amino acids are small organic molecules consisting of a chiral carbon with four substituents. Of those only the fourth the side chain is different among amino acids.

Secondary Structure of Protein

Secondary structure of protein refers to local folded structures that form within a polypeptide due to interactions between atoms of the backbone.

- The proteins do not exist in just simple chains of polypeptides.
- These polypeptide chains usually fold due to the interaction between the amine and carboxyl group of the peptide link.
- The structure refers to the shape in which a long polypeptide chain can exist.
- They are found to exist in two different types of structures α–helix and β–pleated sheet structures.
- This structure arises due to the regular folding of the backbone of the polypeptide chain due to hydrogen bonding between -CO group and -NH groups of the peptide bond.
- However, segments of the protein chain may acquire their own local fold, which is much simpler and usually takes the shape of a spiral an extended shape or a loop. These local folds are termed secondary elements and form the proteins secondary structure.

(a) a-Helix:

α–Helix is one of the most common ways in which a polypeptide chain forms all possible hydrogen bonds by twisting into a right-handed screw with the -NH group of each amino acid residue hydrogen-bonded to the -CO of the adjacent turn of the helix. The polypeptide chains twisted into a right-handed screw.

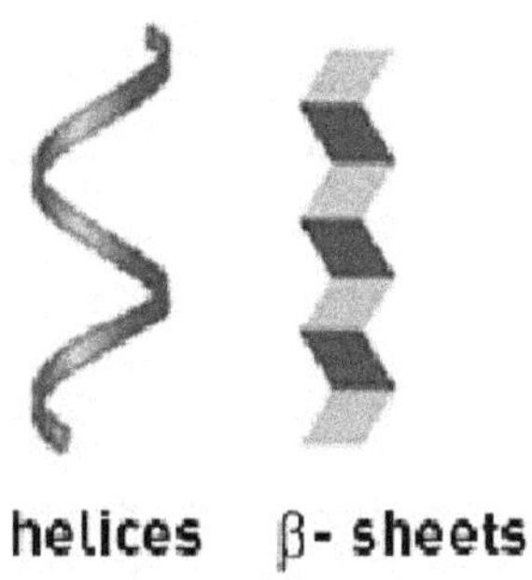

Fig. 12. Secondary Structure of Protein

(b) b-pleated sheet:

In this arrangement, the polypeptide chains are stretched out beside one another and then bonded by intermolecular H-bonds. In this structure, all peptide chains are stretched out to nearly maximum extension and then laid side by side which is held together by intermolecular hydrogen bonds. The structure resembles the pleated folds of drapery and therefore is known as β–pleated sheet

Tertiary Structure of Protein

- This structure arises from further folding of the secondary structure of the protein.
- H-bonds, electrostatic forces, disulphide linkages, and Vander Waals forces stabilize this structure.
- The tertiary structure of proteins represents overall folding of the polypeptide chains, further folding of the secondary structure.

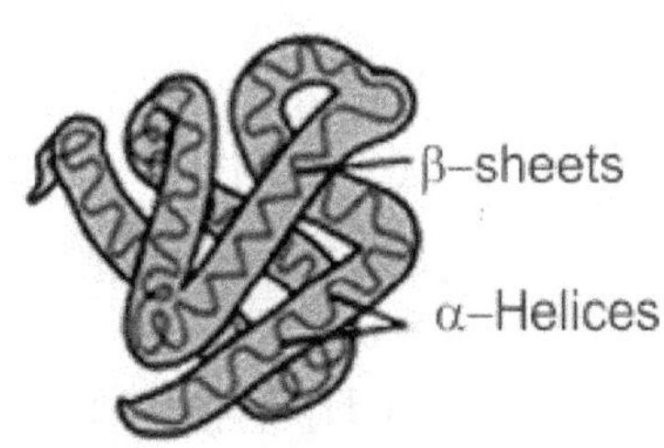

Fig. 13. Tertiary Structure of Protein

- It gives rise to two major molecular shapes called fibrous and globular.
- The main forces which stabilize the secondary and tertiary structures of proteins are hydrogen bonds, disulphide linkages, van der Waals and electrostatic forces of attraction.

Quaternary Structure of Protein

The spatial arrangement of various tertiary structures gives rise to the quaternary structure. Some of the proteins are composed of two or more polypeptide chains referred to as sub-units. The spatial arrangement of these subunits with respect to each other is known as quaternary structure.

The exact amino acid sequence of each protein drives it to fold into its own unique and biologically active three-dimensional fold also known as the tertiary structure. Proteins consist of different combinations of secondary elements some of which are simple whereas others are more complex. Parts of the protein chain, which have their own three-dimensional fold and can be attributed to some function are called ***"domains"***. These are considered today as the evolutionary and functional building blocks of proteins.

Many proteins, most of which are enzymes contain organic or elemental components needed for their activity and stability. Thus the study of protein evolution not only gives structural insight but also connects proteins of quite different parts of the metabolism.

Rules of Protein Structure

- The type determines the function of a protein.
- A protein's shape is determined by its primary structure (the amino acid sequence).
- The amino acid sequence within a protein is determined by the encoding sequence of nucleotides in the gene (DNA).

Summary of Protein Structure

Linderstrom-Lang in particular first suggested a hierarchy of protein structure with four levels: central, secondary, tertiary , and quaternary. You are already familiar with this hierarchy, because the most useful starting point for teaching basic protein structure is this structural grouping.

- The primary structure of protein is the hierarchy's basic level, and is the particular linear sequence of amino acids comprising one polypeptide chain.
- Secondary structure is the next level up from the primary structure, and is the regular folding of regions into specific structural patterns within one polypeptide chain. Hydrogen bonds between the carbonyl oxygen and the peptide bond amide hydrogen are normally held together by secondary structures.
- Tertiary structure is the next level up from the secondary structure, and is the particular three-dimensional arrangement of all the amino acids in a single

polypeptide chain. This structure is usually conformational, native, and active, and is held together by multiple noncovalent interactions.

- Quaternary structure is the next 'step up' between two or more polypeptide chains from the tertiary structure and is the specific spatial arrangement and interactions.

Fig. 13. Tertiary Structure of Protein

Protein Composition of fish Muscle

Table 1 Protein Composition of Fish, Shellfish and Meat

Type of Meat		% of total proteins	
	Sarcoplasmicproteins	Myofibrillar proteins	Stromaprotein
Cod	21	76	3
Carp	23-25	70-72	5
Flatfish	18-24	72-79	3
Mackerel	32	58	7
Lactarius sp.	26	77	7
Prawn (Penaeus indicus)	29.85	65-70	70
Meat	16-28	39-68	16-28

The protein of fish muscle tissue can be divided into the following three groups based on solubility to sarcoplasmic, myofibrillar, and stroma protein. The sarcoplasmic protein is located in the muscle plasma. The myofibrillar protein is present in the myofibrils and the stroma protein is present in the connective tissue. A similar composition is found in fish and livestock meat. The livestock meat contains more stroma than fish meat. The protein compositions of fish, shellfish and animal muscle tissue are presented in Table 1.

CHROMOSOMES

In the nucleus of each cell, the DNA molecule is packaged into thread-like structures called chromosomes. Each chromosome is made up of DNA tightly coiled many times around proteins called histones that support its structure.

Chromosomes are not visible in the cell's nucleus—not even under a microscope—when the cell is not dividing. However, the DNA that makes up chromosomes becomes more tightly packed during cell division and is then visible under a microscope. Most of what researchers know about chromosomes was learned by observing chromosomes during cell division.

Each chromosome has a constriction point called the centromere, which divides the chromosome into two sections, or "arms." The short arm of the chromosome is labeled the "p arm." The long arm of the chromosome is labeled the "q arm." The location of the centromere on each chromosome gives the chromosome its characteristic shape, and can be used to help describe the location of specific genes.

The unique structure of chromosomes keeps DNA tightly wrapped around spool-like proteins, called histones. Without such packaging, DNA molecules would be too long to fit inside cells. For example, if all of the DNA molecules in a single human cell were unwound from their histones and placed end-to-end, they would stretch 6 feet.

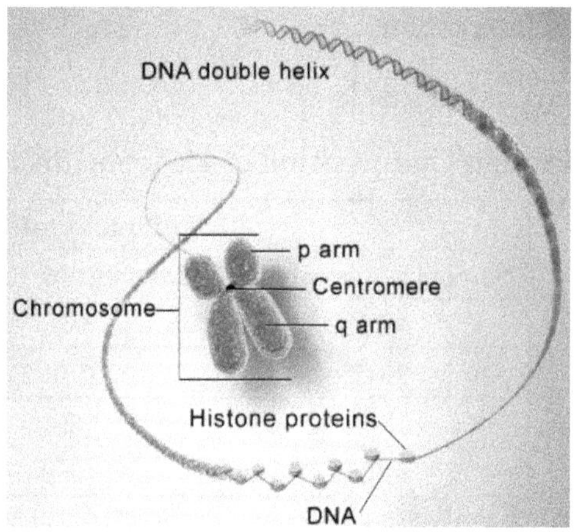

Fig. 15. DNA and histone proteins are packaged into structures called chromosomes.

For an organism to grow and function properly, cells must constantly divide to produce new cells to replace old, worn-out cells. During cell division, it is essential that DNA remains intact and evenly distributed among cells. Chromosomes are a key part of the process that ensures DNA is accurately copied and distributed in the vast majority of cell divisions. Still, mistakes do occur on rare occasions.

Changes in the number or structure of chromosomes in new cells may lead to serious problems. For example, in humans, one type of leukemia and some other cancers are caused by defective chromosomes made up of joined pieces of broken chromosomes.

It is also crucial that reproductive cells, such as eggs and sperm, contain the right number of chromosomes and that those chromosomes have the correct structure. If not, the resulting offspring may fail to develop properly. For example, people with Down syndrome have three copies of chromosome 21, instead of the two copies found in other people.

Chromosomes vary in number and shape among living things. Most bacteria have one or two circular chromosomes. Humans, along with other animals and plants, have linear chromosomes that are arranged in pairs within the nucleus of the cell.

The only human cells that do not contain pairs of chromosomes are reproductive cells, or gametes, which carry just one copy of each chromosome. When two reproductive cells unite, they become a single cell that contains two copies of each chromosome. This cell then divides and its successors divide numerous times, eventually producing a mature individual with a full set of paired chromosomes in virtually all of its cells.

Besides the linear chromosomes found in the nucleus, the cells of humans and other complex organisms carry a much smaller type of chromosome similar to those seen in bacteria. This circular chromosome is found in mitochondria, which are structures located outside the nucleus that serve as the cell's powerhouses.

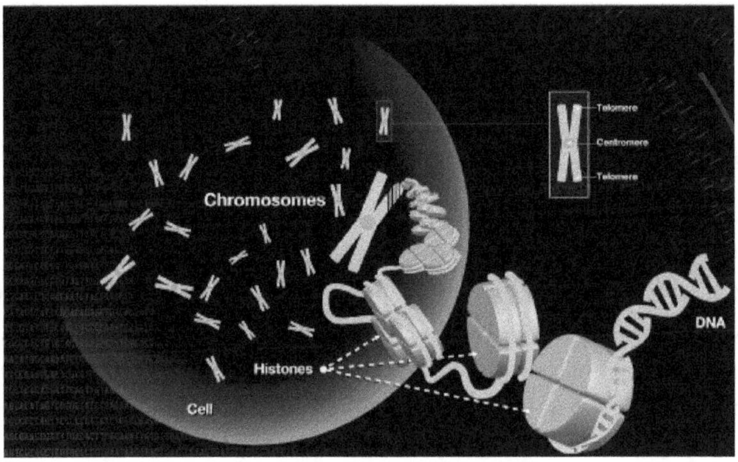

Fig. 16. Chromosomes

Scientists think that, in the past, mitochondria were free-living bacteria with the ability to convert oxygen into energy. When these bacteria invaded cells lacking the power to tap into oxygen's power, the cells retained them, and, over time, the bacteria evolved into modern-day mitochondria.

HOW DOES SEXUAL REPRODUCTION PRODUCE VARIATION?

Genetic variation. It is this variation that is the essence of evolution. Without genetic differences among individuals, "survival of the fittest" would not be likely. Either all survive, or all perish.

Sexual reproduction results in infinite possibilities of genetic variation. In other words, sexual reproduction results in offspring that are genetically unique. They differ from both parents and also from each other. This occurs for a number of reasons.

- When homologous chromosomes form pairs during prophase I of meiosis I, crossing-over can occur. Crossing-over is the exchange of genetic material between homologous chromosomes. It results in new combinations of genes on each chromosome.
- When cells divide during meiosis, homologous chromosomes are randomly distributed to daughter cells, and different chromosomes segregate independently

of each other. This called is called independent assortment. It results in gametes that have unique combinations of chromosomes.

- In sexual reproduction, two gametes unite to produce an offspring. But which two of the millions of possible gametes will it be? This is likely to be a matter of chance. It is obviously another source of genetic variation in offspring. This is known as random fertilization.

All of these mechanisms working together result in an amazing amount of potential variation. Each human couple, for example, has the potential to produce more than 64 trillion genetically unique children. No wonder we are all different!

Crossing-Over

Crossing-over occurs during prophase I, and it is the exchange of genetic material between non-sister chromatids of homologous chromosomes. Recall during prophase I, homologous chromosomes line up in pairs, gene-for-gene down their entire length, forming a configuration with four chromatids, known as a **tetrad**. At this point, the chromatids are very close to each other and some material from two chromatids switch chromosomes, that is, the material breaks off and reattaches at the same position on the homologous chromosome (Fig. 17). This exchange of genetic material can happen many times within the same pair of homologous chromosomes, creating unique combinations of genes. This process is also known as recombination.

Sexual reproduction and evolution

Sexual reproduction involves the fusion of the nuclei of a male and female sex cell during fertilisation. The offspring inherit a mixture of alleles from both parents.

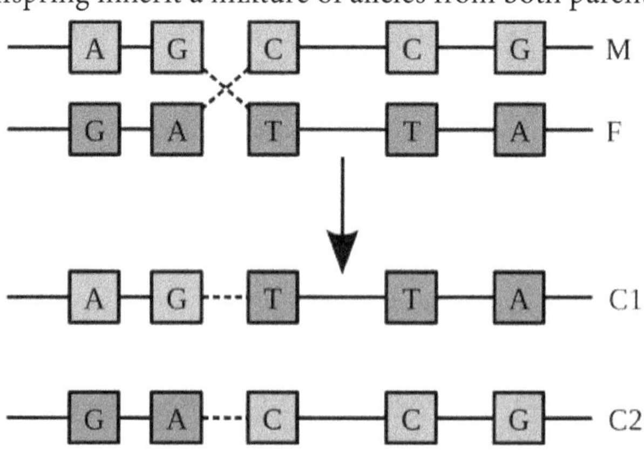

Fig. 17. Crossing-over. A maternal strand of DNA is shown in red. A paternal strand of DNA is shown in blue. Crossing over produces two chromosomes that have not previously existed. The process of recombination involves the breakage and rejoining of parental chromosomes (M, F). This results in the generation of novel chromosomes (C1, C2) that share DNA from both parents.

Advantages of sexual reproduction

- Produces genetic variation in the offspring.
- The species can adapt to new environments due to variation, which gives them a survival advantage.
- A disease is less likely to affect all the individuals in a population.

Disadvantages of sexual reproduction

- Time and energy are needed to find a mate.
- It is not possible for an isolated individual to reproduce.

Asexual reproduction and evolution

In asexual reproduction an exact genetic copy of the parent organism is produced (a clone).

Unlike sexual reproduction, asexual reproduction only introduces genetic variation into the population if a random mutation in the organism's DNA is passed on to the offspring.

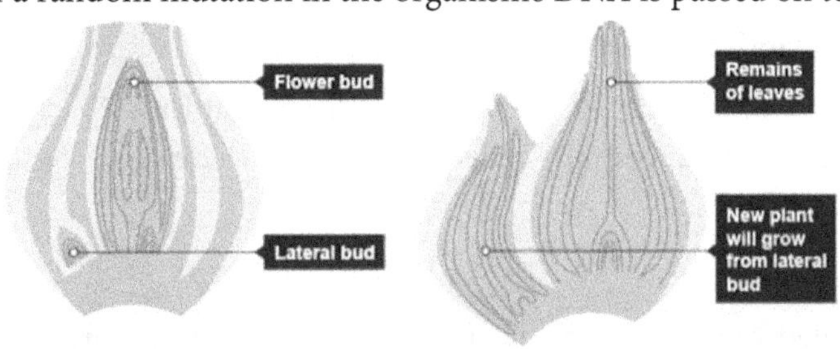

Fig. 18. Daffodil bulb at the start and end of spring

Advantages of asexual reproduction

- The population can increase rapidly when the conditions are favourable.
- Only one parent is needed.
- It is more time and energy efficient as you don't need a mate.
- It is faster than sexual reproduction.

Disadvantages of asexual reproduction

- It does not lead to genetic variation in a population.
- Disease may affect all the individuals in a population.

Animals produce offspring through asexual and/or sexual reproduction. Both methods have advantages and disadvantages.

Asexual reproduction produces offspring that are genetically identical to the parent because the offspring are all clones of the original parent. A single individual can produce offspring asexually and large numbers of offspring can be produced quickly. In a stable or predictable environment, asexual reproduction is an effective means of reproduction because all the offspring will be adapted to that environment.

In an unstable or unpredictable environment asexually-reproducing species may be at a disadvantage because all the offspring are genetically identical and may not have the genetic variation to survive in new or different conditions. On the other hand, the rapid rates of asexual reproduction may allow for a speedy response to environmental changes if individuals have mutations. An additional advantage of asexual reproduction is that colonization of new habitats may be easier when an individual does not need to find a mate to reproduce.

During sexual reproduction the genetic material of two individuals is combined to produce genetically diverse offspring that differ from their parents. The genetic diversity of sexually produced offspring is thought to give species a better chance of surviving in an unpredictable or changing environment. Species that reproduce sexually must maintain two different types of individuals, males and females, which can limit the ability to colonize new habitats as both sexes must be present.

Asexual Reproduction

Asexual reproduction occurs in prokaryotic microorganisms (bacteria) and in some eukaryotic single-celled and multi-celled organisms. There are a number of ways that animals reproduce asexuall

Fission

Fission, also called binary fission, occurs in prokaryotic microorganisms and in some invertebrate, multi-celled organisms. After a period of growth, an organism splits into two separate organisms. Some unicellular eukaryotic organisms undergo binary fission by mitosis. In other organisms, part of the individual separates and forms a second individual. This process occurs, for example, in many asteroid echinoderms through splitting of the central disk. Some sea anemones and some coral polyps (Fig. 19) also reproduce through fission.

Budding

Budding is a form of asexual reproduction that results from the outgrowth of a part of a cell or body region leading to a separation from the original organism into two individuals. Budding occurs commonly in some invertebrate animals such as corals and hydras. In hydras, a bud forms that develops into an adult and breaks away from the main body, whereas in coral budding, the bud does not detach and multiplies as part of a new colony.

Fragmentation

Fragmentation is the breaking of the body into two parts with subsequent regeneration. If the animal is capable of fragmentation, and the part is big enough, a separate individual will regrow.

For example, in many sea stars, asexual reproduction is accomplished by fragmentation. Figure 20 illustrates a sea star for which an arm of the individual is broken off and regenerates a new sea star. Fisheries workers have been known to try to kill the sea stars eating their clam or oyster beds by cutting them in half and throwing them back into the ocean. Unfortunately for the workers, the two parts can each regenerate a new half, resulting in twice as many sea stars to prey upon the oysters and clams. Fragmentation also occurs in annelid worms, turbellarians, and poriferans.

Fig. 19. Coral polyps reproduce asexually by fission.

Note that in fragmentation, there is generally a noticeable difference in the size of the individuals, whereas in fission, two individuals of approximate size are formed.

Parthenogenesis

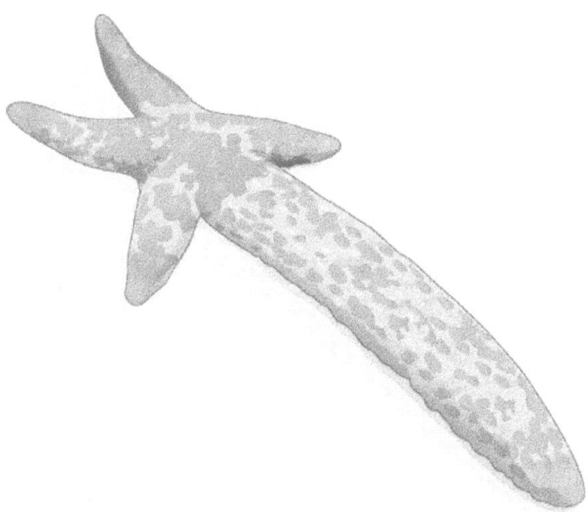

Fig. 20. Sea stars can reproduce through fragmentation. The large arm, a fragment from another sea star, is developing into a new individual.

Parthenogenesis is a form of asexual reproduction where an egg develops into a complete individual without being fertilized. The resulting offspring can be either haploid or diploid, depending on the process and the species. Parthenogenesis occurs in invertebrates such as water flees, rotifers, aphids, stick insects, some ants, wasps, and bees. Bees use parthenogenesis to produce haploid males (drones) and diploid females (workers). If an egg is fertilized, a queen is produced. The queen bee controls the reproduction of the hive bees to regulate the type of bee produced.

Some vertebrate animals—such as certain reptiles, amphibians, and fish—also reproduce through parthenogenesis. Although more common in plants, parthenogenesis has been observed in animal species that were segregated by sex in terrestrial or marine zoos. Two female Komodo dragons, a hammerhead shark, and a blacktop shark have produced parthenogenic young when the females have been isolated from males.

Sexual Reproduction

Sexual reproduction is the combination of (usually haploid) reproductive cells from two individuals to form a third (usually diploid) unique offspring. Sexual reproduction produces offspring with novel combinations of genes. This can be an adaptive advantage in unstable or unpredictable environments. As humans, we are used to thinking of animals as having two separate sexes—male and female—determined at conception. However, in the animal kingdom, there are many variations on this theme.

Hermaphroditism

Hermaphroditism occurs in animals where one individual has both male and female reproductive parts. Invertebrates such as earthworms, slugs, tapeworms and snails, shown in Fig. 21, are often hermaphroditic. Hermaphrodites may self-fertilize or may mate with another of their species, fertilizing each other and both producing offspring. Self fertilization is common in animals that have limited mobility or are not motile, such as barnacles and clams.

Sex Determination

Mammalian sex determination is determined genetically by the presence of X and Y chromosomes. Individuals homozygous for X (XX) are female and heterozygous individuals (XY) are male. The presence of a Y chromosome causes the development of male characteristics and its absence results in female characteristics. The XY system is also found in some insects and plants.

Avian sex determination is dependent on the presence of Z and W chromosomes. Homozygous for Z (ZZ) results in a male and heterozygous (ZW) results in a female. The W appears to be essential in determining the sex of the individual, similar to the Y chromosome in mammals. Some fish, crustaceans, insects (such as butterflies and moths), and reptiles use this system.

Fig. 21. Many snails are hermaphrodites. When two individuals mate, they can produce up to one hundred eggs each

The sex of some species is not determined by genetics but by some aspect of the environment. Sex determination in some crocodiles and turtles, for example, is often dependent on the temperature during critical periods of egg development.

This is referred to as environmental sex determination, or more specifically as temperature-dependent sex determination. In many turtles, cooler temperatures during egg incubation produce males and warm temperatures produce females. In some crocodiles, moderate temperatures produce males and both warm and cool temperatures produce females. In some species, sex is both genetic- and temperature-dependent.

Individuals of some species change their sex during their lives, alternating between male and female. If the individual is female first, it is termed protogyny or "first female," if it is male first, its termed protandry or "first male." Oysters, for example, are born male, grow, and become female and lay eggs; some oyster species change sex multiple times.

INHERITANCE OF MITOCHONDRIAL AND CHLOROPLAST DNA

If you were asked to name the organelle that contains DNA, what would you say? If you said the nucleus, you'd definitely get full points, but the nucleus is not the only source of DNA in most cells.

Instead, DNA is also found in the mitochondria present in most plant and animals cells, as well as in the chloroplasts of plant cells. Here, we'll explore how mitochondrial and chloroplast DNA are inherited.

MITOCHONDRIAL AND CHLOROPLAST DNA

The DNA molecules found in mitochondria and chloroplasts are small and circular, much like the DNA of a typical bacterium. There are usually many copies of DNA in a single mitochondrion or chloroplasts.

Similarities between the DNA of mitochondria and chloroplasts and the DNA of bacteria are an important line of evidence supporting the endosymbiont theory, which suggests that mitochondria and chloroplasts originated as free-living prokaryotic cells.

How is non-nuclear DNA inherited?

Here are some ways that mitochondrial and chloroplast DNA differ from the DNA found in the nucleus:

- **High copy number.** A mitochondrion or chloroplast has multiple copies of its DNA, and a typical cell has many mitochondria (and, in the case of a plant cell, chloroplasts). As a result, cells usually have many copies – often thousands – of mitochondrial and chloroplast DNA.

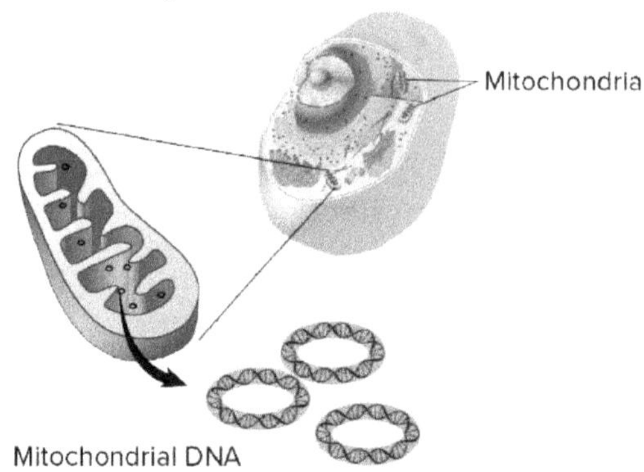

Fig. 22. Mitochondrial DNA

- **Random segregation.** Mitochondria and chloroplasts (and the genes they carry) are randomly distributed to daughter cells during mitosis and meiosis. When the cell divides, the organelles that happen to be on opposite sides of the cleavage furrow or cell plate will end up in different daughter cells.
- **Single-parent inheritance.** Non-nuclear DNA is often inherited uniparentally, meaning that offspring get DNA only from the male or the female parent, not both. In humans, for example, children get mitochondrial DNA from their mother (but not their father).

Chloroplast inheritance: Early experiments

At the turn of the 20th century, Carl Correns, a German botanist, did a series of genetic experiments using four o'clock plants (*Mirabilis jalapa*). We now know that his work demonstrated how chloroplast DNA is passed on from cell to cell and from parent to offspring.

Correns' experiments

The *Mirabilis* plants that Correns worked with came in three types: pure green, pure white, or variegated (mottled green and white). Green and white branches could appear on variegated plants, but variegated branches did not appear on green or white plants.

Correns was curious about this coloration trait, and he carried out a number of crosses between plants of different colors. He found that:

- The color of the egg cell-donating branch (female parent) determined the color of the offspring.
- Female parent branches that were pure green or pure white produced only pure green or pure white offspring, respectively.
- Female parent branches that were variegated could produce all three types of offspring, but not in any predictable ratios.

Correns speculated that some factor in the cytoplasm of the egg cell must determine the color of the offspring. It was actually a different German botanist, Erwin Baur, who suggested that the chloroplasts in the cytoplasm might carry hereditary factors (genes).

Baur thought that, in variegated plants, some of the chloroplasts must have mutations that made them unable to turn green (produce pigment). Today, we know that this hypothesis was exactly right!

Explaining Correns' results

How can the idea of chloroplast inheritance make variegated plants variegated? Let's follow a zygote (1-celled embryo) with mixture of chloroplasts inherited from the egg cell. Some of the chloroplasts are green, while others are white. As the zygote undergoes many rounds of mitosis to form an embryo and then a plant, the chloroplasts also divide and are distributed randomly to daughter cells at each division.

Chloroplast DNA

Chloroplasts are the organelles that carry out photosynthesis in plants. These organelles contain photosynthetic pigments called chlorophylls. Similar to mitochondria, chloroplasts also contain their own DNA (plastid DNA). This chloroplast DNA is present in the chloroplast stroma. cpDNA and plastome are synonyms for chloroplast DNA. cpDNA is double-stranded circular DNA.

Similarities Between Mitochondrial DNA and Chloroplast DNA

- Mitochondrial DNA and chloroplast DNA are single, circular chromosomes.
- Both DNA types are double-stranded.
- Also, both exist as multiple copies.
- moreover, both are randomly distributed to daughter cells, unlike nuclear DNA.
- Furthermore, they are organelle DNA; hence, they are non-nuclear DNA.
- They are devoid of histone proteins and are not enclosed with a membrane.

- Additionally, both mtDNA and cpDNA do not have introns.
- They are rich in AT regions.
- Besides, both mitochondrial and chloroplast DNA contain genes that are important for their functioning.
- In comparison to nuclear DNA, both mtDNA and cpDNA are small in size.

Difference Between Mitochondrial DNA and Chloroplast DNA

Mitochondrial DNA is present inside the mitochondria while chloroplast DNA is present in chloroplasts. So, we can consider this as the key difference between mitochondrial DNA and chloroplast DNA. Furthermore, human mitochondrial DNA contains 16,569 base pairs while chloroplast DNA contains 120,000 to 170,000 base pairs. Thus, this is also a significant difference between mitochondrial DNA and chloroplast DNA. Additionally, the mitochondrial genome contains 37 genes while chloroplast genome contains about 200 genes.

3

HOW CAN GENETIC VARIATION BE MEASURED?

INTRODUCTION

Aquaculture was any form of activity to produce biota or aquatic organisms in a controlled environment and in nature in order to obtain production or output and benefits or outcomes. Activities carried out in the sea, brackish waters, fresh water, including the general waters of the lake, reservoirs and rivers. In principle, aquaculture activities were to control growth, death, and reproduction.

Observing the status and potential of our nation's aquaculture, although overall national fishery production was still dominated by capture fisheries, the contribution of aquaculture has a much higher production growth compared to capture fisheries. Data from the Ministry of Marine Affairs and Fisheries 2018 shows that the contribution of aquaculture to national production increased from 70.07 percent in 2015 to 74.03 percent in 2017. In contrast, the contribution of capture fisheries decreased from 29.93 percent in 2015 to about 25.97 percent in 2017. This gives the impression that aquaculture in the future will play an increasingly important role, but it certainly must be accompanied by some notes of improvement of the weakness that has been found.

Domestic and world demand for fishery products continues to increase as population increases and human awareness of the benefits of healthful and educated fish increases. The capability of producing fishery products from capture fisheries activities at the global level was maximum of 100 million tons per year. The FAO data in 2018, the total volume of capture fisheries production was estimated at 94 million tons in 2025, up from about 93 million tons on average between 2013 and 2015. The main reason for the stagnation in this capture fishery production was that most fish stocks waters have been fully exploited, and now its catch quantity tends to decrease.

The potential of aquaculture production owned by Indonesia was about 57.7 million tons per year, the largest in the world. While in 2017 the total production of national aquaculture was only 17.22 million tons or 29.8 percent. Thus, if aquaculture was more excavated and empowered through change and improvement, it was not impossible that the future will become a mainstay of the economy and increase the effect of a very large domino again. In the sense of providing opportunities for economic growth, job creation, and reducing poverty.

More than 90 percent of the world's aquaculture production comes from Asia, so it was only natural that Asian countries were the most likely to attract foreign exchange from aquaculture business. Indonesia has the opportunity to become a world leader in aquaculture as well as a mainstay of the economy. Moreover, the government's attention to this sector was increasingly evident with the big vision of making Indonesia sovereign in the maritime field by making Indonesia as the world maritime pivot.

Successful development of aquaculture is highly determined, among others: by the supply of seeds that include the quality and quantity factors of seed. Seed quality was influenced by the quality of the mother as genetic factors and environmental factors, i.e., water quality, food, and disease. The quality of the parent as a genetic quality factor has an influence on the quality of the seeds in terms of increasing the number of seeds, rapid growth, and resistance to environmental changes. Improving the quality of the mother through the improvement of both qualitative and quantitative genetic traits that chould be done through selection of traits or characters of prospective mothers. In addition, the management of the parent to avoid the occurrence of inbreeding that causes decreased heritability the parent.

This manuscript aimed to review the results of genetic studies that have been applied in the field of aquaculture. In this article, the study includes basic genetic concepts and theories, genetic research methods, and the role and function of genetic analysis in aquaculture.

THE BASIC GENETIC CONCEPTS AND THEORIES

Although people usually establish genetics beginning with the rediscovery of the manuscript of an article written by Gregor Mendel in 1900, genetics as "the science of inheritance" or heredity has been known since prehistoric times, such as the domestication and development of various breeds of cattle and cultivars. People have also known the effects of crossing and close marriage and made a number of procedures and regulations about it since before genetics stood as an independent science. Genealogy of the disease in the family, for example, has been reviewed before. However, this practical knowledge does not explain the cause of the symptoms.

The popular theory of inheritance adhered to in that period was the theory of inheritance: one inherits a flat mixture of attributes brought by the elderly, especially of the males for carrying sperm. Mendel's results show that this theory was not applicable because the properties were carried in combinations carried by typical alleles, rather than mixed blends. Another related opinion was Lamarck's theory: the traits that the elder inherited in his life be passed on to his son. This theory was also broken with Mendel's explanation that the nature brought by the gene was not influenced by the experience of the individual who inherits that trait.

Charles Darwin also gave an explanation of the pangenesis hypothesis and later modified by Charlesworth and Charlesworth. In this view, the body's cells produce particles called gemmules that will be collected in the reproductive organs before fertilization occurs. Thus, every cell in the body has a contribution to the properties that its offspring.

Gene, from the Dutch: genes, were the inheritance units of nature for living organisms. The physical form was a sequence of DNA encoding a protein, polypeptide, or an RNA that has a function for the organism that has it.

Modern gene boundaries were a particular location in the genome that was related to inheritance properties and chould be associated with functions as regulators or controllers, transcription goals or other functional roles. The use of "genes" in everyday conversations, eg "smart genes" or "hair genes", were often meant for alleles: the choice of variations available by a gene (Fig. 1).

Although the allele's expression chould be similar, people more often use the term allele for expression of genes that were phenotypically different. The genes were inherited from one individual to their offspring through a process of reproduction, together with the DNA that carries it. Thus, information that maintains the integrity of the form and function of an organism's life chould be maintained.

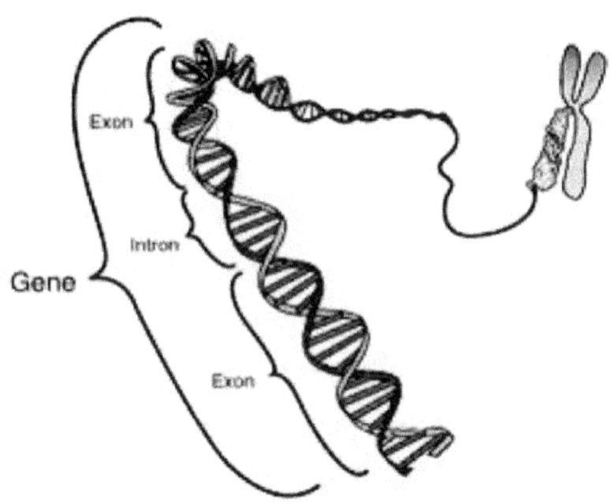

Fig. 1: Gene structure

At that time DNA has been discovered and was known to reside only on chromosomes, but people have not yet realized that DNA was linked to genes. Through Oswald Avery's study of Pneumococcus bacteria, and Alfred Hershey and Martha Chase with T2 bacteriophage virus, did people know that DNA was a genetic material. In the 1940s, George Beadle and Edward Tatum experimented with *Neurospora crassa*. From these experiments, Beadle and Tatum chould draw hypotheses that genes encode enzymes and they conclude that one gene synthesizes one enzyme or one gene-one enzyme theory.

A few decades later, it was discovered that genes encode proteins that not only function as enzymes and some proteins were composed of two or more polypeptides. With these findings, Beadle and Tatum's opinion, one gene-one enzyme theory, has been modified into the theory of one gene-one polypeptide or one gene-one polypeptide theory.

In eukaryotic cells, the gene consists of the transcriptional initiation regulatory domain, comprising the following: GCCACACCC series, ATGCAAAT, GC boxes, CCAAT boxes and TATA boxes, introns and exons that were encoded protein areas that chould be transcribed overlapping or nonoverlapping.

For example, in a code with three series of nucleotides, as triplet codons, i.e., AUU GCU CAG, it chould be read nonoverlapping as AUU GCU CAG or read overlapping as UGC UUG UGC GCU CUC CAG. Although in about 1961, it has been found that amino acids coded by codons by nonoverlapping have found different proteins transcribed with shifting overlapping codons, as well as the final regulatory domain of the transcription. Gene expression was the process by which the information codes present in the gene were converted into proteins that operate only within the cell.

Gene expression consists of two stages: the process of making RNA copies or transcription and polypeptide synthesis processes specifically in the ribosome or translational. The process of transcribing DNA into mRNA and translating mRNA into a polypeptide was called the central dogma. The central dogma applies to prokaryotes and eukaryotes.

However, in eukaryotes, there was an additional stage that occurs between transcription and translation called the premRNA stage. The pre-mRNA stage was to select the mRNA to be sent out of the nucleus to translate in the ribosome. The exon was the mRNA to be sent out of the nucleus to be translated, while the introns were the mRNAs that will remain in the nucleus because it was likely that the mRNA will form a protein that was not functional if it was translated. Intron will then break down again to form a new mRNA chain. It was also known that some of the so-called mutations chould occur in the process of expression of this gene (Fig. 2).

Genotype, literally "gene type", was the term used to express the genetic state of an individual or a group of individual populations. Genotypes may refer to the genetic state of a locus as well as the entire genetic material carried by the chromosome or genome. Genotypes chould be homozygous or heterozygous. Once people chould transfer gene, there was also the use of the term hemizygote. In Mendel's genetics, as classical genetics, genotypes were often denoted by pairs of letters; such as AA, Aa, or B1B1.

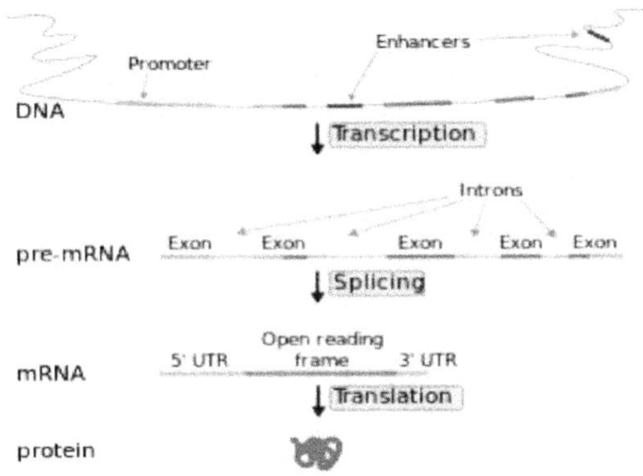

Fig. 2: Illustration of gene expression.

The same pair of pairs indicates that the individual symbolized was homozygous, i.e., AA and B1B1, while the different letter pairs represent the heterozygous individual. A pair of letters indicates that this symbolized individual was diploid or 2n. As a consequence, for example, the individual tetraploid or 4n homozygote was denoted by AAAA.

The phenotype was a characteristic, both structural, biochemical, physiological, and behavioral, that chould be observed from an organism regulated by genotype and its environment and interaction. The phenotypic definition includes the various levels of the gene expression of an organism. At the organism level, a phenotype was something that chould be seen or observed, something of nature or character. In this stage, phenotypes such as eye color, weight or resistance to a particular disease. At the biochemical level, the phenotype may be the content of certain chemical substances in the body.

For example, blood sugar or protein content in rice. At the molecular level, the phenotype may be the amount of RNA produced or the detection of DNA or RNA bands in electrophoresis. Phenotypes were determined in part by individual genotypes, in part by the environment in which the individual lives, time, and, in some traits, the interaction between genotype and environment. Phenotypic observations can be simple - usually flower color - or very complicated to require special tools and methods. However, due to the genetic expression of a gradual genotype from the molecular level to the individual level, there was often a link between a number of phenotypes at different levels.So far the discussion of a phenotype was assumed to describe its genotype. Phenotypes of such traits were easy to distinguish, for example, the color of the bean husk was alternatively white or gray, the plant was high or low, so it was easy to distinguish. Such properties were known as discontinuous traits. In such traits, there was a simple relationship between genotype and phenotype. In many cases a genotype produces only one kind of phenotype, otherwise, a phenotype was the result of the activity of a genotype.

Nevertheless, we have come to understand that the relationship between genotype and phenotype was influenced by phenomena, such as expressivity, penetration, and pleiotropy. In other words, a genotype chould produce different phenotypes because the genotype interacts with its environment during the growth and development process. The properties such as fish weight, fish length, and protein content show a wide range of phenotypes.

The properties that indicate such a range of phenotypic diversity were often called continuous traits. On discontinuous properties have a discrete distribution, whereas continuous properties have a continuous distribution, so it must be expressed quantitatively. Such properties were often called quantitative properties. Continuous properties have a continuous range of phenotypes as well. To examine the inheritance of the continuous or quantitative traits it was necessary to see why a trait has various phenotypes.

Quantitative properties chould occur in various ways. Generally, the phenotype range occurs because different genotypes exist in an individual group or population. This was common when a character was controlled by many loci. For example, if a locus with two alleles per locus controls a trait, then there were three possible genotypes namely AA, Aa, aa. For two loci with two alleles per locus, there will be 9 genotypes, eg AABB, AABb, Aabb, AaBB, AaBb, Aabb, aaBB, aaBb and aabb. Similarly, the increase in the number of loci as much as n, then the genotype formed as much as 3n. if there were more than two alleles for each locus it will produce much more genotype.

Thus the more the locus that controls a trait, the more likely genotypes may be formed. The properties controlled by many such loci were called polygenic or quantitative properties. polygenic or quantitative properties. The basic difference between qualitative and quantitative properties involves the number of genes that contribute to the phenotype variability and the degree to which the phenotype chould be modified by environmental factors. Quantitative properties chould be governed by many genes (perhaps 100 to 100 or more), each contributing to the phenotype so little that its individual influences chouldnot be detected by Mendel's methods. Such genes were called polygenic.

Below was presented a summary of some major differences between quantitative and qualitative genetics (Table 1). In the inheritance of quantitative properties, the concept of polygenic was used to explain the formation of quantitative traits. Ronald Fisher chould explain that the quantitative traits was formed of many genes with little influence, each of which segregates according to Mendel's theory.

Table 1. Main differences in inheritance of quantitative and qualitative traits

	Quantitative Traits	**Qualitative traits**
1	Characteristics of the degree	The characteristics of the type.
2	Continuous variation; phenotype measurement	Discontinuous variations; clear phenotype
	is a spectrum	classes
3	Polygenic control; the influence of single genes	The single gene gives a distinctly differ-
	is too small to be detectable	entiated influence
4	Questioning a population of organisms comp-	Questioning the marriages of individuals and

	rising all kinds of marriages that chould occur	their descendants
5	Statistical analysis provides estimates of popula-	Analyzed by making calculations and ratios
	tion parameters such as averages and standard	
	deviations	

Because of its small effect, the phenotype regulated by these genes chould be affected by the environment. Nevertheless, Fisher's explanation still places the "genes" that govern quantitative properties as abstract because they were only concepts. The proof of the existence of genes that regulate quantitative properties begins to open up after the availability of many genetic markers to enable people to create genetic maps that chould reach most of the chromosomes. Genetic markers were used to indicate the allelic situation in a particular chromosome part. Allele variations on a marker become genotypes for chromosomes or clusters.

Heritability was the proportion of genetic variation to the total amount of genetic variation coupled with the variety of environment, in other words, heritability was the proportion of the genetic variation to the phenotype variety for a particular character. There were two heritability values known in plant breeding that were heritability in the broad sense and heritability in the narrow sense.

The value of heritability in the broad sense takes into account the total genetic variation in relation to phenotype diversity. In this case, the genotype was considered a unit in relation to the environment. While narrow heritability was the focus of attention, the diversity attributed to the role of additive genes was part of the total genetic diversity.

Based on this explanation it chould be understood that the value of heritability in the narrow sense will never be greater than the value of heritability in a broad sense for a particular character.

Heritability parameters involve all types of gene action and therefore form a broad heritability estimate. In the case of perfect dominance, if a gamete containing the dominant active allele A2 combines with a gamete containing a zero allele A1, the resulting phenotype may consist of two units.

When two A2 gametes combine, the phenotype results will consist of two units. Conversely, if genes that have no dominance or additive genes were involved, then the A2 gametes will add one unit to the phenotype of the resulting zygote, regardless of the contribution of the allele of the gamete to which it combines. Thus only the genetic additive component of the variance has predictable quality, which was in the formulation of breeding roles. Heritability in this narrower sense was the ratio of genetic additive variance to the phenotype variance.

GENETIC DIVERSITY IN FISHES

Fishes are the most speciose group of vertebrates (over 33,000 species described), inhabit nearly all major aquatic habitat types, and perform a diverse set of biological functions in ecosystems. However, fishes are increasingly faced with altered environmental conditions and disturbance resulting from human activity. In particular, overharvesting, habitat destruction, pollution, and the introduction of non native species have led to a global decline in marine and freshwater fish biodiversity.

Furthermore, fishes increasingly need to contend with the effects of global climate change, which is driving ocean acidification and increases in aquatic temperatures and is expected to impose regional changes to salinity, dissolved oxygen availability, and circulation patterns in aquatic environments. Thus, disentangling species that are likely to adapt to future environmental changes from those that will require intervention remains a fundamental challenge for successful conservation and management of fishes.

One metric that may help predict which species are most likely to adapt to future conditions is genetic diversity. Broadly speaking, genetic diversity is any measure that quantifies within population variability in alternative forms of genes or noncoding loci. The ability of a population to evolve and adapt may be related to both heterozygosity (i.e., the proportion of diploid individuals that have two different alleles at a single locus) and the total number of alleles present within a population.

Additionally, reduced genetic diversity may result in decreased population viability and increased extinction likelihood, particularly for populations faced with stressful environmental conditions.

Understanding how patterns of genetic diversity vary across fishes could help inform predictions regarding which species are likely to adapt in response to future disturbance while simultaneously identifying species that might be particularly susceptible to extinction.

A landmark study by DeWoody and Avise analyzed 32 fish species and provided one of the first quantitative comparisons of genetic diversity in fishes occupying different habitats. The study suggested that genetic diversity was higher in marine fishes relative to freshwater fishes. We have since identified hundreds of additional studies that directly estimate genetic diversity in 463 distinct fish species, providing a much broader taxonomic survey and increased power and precision for statistical analyses.

Furthermore, improvements to phylogenies and statistical methods that consider evolutionary relatedness among species now allow us to account for the confounding effect of shared evolutionary histories and ask new questions about potential drivers of patterns of genetic diversity in fishes.

In this study, we ask how genetic diversity relates to species habitat needs, conservation status, and life history characteristics. We have three predictions. First, following

DeWoody and Avise, we predict that marine fishes should generally have higher genetic diversity than freshwater fishes even after accounting for species relatedness. Second, increased conservation need, as determined by the International Union for Conservation of Nature (IUCN), is determined in part by a reduction in census population size (IUCN, 2018), which is often associated with a reduction in genetic diversity.

Therefore, we predict that threatened species (i.e., species of high conservation concern) are likely to have less genetic diversity compared to not threatened species. Finally, certain life history characteristics can influence genetic diversity across a wide variety of taxa. Researchers showed that species exhibiting life history characteristics typically associated with "r strategists," including both early age at maturity and high fecundity, tend to have more genetic diversity than species that mature later and have fewer offspring (hereafter referred to as "K strategists"). As a result, we predict that genetic diversity should decrease as age at maturity increases in fishes.

Additionally, given that r strategists tend to exhibit high fecundity, we predict that genetic diversity will increase with fecundity in fishes. To identify patterns and drivers of genetic diversity in fishes, we test each of these predictions by determining the relationship between genetic diversity and (i) habitat, (ii) conservation status, and (iii) life history characteristics by performing a quantitative review of 463 globally distributed fish species.

DNA sequence variation

The crude extraction of DNA from animal or plant tissue is a simple process which involves mechanically or chemically breaking down the insoluble cellular structures and removing them by centrifugation. Soluble cellular proteins, and the proteins which bind the DNA into the chromosomes, can then be broken down using a strong protease enzyme and removed, usually using solvents such as phenol-chloroform. The DNA is present in the water-soluble component and can then be precipitated using an alcohol. There are a number of commercial kits on the market which enable further purification of DNA. The next problem is to produce multiple copies of specific fragments of DNA and this can be done either by cloning the fragment or by the use of the polymerase chain reaction (PCR). In the process of cloning, the target DNA is inserted into a vector molecule which is taken up or inserted into host cells. Subsequent rapid replication of these host cells and the vector molecules inside them results in the production of millions of copies of the target DNA.

As far as most DNA markers are concerned, cloning is usually only needed during the development phase – once the DNA sequences flanking the markers have been found from the cloned fragments, PCR can be used to produce millions of copies of the target sequence within a few hours. The PCR method relies on the fact that double-stranded DNA becomes denatured and separates into single strands when heated above 90°C. Once denatured, the temperature is lowered to a predetermined annealing temperature which allows short manufactured lengths of single-stranded DNA of known sequence (primers), designed to be complementary to the regions flanking the target DNA, to attach (anneal) to these flanking regions.

Raising the temperature to 72°C in the presence of a DNA polymerase enzyme and the building blocks of DNA results in two copies of the double-stranded target DNA. Each time the cycle is repeated the number of copies is doubled and, since each cycle takes only a minute or two, millions of copies can be produced within a few hours by this method.

DNA FRAGMENT SIZE VARIATION MOLECULAR MARKERS

All organisms are subject to mutations because of normal cellular operations or interactions with the environment, leading to genetic variation (polymorphism). Genetic variation in a species enhances the capability of organism to adapt to changing environment and is necessary for survival of the species. In conjunction with other evolutionary forces like selection and genetic drift, genetic variation arises between individuals leading to differentiation at the level of population, species and higher order taxonomic groups.

Molecular genetic markers are powerful tools to detect genetic uniqueness of individuals, populations or species. These markers have revolutionized the analytical power, necessary to explore the genetic diversity. The conclusion from genetic diversity data has varied application in research on evolution, conservation and management of natural resources and genetic improvement programmes, etc. In addition to protein markers, application of DNA markers is finding wide acceptance in population genetics.

With DNA markers, it is theoretically possible to observe and exploit genetic variation in the entire genome. Both genomic and mitochondrial DNA is used for varied applications. The commonly used technique are allozyme analysis, types of restriction fragment length polymorphism (RFLP), randomly amplified polymorphic DNA (RAPD), amplified fragment length polymorphism (AFLP), microsatellite typing, single nucleotide polymorphism (SNP), and expressed sequence tag (EST) markers, etc. Molecular markers can be classified into type I and type II markers.

Type I markers are associated with genes of known function, while type II markers are associated with anonymous genomic regions. Under this classification, allozyme markers are type I markers because the protein they encode has known function. RAPD markers are type II markers because RAPD bands are amplified from anonymous genomic regions via the polymerase chain reaction (PCR).

Microsatellite markers are also type II markers unless they are associated with genes of known function. The significance of type I markers is becoming extremely important for aquaculture genetics. Type I markers serve as a bridge for comparison and transfer of genomic information from a map-rich species into a relatively map-poor species. In general, type II markers such as RAPDs, microsatellites, and AFLPs are considered non-coding and therefore selectively neutral. Such markers have found widespread use in population genetic studies to characterize genetic divergence within and among the populations or species.

Molecular markers for population genetics and conservation

The definition of "marker" in wide sense is "something that serves to identify, predict, or characterize". In biology, the markers refer to any stable variation, which is heritable and can be measured or detected by an appropriate method. Such variations are produced at different levels for example, morphological, gene, chromosomal, biochemical or genomic. The markers that represent variations that appear at the DNA level are so-called *molecular markers*. The molecular markers occupy specific places in the genome and possess the role to "mark" the position of a specific gene or the inheritance of a particular character.

All organisms are subject to mutations that appear when genetic material fails to copy accurately or as a consequence of interactions with the environment, leading to genetic variation or polymorphism. Individuals of a species are distributed into more or less separate groups called populations, distributed over the species range. The genetic variation of a species is distributed both within and between populations.

The diversity at the gene level is often referred to as intra-specific variability as it represents biological diversity within a single species. The researchers are very interested in assessing genetic variation within and between populations and detecting similarities as well as differences between individuals/ populations in order to establish optimum conservation strategies.

For this variation to be useful in different genetic studies it is necessary to be inherited and to be perceptible for the researchers, whether is recognizable as phenotypic variation or as a genetic mutation detectable by different molecular techniques. At the DNA level the genetic variation is represented by point mutations called also single nucleotide polymorphisms (SNPs), indels (insertions or deletions of nucleotide sequences), inversion of a segment of DNA within a locus, and rearrangement of DNA segments around a locus of interest.

As the mutation rate is very low the evolution through mutation is extremely slow. The process of mutation is the only way in which genetic variation is created and in the lack of mutations there would be no biological diversity. DNA marker technology can be applied to reveal these mutations.

Large deletions and insertions determine shifts in the size of DNA fragments resulted consequently digestion by restriction enzymes, and are among the easiest type of mutations to detect by electrophoresis on agarose gel; smaller indels require DNA sequencing or more elaborate electrophoretic techniques, while the SNPs can be easily detected by DNA sequencing.

By using molecular markers is possible to observe and exploit genetic variation across the entire genome. Thus, the application of molecular markers in fish allowed recording rapid progress regarding the study of genetic variability and inbreeding, parentage determination, species identification, genetic linkage map construction for aquaculture species, identifying Quantitative Trait Loci (QTL) related to specific traits for marker assisted selection.

One of the classifications of the molecular markers refers to markers of type I as markers associated with genes having a known function and markers of type II as markers associated with unknown genomic segments. The type I of DNA markers were not initially consider to be appropriate for genetic studies in fishes, but over time it became clear that these markers are very important both for the study of wild populations and aquaculture.

Thus, this of markers has become very important for studying the phenomenon of linkage and for QTL mapping, being of great use in comparative genomic studies and for identification of candidate genes for quantitative traits in different fish species raised in aquaculture. An important number of studies were focused on the elucidation of the molecular basis of economically important traits in different species of salmonids. For example in the case of Atlantic salmon the mapping the QTLs was performed for the loci correlated with viral disease resistance, flesh color and growth traits, salinity tolerance, late sexual maturation, etc.

The type II of molecular markers (RAPD, AFLP, microsatellites) is considered to be non-coding. Such markers are used in population genetic studies aiming to characterize the genetic diversity. The type II of DNA markers proved useful in identifying species, populations and subpopulation, but also in identifying interspecific hybrids.

A second classification of markers is the one regarding their position in the cell. Thus we can distinguish between nuclear and mitochondrial DNA (mtDNA) markers, depending on their localization in the nuclear or mitochondrion genome.

The molecular markers might be highlighted by a variety of techniques that differ by difficulty, repeatability, cost and nature of polymorphism that is detected. The DNA markers detected by Polymerase Chain Reaction (PCR) are the most frequent used in assessing the genetic diversity and have a particular role in conservation. By PCR is possible to amplify DNA sequences up to several million times, so these markers present the advantage of nonlethal sampling. Small amount of tissue (fin clips or scales) are sufficient for analysis and is not necessary that the individuals to be sacrificed for sampling.

This can be an important feature when evaluating genetic change in protected or declining populations and for providing access to DNA of ancient or archived tissue samples. It can provide information about genetic diversity over extensive temporal and spatial scales, especially for populations that no longer exist. Taken together, these characteristics suggest capability to monitoring populations that are small, exploited or declining.

Depending on the primers used for PCR amplification these markers can be divided in two groups: (i) PCR markers for target sequences – in this case the fragment of interest is amplified with two specific primers and (ii) PCR markers for arbitrary sequences – one primer with a arbitrary nucleotide sequence is used; the primer binds to randomly in the genome resulting unknown DNA fragments. The primer used in this type of techniques is usually short (of 8-10 nucleotides) and so the probability for it to bind at multiple sites in genome increase.

In the first category are found molecular markers like PCR-RFLP (PCR Restriction Fragment Length Polymorphism), PASA (PCR Amplification of Specific Alleles), SNP (Single Nucleotide Polymorphism), repetitive DNA sequences (minisatellites, microsatellites, etc.). The second category includes markers like RAPD (Randomly Amplified Polymorphic DNA), AP-PCR (Arbitrary Primed-PCR) and AFLP (Amplified Fragment Length Polymorphism), but these are less preferred comparing with PCR markers for target sequences due to the difficulty of analysis and lack of results accuracy and reproducibility.

Allozyme Markers

Analysis of allozyme loci remained one of the most popular approaches in examining population genetics and stock structure questions in fishes. The technique is rapid, relatively inexpensive and provides an independent estimate of level of variation within a population without an extensive morphological and quantitative survey. Isohyets are structurally different molecular forms of an enzyme system with qualitatively the same catalytic function encoded by one or more loci. Isohyets, which are encoded by different alleles of the same gene locus, are designated as "allozymes" or "alloenzymes". Amino acid differences in the polypeptide chain of the different allelic forms of an enzyme reflect changes in the underlying DNA sequence. Depending on the nature of the amino acid changes, the resulting protein products may migrate at different rates (due to charge and size differences) when run through a gel subjected to an electrical field. Differences in the relative frequencies of alleles are used to quantify genetic variation and distinguish among genetic units at the levels of populations, species, and higher taxonomic designations.

Disadvantages associated with allozymes include occasional heterozygote deficiencies due to null (enzymatically inactive) alleles and sensitive to the amount as well as quality of tissue samples. In addition, some changes in DNA sequence are masked at the protein level, reducing the level of detectable variation.

Some changes in nucleotide sequence do not change the encoded polypeptide, and some polypeptide changes do not alter the mobility of the protein in an electrophoretic gel. At present 75 isozyme systems representing several hundred genetic loci are known.

With the strength as codominant marker, ease of use, and low cost, the allozyme markers are popular in population structure and phylogenetic studies, though has limited role in aquaculture genetics.

Mitochondrial DNA Markers

Mitochondrial DNA (mtDNA) analysis is being increasingly used in recent population and phylogenetic surveys of organisms. Studies of vertebrate species generally have shown that sequence divergence accumulates more rapidly in mitochondrial than in nuclear DNA. This has been attributed to a faster mutation rate in mtDNA that may result from a lack of repair mechanisms during replication and smaller effective population size due to the strict maternal inheritance of the haploid mitochondrial genome.

Due to its rapid rate of evolution, mtDNA analysis has proven useful in clarifying relationships among closely related species. Different parts of the mitochondrial genome are known to evolve at different rates. Almost the entire mtDNA molecule is transcribed except for the approximately 1-kb control region (D-loop), where replication and transcription of the molecule is initiated. In general, non-coding segments like the D-loop exhibit elevated levels of variation relative to coding sequences such as the cytochrome b gene, presumably due to reduced functional constraints and relaxed selection pressure. The 16S rRNA gene in the mitochondrial genome is one of the slowest evolving genes whereas rapidly evolving regions are control regions.

Due to non-Mendelian mode of inheritance, the mtDNA molecule is considered as a single locus. In addition, because mtDNA is maternally inherited, the phylogenies and population structures derived from mtDNA data may not reflect complete picture of the nuclear genome if gender-biased migration or selection or introgression exists.

Analyses of mtDNA markers have been used extensively to investigate stock structure in a variety of vertebrates including fishes, birds, mammals and reptiles.

Random Amplified Polymorphic DNA (RAPD) Markers

RAPD markers are the amplified products of less functional part of the genome that do not strongly respond to selection on the phenotypic level. Such DNA regions may accumulate more nucleotide mutations with potential to assess inter-population genetic differentiation. The amplification of genomic DNA by PCR with arbitrary nucleotide sequence primers, RAPD can detect high levels of DNA polymorphisms.

The technique detects coding as well as non-coding DNA sequences, and many of the most informative polymorphic sequences are those derived from repetitive (non-coding) DNA sequences in the genome. Because 90% of the vertebrate nuclear genome is non-coding, it is presumed that most of the amplified loci will be selectively neutral. RAPD loci are inherited as Mendelian markers in a dominant fashion and scored as present/absent. RAPDs have all the advantages of a PCR-based marker, with the added benefit that primers are commercially available and do not require prior knowledge of the target

DNA sequence or genome organization. Other advantages of RAPDs include the ease with which a large number of loci and individuals can be screened simultaneously.

Shortcomings of this type of marker include the difficulty of demonstrating Mendelian inheritance of the loci and the inability to distinguish between homozygotes and heterozygotes. In addition, the presence of paralogous PCR product (different DNA regions which have the same lengths and thus appear to be a single locus), low reproducibility due to the low annealing temperature used in the PCR amplification, have limited the application of this marker in fisheries science.

Single Nucleotide Polymorphism (SNP)

Single nucleotide polymorphism (SNP) describes polymorphisms caused by point mutations that give rise to different alleles containing alternative bases at a given nucleotide position within a locus. SNPs are becoming a focal point in molecular marker development since they represent the most abundant polymorphism in any organism's genome (coding and non-coding regions), adaptable to automation, and reveal hidden polymorphism not detected with other markers and methods.

Theoretically, a SNP within a locus can produce as many as two alleles, each containing one of two possible base pairs at the SNP site. Therefore, SNPs have been regarded as bi-allelic. SNP markers are inherited as co-dominant markers. Several approaches have been used for SNP discovery including SSCP analysis, heteroduplex analysis, and direct DNA sequencing. DNA sequencing has been the most accurate and most used approach for SNP discovery. SNPs are not without their limitations, however, might provide marginal additional, or even less, utility in some applications (e.g. relatedness).

Microsatellite Markers

Microsatellites consist of multiple copies of tandemly arranged simple sequence repeats (SSRs) that range in size from 1 to 6 base pairs [e.g., ACA or GATA]. Abundant in all species studied to date, microsatellite motifs have been estimated to occur as often as once every 10 kb in fishes. Microsatellites tend to be evenly distributed in the genome on all chromosomes and all regions of the chromosome. However, data from whole genome sequencing has somewhat contradicted this statement. They have been found inside gene coding regions, introns, and in the non-gene sequences.

Most microsatellite loci are relatively small, ranging from a few to a few hundred repeats. Regardless of specific mechanisms, changes in numbers of repeat units can result in a large number of alleles at each microsatellite locus in a population. Microsatellites have been inherited in a Mendelian fashion as codominant markers. Microsatellites were found to be informative in several species, which showed almost no variation at other markers.

However, use of microsatellite markers involves a large amount of up-front investment and effort. Each microsatellite locus has to be identified and its flanking region sequenced to design of PCR primers. Due to polymerase slippage during replication, small size differences between alleles of a given microsatellite locus (as little as 2 bp in a locus comprised of di-nucleotide repeats) are possible. Microsatellites recently have become an extremely popular marker type in a wide variety of genetic investigations.

NEW DEVELOPING MARKERS IN FISHERIES AND AQUACULTURE

Various type of DNA markers have been developed, including Allozymes, microsatellites, RAPDs, mt-DNA and SNPs. These markers in fish populations have revealed high levels of genetic variation distributed throughout the fish genome. A recent initiative has been made to accelerate efforts of DNA marker development, genome mapping and species identification. Major progress has been made toward Expressed Sequence Tags (EST) and DNA barcode development in several aquaculture species.

Expressed Sequence Tags (ESTs)

Expressed sequence tags (ESTs) are single-pass sequences generated from random sequencing of cDNA clones. The EST is use to identify genes and analyze their expression by means of expression profiling. It helps for rapid and valuable analysis of genes expressed in specific tissue types, under specific physiological conditions, or during specific developmental stages. ESTs offer the development of cDNA microarrays that allow analysis of differentially expressed genes to be determined in a systematic way, in addition to their great value in genome mapping. For genome mapping, ESTs are most useful for linkage mapping and physical mapping in animal genomics such as those of cattle and swine, where radiation hybrid panels are available for mapping non-polymorphic DNA markers.

A radiation panel is composed of lines of hybrid cells, with each hybrid cell containing small fragments of irradiated chromosomes of the species of interest. Typically, the cells from species of interest are radiated to break chromosomes into small fragments. The radiated cells are unable to survive by themselves. However, the radiated cells can be fused with recipient cells to form hybrid cells retaining a short segment of the radiated chromosome.

Characterization of the chromosomal break points within many hybrid cell lines would allow linkage and physical mapping of markers and genes. In spite of its popularity in mammalian genome mapping, radiation hybrid panels are not yet available for any aquaculture species. Development of radiation hybrid panels from aquaculture species is not expected in the near future, given the fact that physical mapping using BAC libraries can provide even higher resolution and the fact that BAC libraries are already available from several aquaculture species. Therefore, ESTs are useful for mapping in aquaculture species only if polymorphic ESTs are identified.

The value of EST resources and applications of bioinformatics in aquaculture genetics/genomics is inevitable, and it is expected that various EST databases will serve as rich sources of genomic information not only for aquaculture geneticists, but also for aquaculture physiologists, immunologists and biotechnologists.

DNA Barcoding

The principle of conservation biology is the preservation and management of biodiversity. The two major problems to such an endeavor are the difficulty of developing an assessment of this diversity for prioritization of hotspots of species richness and the identification of lineages particularly worthy, or in need, of preservation. Understudied taxa are greatly susceptible to extinction, suggesting there is a conservation penalty for our ignorance.

Even there are millions of unidentified and unknown species. DNA barcodes, segments of approximately 600 base pairs of the mitochondrial gene cytochrome oxidase I (COI), have been proposed as a fast, efficient, and inexpensive technique to catalogue all biodiversity. Barcoding is the use of universal polymerase chain reaction (PCR) primers to amplify and sequence an approximately 600-basepair fragment of the COI gene.

That portion of sequence is then compared using distance-based algorithms with an existing database of "known" sequences from specimens previously identified by taxonomists. DNA barcodes from a small portion of the mitochondrial genome might seem like an effective and rapid way to assess at least some, perhaps minimal, level of biodiversity. And for groups that are already relatively well known, especially birds and mammals, molecular studies based on barcode sized sequences have revealed cryptic DNA lineages and may be helpful.

Application Of Molecular Markers Species Identification

The inter-specific genetic divergence established through species specific diagnostic molecular markers provides precise knowledge on phylogenetic relationships and also resolve taxonomic ambiguities. These markers can be used to detect hybrid and introgressed or backcrossed individuals, distinguish early life history stage of morphologically close species both in hatchery and in natural populations.

Species-specific allozyme markers have been identified in many fishes [Tilapia; Sciaenid; *Anguilla* sp; Mugilidae] Specific diagnostic allozyme loci were used for different species: apache trout (*Oncorhynchus apache*), cutthroat (*Oncorhynchus clarki*) and rainbow trout (*Oncorhynchus mykiss*) and *Gambusia affinis* and *G. holbrooki*. Allozyme markers have also been used for individual classification in cyprinid species *Zacco pachycephalus* and *Z. platypus*, in cyprinodontid species *V. letourneuxi* and *V. hispanica*, in mullets *Mullus barbatus* and *M. surmuletus* and hake species *Merluccius australis* and *M. hubbsi*.

Species-specific diagnostic RAPD fingerprints were generated in several fish species and their taxonomic relationship has been analyzed. The RAPD-PCR technique was employed to identify three endemic morphologically similar Spanish species of Barbus: *Barbus bocagei, B. graellsii and B. sclateri* that have similar morphologies. RAPD markers were characterized to identify five species of family Cyprinidae: *Chondrostoma lemmingii, Leuciscus pyrenaicus, Barbus bocagei, Barbus comizo*, all endemic in the Iberian Peninsula, and introduced *Alburnus alburnus*, for studying genetic relationship and diversities in four species of Indian Major carps (family Cyprinidae): rohu (*Labeo rohita*), kalbasu (*L. calbasu*), catla (*Catla catla*) and mrigal (*Cirrhinus mrigala*), for identification of three eel species, *A. japonica, A. australis and A. bicolor* and to estimate the population structure and phylogenetic relationships among the eight species of the genus Barbus.

Large variation in mtDNA sequences among species can be utilized to produce species-specific markers. Since the structures of mitochondrial RNA genes (tRNA and rRNA) and the functional molecule of the 16S rRNA are highly conserved among the animal taxa that are related even distantly, change of even few nucleotides in such a gene between closely related taxa might indicate a substantial degree of genetic divergence. Mt-DNA sequences have been used as useful marker for species-specific identification in many fishes [Tuna; Billfish; Snappers; Myctophidae; Grey mullets].

Comparable levels of divergence based on 12S rRNA and 16S rRNA sequences have been reported for several recently diverged fish species [genus Sternoptyx; *Cyclothone* sp]. Sequence variation in the control region (D-loop) of the mitochondrial DNA (mtDNA) was examined to assess the genetic distinctiveness of the short-jaw cisco, *Coregonus zenithicus* and revealed high similarity of *C. zenithicus* and the related species *C. artedi, C. hoyi, C. kiyi*, and *C. clupeaformis*. Identification of *Astyanax altiparanae* (Teleostei, Characidae) in the Iguacu River, Brazil, was done on the basis of mitochondrial DNA and RAPD markers. Two species, *Acipenser baeri*, and *A. stellatus*, was studied using mitochondrial DNA (D-loop, cytochrome b (cyt-b) and ND5/6 genes) sequencing to determine whether traditionally defined subspecies correspond to taxonomic entities and conservation management units.

GENETIC VARIATION AND POPULATION STRUCTURE STUDY IN NATURAL POPULATIONS

Molecular markers provide direct assessment of pattern and distribution of genetic variation thus helping in answering, "if the population is single unit or composed of subunits". Several evolutionary forces affect the amount and distribution of genetic variation among populations and thereby population differentiation. Geographic distance and physical barriers enhance reproductive isolation by limiting the migration and increase genetic differentiation between populations.

Impact of migration and gene flow on genetic differentiation also depends upon effective size of receiving population and number of migrants. Increased computational power and mathematical models have enhanced the scope of conclusions that can be drawn out of genotype data generated through molecular markers. Some of the possibilities are assignment of migrants, determination of genetic bottlenecks, effective breeding population estimates besides genetic variation and differentiation estimations.

These markers have been extensively employed across various taxonomic groups [mosquito; turtle; amphibians; panda; five vertebrate classes including fish, amphibian, reptiles, birds and mammals]. Experiments on fish populations have significantly contributed towards development of science of population genetics, models and analytical softwares.

Population genetic structure has been investigated using allozyme markers in many fish species [*Oncorhynchus gorbuscha*; *Tenualosa ilisha*; *Pagrus auratus*].

Fifteen random primers were used to analyze the genome DNA of Jian carp (*Cyprinus carpio* var *jian*) by the RAPD technique. Study on cold tolerant traits for common carp *Cyprinus carpio* was conducted and nine RAPD-PCR markers associated with cold tolerance of common carp were identified. The genetic diversity has been studied using RAPD markers in *Carassius auratus, Epinephelus merra* population and *Solea solea*.

Genetic variation have been assessed with Allozyme and RAPD markers on *Mullus surmuletus* L., and three species of Pimelodidae catfish.

Population structure has been examined using microsatellite markers of sockeye salmon, Chinook salmon and Arctic charr populations. Genetic variation have been assessed using microsatellite genetic markers to identify the population structure of brook charr, *Salvelinus fontinalis* and 14 populations of northern pike (*Esox lucius*) in the North Central United States and in six populations from Quebec, Alaska, Siberia, and Finland.

Based on five microsatellite loci, the genetic structure of endangered fish species Anaecypris hispanica was studied in eight distinct populations in the Portuguese Guadiana drainage to determine levels of genetic variation within and among populations and suggested implications for conservation of the species.

Combination of allozyme and microsatellites was used to investigate genetic divergence in Salmo trutta and Salmo salar.

Researchers represents population genetic analysis of gilthead sea bream (Sparus aurata) and Kanda, Kanda and Allendorf examine population genetic structure of bull trout Salvelinus confluentus using a combination of allozyme, microsatellite and mtDNA variation.

Genetic variability of *Salmo trutta* and *Sparus aurata* was evaluated on the basis of Allozyme, Microsatellites and RAPD markers.

Patterns of population subdivision and the relationship between gene flow and geographical distance in the tropical estuarine fish Lates calcarifer (Centropomidae) were investigated using mtDNA control region sequences.

Allozymes and mtDNA sequences were assessed to evaluate the genetic variability in small marine fish Pomatoschistus microps, brown trout and Macquaria novemaculeata.

Comparison of Genetic Variation Between Wild and Hatchery Populations

Molecular markers also find application in aquaculture to assess loss of genetic variation in hatcheries through, comparison of variation estimates between hatchery stocks and wild counterparts. The information is useful obtained in monitoring farmed stocks against inbreeding loss and to plan genetic up gradation programmes. A major aspect such studies address is concerned with the assessment of farm escapes into the natural population and introgression of wild genome.

Brook trout *Salvelinus fontinalis* from ustocked waters, naturalized lakes, and hatcheries in New York and Pennsylvania were analyzed electrophoretically for allozyme expression. All wild-unstocked samples were highly differentiated populations and significantly different from each other and from hatchery samples.

Genetic diversity was investigated using microsatellites between farmed and wild populations of Atlantic salmon. Farmed salmon showed less genetic variability than natural source population in terms of allelic diversity.

Variation in allozymes and three microsatellite loci was assessed in populations of wild and cultured stocks of *Sparus aurata* and *Sparius auratus*. The microsatellite heterozygosity values were high in wild, but lower in the cultured samples.

Assessment of Demographic Bottleneck in Natural Population

Demographic bottlenecks occur when populations experience severe, temporary reduction in size. Because bottlenecks may influence the distribution of genetic variation within and among populations, the genetic effects of reductions in population size have been studied extensively by evolutionary biologist.

It may often be necessary to perform genetic analyses of temporal replicates to estimate the significance of spatial variation independently from that of temporal variation in order to ensure the reliability of estimates of a defined population structure. Such estimates provide understanding about changes in genetic variation, effective population size and other historical bottlenecks and can be extrapolated to define evolutionary trends of species.

Today various models are available that can resolve bottlenecks or effective population size changes through use of heterozygosity excess, linkage disequilibrium etc. However, estimates through temporal changes are considered more accurate. Analysis of temporal changes is limited due to lack of historical data as well as samples. Therefore, such studies are limited and mostly use archived samples, wherever available. In vertebrates, a limited number of studies have specifically assessed the temporal changes in genetic variation for more than one generation.

Researchers recommended close monitoring of negative effects on sea otter population based on the conclusion from mtDNA, D-loop, microsatellite variability comparison between prefur trade and present population. Prefur trade DNA samples were obtained from excavated bones.

Propogation Assisted Rehabilitation Programmes

Habitat alterations and over harvesting have contributed to the decline or disappearance of numerous natural populations. In addition, reinforcement programs of wild populations based on releases of hatchery reared fish of non-native origin compromise the conservation of remnant native trout resources. Effect of these programmes through releases in natural populations has been studies in many fishes through molecular markers.

Researchers found through allozyme polymorphism that brown trout (*Salmo trutta* L.) in the Abatesco river basin on the eastern coast of Corsica restoration was mainly due to the populations of the tributaries, which had been less disturbed by the spate. This study has shown that the wild population was primarily restored by the surviving individuals, particularly those from the tributaries that escaped the spate.

To assess the levels of gene introgression from cultured to wild brown trout populations, four officially stocked locations and four non-stocked locations were sampled for one to three consecutive years and compared to the hatchery strain used for stocking. Allozyme analysis for 25 loci included providing allelic markers distinguishing hatchery stocks and native populations. Different levels of hybridization and introgression with hatchery individuals were detected in stocked drainages as well as in protected locations.

GENETIC RESEARCH METHODS

Phenotypic performance analysis

Phenotypic performance analysis includes morphological and morphometric data. Measured morphometric data include weight, length, width, height, shell thickness, area and circumference. The morphometric ratio or index analysis was performed to obtain the coefficient of interpersonal performance within a population by calculating the distance coefficient or Eucleudian value, the coefficient of similarity and coefficient of variation. The coefficient values were mapped in the dendrogram to see whether there were similarities and differences between individuals in the ring snail population.

Biochemical analysis method

The measured biochemical performance was the analysis of enzyme or allozyme, protein, amino acids using starch gel electrophoresis. The analysis of the buffer migration rate and the staining procedure. Fitness test for Hardy- Weinberg equilibrium and differences between populations using GENEPOP. The interpretation of locus and allele identification follows the Allendorf and Utler method. The allele's naming was based on the relative distance of the migration of the allele to the most commonly encountered allele coded

100. The data was interpreted to obtain allelic frequency, the degree of polymorphism, heterozygosity and genetic distance.

Discriminant analysis was used to see the significant differences between groups determined by observed morphometric. The measured results were relativized using standard deviations to avoid the diversity of sizes and possibly different ages. Next will be grouped morphological characters that most characterize the difference.

The influence of the interaction of fellow characters measured resulted in the depreciation of these characters. This component was also used to see the most powerful influence on the formation of a population. This analysis uses Minitab program and SPSS program based on Principal Component Analysis or PCA. PCA analysis aims to interpret most of the information contained in a data matrix into graphs and to obtain maps of population distribution and similarity values within and outside the group. The results obtained were shown in dendrogram form.

RNA and DNA ratio analysis

Total DNA and RNA extracts follow the standard method developed. Amplification and visualization of DNA and RNA using PCR and electrophoresis methods with primers tailored to the sample. DNA sequencing and RNA from PCR using Sequenzer ABI377A Applied Biosystem. The DNA sequence and RNA were edited manually using Genetyx software, then aligned using Clustal software. Characterization of DNA sequences and their analysis using the maximum parsimony and Neighbor Joint methods performed using MEGA software. To find out the concentration of DNA and RNA measured by using Gene quant program. Results of DNA and RNA genomic concentrations of each sample were further calculated for determination of RNA-DNA ratio.

DNA MARKER METHODS

Some DNA markers that can be used to analyze genetic variation, the evolution of a plant, the linking of specific genes to specific characters, parent search, analysis of quantitative character loci and revision of plant classification were Restriction Fragment Length Polymorphism, Random Amplified Polymorphic DNA, Amplified Fragment Length Polymorphism and microsatellite or Simple Sequence Repeat.

DNA fragment size variation

At the beginning of this chapter we said that genetic variation can be measured and quantified at several levels. We have shown how we can determine the precise sequence of a length of DNA, and how it varies between individuals. Now we shall progress to see how differences between sizes of DNA fragments can be identified and used to address particular genetic questions. Techniques that fall into this category include those known by the acronyms RFLP, VNTR, DNA fingerprinting, RAPD and AFLP. Of these, VNTR markers (microsatellites in particular) have come to the fore in recent years as being the most generally useful, though the others all have their place in answering particular genetic questions.

Restriction fragment length polymorphisms (RFLPs)

We can make good use of fragments of DNA as genetic markers without going through the procedure of sequencing them. If we have a high copy number of a particular fragment produced by the cloning method or from the PCR machine, this can be incubated with a number of different restriction endonucleases (REs) which will cleave it into a number of lengths depending on the position of the RE recognition sites. The various lengths produced can be separated by size and stained on an agarose gel. The same piece of DNA from different individuals will produce different sets of restricted fragments if there have been point mutations affecting the RE recognition sequences. In this way, polymorphisms can be identified based on the pattern of the size fragments on the agarose gel.

Variable number tandem repeats (VNTR)

Variation in the sequence of DNA can occur at certain sites by a method which is not point mutation. Spread throughout the genome are regions called variable number tandem repeats (VNTR), also known as simple tandem repeats (STR) or simple sequence length polymorphisms (SSLPs), which contain tandem (i.e. linked in chains) repeats of DNA sequences. The sequences may be very short (from 1 to 10 bp) or much longer, but the key feature of these tandem repeats is that the number of repeats can vary between individuals. It is thought that increases or decreases in the number of the repeats occur during copying by recombination or replication slippage and that these processes are not only independent of point mutations, but also occur at a much faster rates. Variation in the number of repeats at these satellite (repeated units 100 to 5000 bp), minisatellite (repeated units 5 to 100 bp) or microsatellite (repeated units 2 to 4 bp) loci can be very extensive in populations and provides a valuable tool for investigation of population genetic changes in the recent past. Microsatellite markers in particular are now used extensively for a number of reasons: because they are co-dominant (both alleles can be identified) and therefore can be analysed under the standard Hardy-Weinberg model; because, as 'junk DNA' they can usually be considered to be free of selective pressures; because of the high number of both loci and alleles at each locus; and, not least, because automatic DNA sequencers can be used for automated genotyping at microsatellite loci, vastly increasing the rate at which samples can be processed.

DNA fingerprinting

DNA fingerprinting can be thought of as a combination of RFLP and VNTR. First, genomic DNA is cut with a particular suite of restriction enzymes and differently sized fragments are separated by electrophoresis. The DNA is transferred from the fragile gel to a nylon membrane by the technique known as Southern blotting and the membrane is then probed with a particular satellite repeat sequence which is common throughout the genome.

Fragments that contain the repeat show up as a number of discrete bands after autoradiography. These banding patterns are so variable as to be in practice unique to each individual (the chances of a match between unrelated individuals are millions to one). Since the bands are inherited in a predictable fashion, DNA fingerprinting is a very accurate way of determining parentage and the forensic uses of this method are now well known.

Restriction fragment length polymorphism (RFLP)

Research to exploit DNA polymorphisms in plant genomes by utilizing the technology of molecular markers continues to increase. RFLP was a molecular technique based on polymorphism caused by the substitution of the base or nucleotide, insertion, dilation, or translocation that may occur in the past. This technique utilizes specific restriction sites in the genome of an organism. Restriction enzyme or restriction endonuclease was an enzyme that cuts DNA strands on a sugar-phosphate framework without damaging a base, in the sequence it recognizes. The recognition sequence was often called the introduction site was a DNA sequence that places the restriction enzyme and cuts on the sequence. These enzymes can be ordered through companies engaged in biotechnology and molecular biology.

The result of genomic cutting using certain restriction enzymes will result in differences in the length of the DNA fragment, which shows the distance from the enzyme restriction sites in an organism's genome. The fragmented genome could then be analyzed for the purpose of the study. Through further analysis could be known whether the target sequence has been changed due to the substitution of the nucleotide, insertion, deletion or translocation.

Random amplified polymorphic DNA (RAPD)

RAPD was widely used to analyze the diversity of genetic traits in a variety of studies with the consideration that, among others, it does not require a background knowledge of the genome to be analyzed, primers used universally - could be used for both prokaryotes and eukaryotes-, capable of producing relatively unlimited characters, the materials used were relatively cheaper, the preparation was easier, and yields faster results than other molecular analyzes.

The RAPD method was able to detect the nucleotide sequence by using only one primer. The primer will bind to the single strand of the one genome and to the DNA strand of its partner in the opposite direction. As long as the primary attachment site remains at an amplified range in general no more than 5000 bp, it will obtain amplification DNA products. With the use of RAPD that was relatively easy, inexpensive and analyzed in DNA level, early selection could be performed on the properties of resistance to biotic and abiotic stress, quantitative characteristics and duplication of collection crops could be avoided.

Random Amplified Polymorphism DNA was one of the most widely used molecular techniques used in the study of the population's genetic structure and measurement of the genetic variation of a population of organisms. The number and distribution of genetic variations reflect the interaction between the processes involved in evolution such as genes, mutations and genetic drift.

Microsatellite (Simple sequence repeat, SSR)

Repeated recurrent DNA that has the highest variation in the plant genome was a recurrent sequence with a simple or short repetitive fragment. This fragment was known by the name minisatellite and microsatellite. Minisatellite was a repetitive DNA usually between 10-60 bp, which at the beginning of its discovery was widely applied to the human genome. While SSR has fewer recurring units ranging from 1 to 6 bp, contained in very large quantities and spread in the genome and was widely used in plants. Variations of these fragments are usually the result of changes in the number of copies of the original loop and were often categorized as a variable number of tandem repeats or VNTR. Because very high levels of polymorphism could be detected with this fragment, VNTR was recognized as a potent tool for fingerprinting and cultivar identification. This fragment could also be used to study the diversity between and intra-population, ecological studies, calculating genetic distance and studying plant evolution.

The short sequence of DNA microsatellites with the clamp DNA sequences was conserved, allowing the primary design to amplify specific sites using PCR. If these primers were used to amplify certain SSR loci, then each primer will produce polymorphism in the form of a difference in amplification length known as Simple Sequence Length Polymorphism or SSLP. Each length represents one allele of a locus. Long differences occur because of the difference in the number of repeating units at specific SSR loci.

The diversity of the number of replicates in microsatellite could be detected by electrophoresing the amplified DNA products in the standard gel sequence, which could separate the fragments that distinguish each nucleotide. Microsatellite DNA was present in large quantities and spreads in the genome. The general form of repetition of DNA microsatellite was a simple repetition of two bases. Microsatellite DNA with the ease and speed of using PCR technology, co-dominant and easily interpreted makes microsatellite the best marker in gene mapping.

Molecular data analysis

Data generated from the molecular analysis were often used as a basis for the classification of individuals in a population, for phylogenetic construction, parent search, identification of markers of a particular trait, taxonomic revision and others. The dendrogram constructed in the analysis was generally based on the degree of similarity between the individuals used. However, the confidence limits for groupings generated through the dendrogram, usually not acceptable for calculations using common statistical procedures.

Molecular data to be analyzed should be converted into binary data based on the presence or absence of the amplified band. If there are amplified bands scored 1, whereas if none was given a score of 0. Binary data obtained could be used to arrange the dendrogram or phylogenetic of the organism being analyzed. For parent search, genetic mapping, and identification of specific fragments of binary data were converted into genotype data, and with specific programs analyzed according to the destination.

The result of the molecular band pattern that has been converted into binary data form could be processed using NtSys software to obtain the matrix of genetic distance and dendrogram or phylogenetic of the sample. To analyze the data according to research objectives, there was now very much software that could be used to process molecular data. For example, Genotype data composed of binary data could be used to analyze the relationship of the parent to the filial, the linkage map, and the correlation between loci with locus and locus with morphological properties. Tracking the elders using Cervus, cross-based depression in molecular markers using POPGENE, making a linkage map using MAPMAKER software with LOD 3 values, and the correlation between loci with locus and between the locus and morphological properties using statistical software such as Excel Stat, Minitab and SPSS

PROTEIN CONSERVATION AND VARIATION

Proteins do not act in isolation but interact with other proteins to fulfill important cellular functions. Often proteins are organized into pathways, which are tightly controlled cascades of protein binding events (and those of other biomolecules). One important cellular function controlled by pathways is the transmission of extra-cellular signals from the cell membrane to the nucleus to provoke a response to changes in the environment of the cell. Signaling pathways are often active in many different cell types and are conserved at a large evolutionary scale. Therefore, the characterization of mechanisms by which these ubiquitous pathways achieve specificity and fulfil largely different functions in different cell types or organisms is of crucial importance.

One characteristic of signaling pathways is the bow-tie (or hourglass) architecture in which signals sensed by receptors converge onto a core consisting of a smaller number of proteins followed by a diverse response of transcription factors. The bow-tie property has been observed in different human signaling pathways such as those downstream of epidermal growth factor receptor and of toll-like receptor. It is generally believed to confer pathways with robustness and evolvability by buffering input signals and modularizing the response. However, robustness by hierarchy comes to a price: mutations in the central core proteins might easily hijack the behavior of the entire system.

The robustness of pathways needs to be in balance with flexibility allowing pathways to vary their response to similar stimuli at different time points or in different cell types of the same organism. One intuitive though largely unexplored link to the bow tie model comes from investigations of protein-protein interaction (PPI) networks associated with gene expression information: it was observed that tissue-specific proteins tend to bind to core cellular proteins, possibly to modulate housekeeping cellular processes in a cell type-specific manner.

The mechanistic understanding about how activation of the same signaling pathways can lead to cell type-specific responses is rather anecdotal and involves diverse mechanisms such as cell type-specific feedback loops, different abundances of transcriptional cofactors, or cell type-specific chromatin states. However, it is largely unknown if there are functional or (network) topological signal protein classes that preferentially act as tissue-specific modulators of signaling. Therefore, we will explore here if we can adapt the bow-tie model to identify protein classes that show distinct evolutionary and abundance variability patterns.

Recently a mass spectroscopy analysis accomplished by Mann and co-workers has determined absolute protein copy numbers for 11 common human cancer cell lines with high coverage. The analyzed cell lines covered distinct tissue origins, such as lung carcinoma, hepatoma, osteosarcoma, colon carcinoma, and leukemia. Multiple technical replicates allow to make robust estimates of protein expression variability by contrasting the inter-cell line with the intra-cell line variability and, therefore, make this dataset a perfect choice for quantifying differences in protein expression among different cell types. Using this dataset and defining a measure of protein conservation covering a broad set of species, we systematically investigate patterns of protein expression variability and phylogenetic conservation in human pathways. We observe large differences in protein expression and in phylogenetic conservation between and within different human pathways.

Focusing on human signaling, we identify components of signaling pathways with distinct properties in respect to these features. By incorporating germline and somatic disease mutations, we show how the thereby identified pathway components underlie different selective constraints.

Protein abundance, variation, and conservation of cellular processes

To identify cellular processes that differ in their phylogentic conservation and cross-cell variability, we selected all Reactome pathways that are expected to be general and not restricted to only some cell types (ten pathways): Cell cycle, DNA replication, and chromosome maintenance [DR], Extracellular matrix organization [MO], Gene expression and RNA processing [GE], Membrane trafficking [MT], Metabolism [MB], Signal transduction [ST], Apoptosis [AP], Developmental Biology [DB], Transmembrane transport of small molecules [TM], and Cell-Cell communication [CO]. Several other Reactome pathways are restricted to very specific body cell types (e.g., Neuronal system and Muscle contraction), or are of low coverage (e.g, Circadian clock proteins), and were neglected for further analysis. We associated all members of the ten pathways with mean protein abundance, abundance variability and phylogenetic conservation values.

A fraction of the proteins (18.7% from the 4069 Reactome proteins that we could associate with at least one of the investigated features) participates in more than one pathway. We compared the distributions of mean abundance, abundance variability and phylogenetic conservation among exclusive proteins associated only with one pathway to the distributions associated with proteins involved in several pathways. We observed that exclusive proteins are significantly less conserved ($P<e^{-16}$; Wilcoxon-Mann-Whitney), more variable ($P<e^{-12}$; Wilcoxon-Mann-Whitney) and less abundant ($P<e^{-8}$; Wilcoxon-Mann-Whitney).

Next, to elucidate the common and variable elements in 11 cell types with respect to the ten Reactome pathways, we considered only proteins found exclusively in one pathway. We observed large differences with respect to the three investigated features among human functional pathway classes. In general, we found two opposing groups of behavior. Housekeeping pathways (GE, MB, MT and DR) are enriched in conserved proteins, have low to average variability (with GE being the only pathway class with a significant depletion in variability) and (except for DR) higher abundance (GE and MB are significantly enriched in highly abundant proteins).

Specific pathways (ST, MO, CO, DB) tend to have less conserved proteins (ST and MO show a significant depletion), have higher variability and less abundance (ST and DB are significantly associated with lower abundance). The remaining two pathways (AP and TM) showed a rather average behavior, with exception of the significantly lower abundance of TM proteins. Signal transduction (ST) shows in all three categories (mean abundance, abundance variability and conservation) a significant deviation from random expectation and has a larger than average spread of the distribution of variability values (fourth highest inter-quartile range among the ten variability distributions). This indicates that while average variability of signaling proteins is high, we will also find a large proportion of proteins with low variability in signaling pathways. Hence, we studied the variable and constant parts of signaling pathways.

Protein abundance, variation, and conservation in signaling-related pathways

To elucidate in more depth the protein abundance variability in signal transduction pathways, we investigated whether we can relate the molecular function of proteins in signaling pathways to their abundance, variability and conservation. For this purpose, we chose two complementary protein classification strategies and signaling resources:

a. We assigned, where non-ambiguously possible, signaling proteins in Reactome to one of the following Gene Ontology and UniprotKB categories: membrane-bound receptor, phosphatase, kinase, transcription factor, adaptors, and GTPase binding.
b. We retrieved the full set of proteins and their classification into signaling-related sections (ligand, receptor, mediator, cofactor, transcription factor) from the signaling pathway database.

SignaLink, which is another manually curated resource classifying proteins into pathway sections based on their role in signal transmission and topological properties. For example, SignaLink distinguishes between mediators and cofactors of signal transduction, where mediators are core pathway members and the cofactors merely modulate the function of signaling proteins. We only considered signaling proteins in SignaLink that are uniquely assigned to one class.

Due to different curation strategies, the sets of proteins associated with pathways largely differ between Reactome and SignaLink: we could automatically assign pathway functions to 802 proteins from Reactome while the SignaLink database contains 667 proteins with unique roles in signaling. The overlap between the two sets consists of only 80 proteins. The differences in protein composition and in the way proteins are associated with pathway functions allow us to study the evolution and expression of signaling proteins on two largely independent datasets.

We observed large differences in conservation, mean abundance and abundance variation for different classes within both sets of annotated signaling proteins. Next, we pooled all functional and topological classes into three layers: input (receptors), transmission (SignaLink: mediators and cofactors; Reactome: kinases, phosphatases, adaptors and GTPase binding proteins) and output (transcription factors).

We compared the resulting feature distributions. With respect to protein abundance variability we observed for both the Reactome and the SignaLink proteins significantly higher values associated with the input layer than with the transmission layer ($P<0.01$, Wilcoxon-Mann-Whitney). The difference between the transmission and the output layer was for both data sources not (Reactome) or only marginally ($P = 0.03$; Wilcoxon-Mann-Whitney; SignaLink) significant. The conservation of proteins of the transmission layer was significantly larger than of proteins of both the input and the output layer (for all comparisons: $P<0.00001$; Wilcoxon-Mann-Whitney).

In summary, taking a mechanistic view on signaling pathways where an input layer receives signals from the environment, a transmission layer integrates and proceeds the signal and an output layer orchestrates the transcriptional response, two patterns emerge: (i) In terms of conservation, we see a bell shaped curve with a high conservation of the transmission layer and lower conservation of the input and output layer. (ii) With respect to protein abundance variability, signaling pathways show a gradient with decreasing variability from the input to the output layer. There is a sharp drop in variability between the input and the transmission layer while the transmission and the output layer are rather similar in terms of variability. . The difference in terms of protein abundance between different functional classes was less pronounced. Interestingly, the variability and conservation of mediators and cofactors of signaling is almost the same (Wilcoxon-Mann-Whitney test does not yield $P<0.05$). This suggests that modulators in the transmission layer contribute less to cell type specific differences.

ALLOZYMES

Allozymes are the products of genetic variation at enzyme-encoding loci. They should not be confused with isozymes that are alternative forms of an enzyme produced at different loci. In order to identify and score allozymes, a small piece of tissue is obtained from an organism and is ground up with a buffer solution in a microtube. This grinding releases the soluble proteins from the cells. After centrifugation, these proteins present in the supernatant are subjected to starch or cellulose acetate gel electrophoresis that separates the proteins on the basis of charge and size. This technique has been called the 'find 'em and grind 'em' method. Following staining for a particular enzyme, individual samples are classified according to the alleles present at the enzyme locus being investigated. Alleles are best identified by their mobility relative to a known standard - i.e. a band that moves 90% of the distance of the standard is called allele 90 - but they are sometimes just labelled numerically or alphabetically

Strengths

The strength of allozymes is simplicity. Because allozyme analysis does not require DNA extraction or the availability of sequence information, primers or probes, they are quick and easy to use. Some species, however, can require considerable optimization of techniques for certain enzymes. Simple analytical procedures, allow some allozymes to be applied at relatively low costs, depending on the enzyme staining reagents used. Allozymes are codominant markers that have high reproducibility. Zymograms (the banding pattern of isozymes) can be readily interpreted in terms of loci and alleles, or they may require segregation analysis of progeny of known parental crosses for interpretation. Sometimes, however, zymograms present complex banding profiles arising from polyploidy or duplicated genes and the formation of intergenic heterodimers, which may complicate interpretation.

Weaknesses

The main weakness of allozymes is their relatively low abundance and low level of polymorphism. Moreover, proteins with identical electrophoretic mobility (co-migration) may not be homologous for distantly related germplasm. In addition, their selective neutrality may be in question. Furthermore, often allozymes are considered molecular markers since they represent enzyme variants, and enzymes are molecules. However, allozymes are in fact phenotypic markers, and as such they may be affected by environmental conditions. For example, the banding profile obtained for a particular allozyme marker may change depending on the type of tissue used for the analysis (e.g. root vs. leaf). This is because a gene that is being expressed in one tissue might not be expressed in other tissues. On the contrary, molecular markers, because they are based on differences in the DNA sequence, are not environmentally influenced, which means that the same banding profiles can be expected at all times for the same genotype.

Applications

Allozymes have been applied in many population genetics studies, including measurements of outcrossing rates, (sub)population structure and population divergence. Allozymes are particularly useful at the level of conspecific populations and closely related species, and are therefore useful to study diversity in crops and their relatives. They have been used, often in concert with other markers, for fingerprinting purposes, and diversity studies, to study interspecific relationships, the mode of genetic inheritance, and allelic frequencies in germplasm collections over serial increase cycles in germplasm banks, and to identify parents in hybrids.

IMMUNOLOGICAL IDENTIFICATION OF PROTEINS

Antigen-antibody reactions are most easily studied *in vitro* using preparations of antigens and antisera. Reactions of antigens and antibodies are highly specific. An antigen will react only with antibodies elicited by itself or by a closely related antigen. Because of the high specificity, reactions between antigens and antibodies are suitable for identifying one by using the other. However, cross-reactions between related antigens can occur, and these can limit the use-fulness of the test. The study of antigen-antibody reactions *in vitro* is called serology, and is important in clinical diagnostic microbiology.

Immunodiagnostic tests use an antigen-antibody reaction to detect and identify a specific antigen or antibody associated with a disease-causing organism. The antigen-antibody reaction itself is very specific. Consequently, if the correct antibodies can be obtained, immunodiagnostic techniques have the advantage of being able to identify the presence of a specific pathogen directly in the 138 Health Management in Aquaculture specimens and also can be used to detect the specific antibodies produced as a result of the immune response of the host to the organism. The primary component in the test is the antibody. The specificity, and to some extent the sensitivity, of the assay depends on the quality of the antibodies used in the reagents.

There are two basic methods for identification of unknown organisms by using sera containing various known antibodies; one is agglutination for particulate antigens and another is precipitation for soluble antigens. Other immunoassay methods include, fluorescent antibody technique (FAT), enzymelinked immunosorbent assay (ELISA) and Western blot analysis.

Vertebrates have fairly-well understood systems in their blood for fighting and removing non-self or foreign proteins such as those in viruses and bacteria. Each protein has sites on its surface that are called antigenic determinants. When a non-self protein is detected the blood produces antibodies (immunoglobulin molecules) specific to the antigenic determinants of that protein. However, changes in the DNA coding for a protein can change the amino acid sequence in the protein and this, in turn, can change the nature of the antigenic determinants. This feature of proteins can be used for a number of purposes: to assess relationships between proteins extracted from different individuals or different taxa; to test for the presence of specific disease proteins in an animal; to identify foreign proteins coded for by transgenic DNA in genetically modified organisms.

It is rather complicated, but the method works as follows. Let us assume we are dealing with a comparison of a protein between two related species, A and B. First, a protein from an individual of species A is injected into the blood stream of a rabbit. The rabbit's blood will respond to this foreign protein by forming antibodies specific to the protein. If a sample of the rabbit's blood serum (blood cells removed), called antiserum, is then mixed in a suitable medium with the original protein from species A, a reaction will take place; the antibodies will link to the antigenic determinant sites on the protein to form aggregations which will precipitate from solution. If the antiserum is mixed with protein from species B, the amount of aggregation will depend on how many of the antigenic determinant sites are common between the two species. If they are very similar, there will be much aggregation; if they are very different, there will be very little aggregation. So the principle is that the amount of aggregation is proportional to the difference between the two proteins.

How then is the amount of aggregation measured? It is done indirectly using a method known as complement fixation. In serum there are a group of proteins called collectively 'complement' which bind on to antigen – antibody complexes and the more of these complexes there are, the more complement is bound up or 'fixed'. In the test, a known amount of complement is added to the antigen – antibody mixture and after a specified period of time, the amount of unfixed complement is assessed. Free, unfixed complement has the ability to lyse specially sensitised sheep red blood cells allowing the release of haemoglobin. Following this reaction, remaining blood cells are removed by centrifugation and the colour change produced by the released haemoglobin is measured in a spectrophotometer. So the colour change is inversely proportional to the amount of antigen–antibody aggregation; strong colour means there has been a weak reaction between the antigen and the antibody.

PHENOTYPIC VARIANCE

There are very few examples of easily identifiable phenotypic variation in aquatic organisms controlled by single genes, or even pairs of genes. The best examples are found in the colouring of ornamental fish. It is interesting to consider just how lucky Gregor Mendel (the nineteenth-century discoverer of the method of genetic inheritance) was to have chosen to work with peas, which had a number of easily identifiable characters (round or wrinkled seeds; tall or short plants) each controlled by single genes. Such easily identifiable single-gene phenotypes are rare in most organisms. Because of the extensive development of DNA and allozyme technologies, the search for single-gene phenotypic variation is now uncommon. However, it is important to realise that the visual identification of varieties can be of critical importance to fish farmers without access to a modern genetic laboratory. Phenotypic rarities can provide high-value niche markets, and understanding how to get them to breed true requires genetic knowledge.

Within any given population, there are multiple potential sources of phenotypic variation, and each of these sources reflects a different underlying cause. The particular source of an instance of phenotypic variation determines whether that trait has the ability to respond to natural or artificial selection (i.e., whether the trait has evolutionary potential), as well as whether the trait can respond to environmental changes. Thus, researchers are highly interested in determining the relative importance of both the genetic factors and the environmental factors that lead to particular traits. Moreover, once investigators have determined this information, they can use it to predict the evolutionary dynamics of an entire population with regard to a specific trait.

Calculating Phenotypic Variance

As previously mentioned, all instances of phenotypic variance (V_P) within a population are the result of genetic sources (V_G) and/or environmental sources (V_E). This relationship can be summarized as follows:

$$V_P = V_G + V_E$$

However, in order to determine the values for both V_G and V_E, researchers must consider a number of additional variables, as described in the following sections.

Genetic Sources of Variation

Genetic sources of variation can themselves be divided into several subcategories, including additive variance (V_A), dominance variance (V_D), and epistatic variance (V_I). Together, the values for each of these subcategories yield the total amount of genetic variation (V_G) responsible for a particular phenotypic trait:

$$V_G = V_A + V_D + V_I$$

Defining the Subcategories of Genetic Variance

Of course, in order to fully understand this equation, one must also understand what each subcategory of genetic variance entails. The first subcategory, additive genetic variance, refers to the deviation from the mean phenotype due to inheritance of a particular allele and this allele's relative (to the mean phenotype of the population) effect on phenotype. In contrast, the second subcategory, dominance genetic variance, involves deviation due to interactions between alternative alleles at a specific locus.

For example, say that a plant produces white flowers if its genotype is A_1A_1 and red flowers if its genotype is A_2A_2. If there is no dominance deviation or interaction between these alleles, then the A_1A_2 genotype would lead to pink flowers due to expression of both alleles. However, if there is an interaction between the alleles such that their expression is not equal, then the phenotype of the heterozygote (A_1A_2) would not be an approximate midpoint; rather, it would more closely resemble one of the homozygotes (either A_1A_1 or A_2A_2). For instance, if the A_2 allele was dominant to the A_1 allele, then the heterozygote would produce red flowers.

Finally, like dominance variance, the third category of genetic variance—epistatic variance—involves an interaction between alleles; however, in this case, the alleles are associated with different loci. This concept can be illustrated by returning to the flower color example, but this time assuming that a second locus (denoted B) determines whether flower pigment is produced. In this case, the B_1 allele is dominant, and it codes for the production of pigment; an absence of pigment means that a plant's flowers will be white. Thus, if the plant's genotype is either B_1B_1 or B_1B_2, its flower color will be determined by the genotype at the A locus. On the other hand, if the plant's genotype is B_2B_2, then no pigment is produced, so the flower color is always white and is not influenced by the genotype at the A locus.

Genetic Variance and Trait Heritability

Once researchers have determined the total amount of genetic variation responsible for a trait, they can use this information in calculations of the trait's heritability. Heritability is a measure of the proportion of phenotypic variance attributable to genetic variance, and it is an important predictor of the degree to which a population can respond to artificial or natural selection.

Consider, for example, a study involving the rare plant *Scabiosa canescens* that was conducted by researcher Patrik Waldmann in 2001. Like many rare species, this plant is often found in populations of very small size. Such populations are expected to have a low level of genetic diversity, which would limit their ability to adapt. In his study, Waldmann used genetic lines of *S. canescens* from both a small and a large natural population to estimate both additive and dominance genetic variation. However, contrary to expectations, Waldmann did not find a lower level of additive genetic variance in the offspring from the smaller population. This finding meant that the smaller population still had the ability to evolve, even though the population consisted of only 25 individuals.

Environmental Sources of Variation

Like their genetic counterparts, environmental sources of variation can also be placed into various subcategories; these include specific environmental variance (V_{Es}),general environmental variance (V_{Eg}), and genotype by environment interaction (V_{GxE}).Together, the values for each of these subcategories yield the total amount of environmental variation (V_E) responsible for a particular phenotypic trait:

$$V_E = V_{Eg} + V_{GxE} + V_{Es}$$

Defining the Subcategories of Environmental Variance

Again, a closer understanding of each of the aforementioned subcategories is critical to an accurate interpretation of the concept of environmental variance. Here, specific environmental variance refers to the deviation from the population mean due to the environmental conditions that are uniquely experienced by each individual. Statistically speaking, this is called the error or residual variance. From the perspective of a plant or animal breeder, specific environmental variance is basically random noise in the expression of the true phenotype, and it can interfere with artificial selection of traits of commercial interest by weakening the correlation between genotype and phenotype.

The second subcategory of environmental variance, general environmental variance, is attributed to the nongenetic sources of variation between individuals that are experienced by multiple individuals in a population. This is typically the largest component of variance in populations in natural conditions. While specific environmental variance is determined by the variation within replicated genetic lines, to obtain an estimate of general environmental variance, replicates of each of the genetic lines need to be assessed in each natural or experimental environment of interest. If just one environmental treatment is used, then all nongenetic sources of variation are due to the specific environment.

Estimates of general environmental variance can be determined by growing or raising the same set of genetic lines in all environments of interest. For example, water is required for all plants, but the amount that promotes maximum growth differs among both species and genetic lines of a species. Consider an experiment that grows 10 genetic lines of corn in each of three levels of soil moisture (low, medium, and high). On average, the genetic lines grow slower in the low and high moisture soils as compared to the medium moisture soil. Here, the changes in growth rate due to the different water treatments are equal to the general environmental variance.

Finally, the third subcategory of environmental variance—genotype by environment interaction—involves the unique or different responses of genetic lines to general environmental variation. To better understand this concept, once again consider the corn grown in different water levels. Among the 10 genetic lines under study, the response to water availability may not be parallel. For instance, some genetic lines may be less negatively affected by low or excessive moisture levels. In this case, there is no genotype by environment interaction, as all of the genotypes have a similar response to the same environmental change. However, in Figure 1B, the lines cross, which indicates that the different genotypes have different responses to an identical change in water availability. This differential response to environmental change (or genotype by environment interaction) can contribute to a species' ability to be successful in heterogeneous environments.

Environmental Variance and Phenotypic Plasticity

Both general environmental variance and genotype by environment interaction are aspects of a characteristic known as phenotypic plasticity. Phenotypic plasticity is a change in the phenotype associated with a given genotype in response to different environmental conditions.

Some researchers believe that genotype by environment interaction provides the best estimate of the potential for phenotypic plasticity to evolve. Other researchers argue that the correlation of genotype expression across two environments is a better indicator of this potential.

PHENOTYPIC VARIANCE: A SUMMARY

In summary, the phenotypic differences among the individuals in a population can be attributed to both genetic and environmental sources. Determination of the underlying sources is essential for assessment of the potential of the population to evolve and adapt to heterogeneous or changing environments.

4

POLYPLOIDY

INTRODUCTION

Polyploidy is the process of genome doubling that gives rise to organisms with multiple sets of chromosomes.

In other words, polyploidy is the condition in which a normally diploid cell or organism acquires one or more additional sets of chromosomes.

The fusion between isolated somatic protoplasts under *in vitro* conditions and subsequent development of their product to a hybrid plant is known as somatic hybridization.

Somatic fusion, also called protoplast fusion, is a type of genetic modification in plants by which two distinct species of plants are fused together to form a new hybrid plant with the characteristics of both, a somatic hybrid.

In fact, recent findings in genome research indicate that many species that are currently diploid, including humans, were derived from polyploid ancestors (Van de Peer & Meyer, 2005). These species that have experienced ancient genome duplications and then genome reduction are referred to as paleopolyploids. This article discusses the mechanisms underlying polyploidy, and both the advantages and disadvantages of having multiple sets of chromosomes.

MECHANISMS OF POLYPLOIDY

How does an organism become polyploid? Polyploids arise when a rare mitotic or meiotic catastrophe, such as nondisjunction, causes the formation of gametes that have a complete set of duplicate chromosomes. Diploid gametes are frequently formed in this way. When a diploid gamete fuses with a haploid gamete, a triploid zygote forms, although these triploids are generally unstable and can often be sterile.

If a diploid gamete fuses with another diploid gamete, however, this gives rise to a tetraploid zygote, which is potentially stable. Many types of polyploids are found in nature, including tetraploids (four sets of chromosomes), hexaploids (six sets of chromosomes), and other chromosome-pair multiples

Researchers usually make a distinction between polyploids that arise within a species and those that arise due to the hybridization of two distinct species. The former are known as autopolyploids, while the latter are referred to as allopolyploids. Autopolyploids are essentially homozygous at every locus in the genome.

However, allopolyploids may have varying degrees of heterozygosity depending on the divergence of the parental genomes. Heterozygosity is apparent in the gametes that polyploids produce. Allopolyploids can generally be distinguished from autopolyploids because they produce a more diverse set of gametes.

POLYPLOIDY INDUCTION IN FISH

Chromosome manipulation research has a short history in fish compared to that of crops and animals. Since 1943, early attempts were initiated, and until recently various techniques have been developed to interfere with normal functioning of the metaphase spindle apparatus during nuclear cycles of cell division in fish eggs using several causal agents, both physical and chemical. As a result, individuals with differing genomic status, viz. polyploids (triploid and tetraploid), gynogenetics (both meiotic and mitotic gynogens) and androgenetics, are being produced in fish population.

Among polyploids, triploid individuals are expecting to be functionally and endocrinologically sterile due to their meiotic inhibition of gametogenesis and lack of essential steroid hormone levels to support gonadal growth. Such sterility in triploid fish can be of advantage to aquaculture where control of reproduction and population is desirable. The production of putative tetraplo,ids might have treme.ndous impact, because of promising future of large scale production of genetically sterile fish (triploids) population lies with the mating of normal diploid and tetraploid individuals.

Induction of two types of gynogenesis has its potential in generating nearly all female and all homozygous individuals in a single generation. Probable application of meiotic gynogenetics in aquaculture relates to coupling with sex inversion using androgen hormones to produce all outbred monosex female population where growth rate of females are superior to males particularly in salmonids and cyprinids.

Mitotic gynogenetics have limitations to use them directly for culture but they are potential and valuable as completely homozygous broodstock to produce second generation of clonal lines. Production of viable androgenetic diploids in fish has its great impact on commercial application of genetic masculinization technique to replace hormonal sex-reversal. Present review highlights mai.nly on the various chromosome manipulation techniques which would easily be used to suppress cell divisions in fish eggs and their probable implications to the cultured species to benefit aquaculture.

For the increasing production of carps it is necessary that they should be made to breed in confined water so that increasing demand of good quality fish and their seed could be available.

This can be done by adopting induced breeding technique by which ripe or mature fishes breed in confined water when stimulated by injection of pituitary hormone. The pituitary hormone is an important gonadotropin, which is extracted from the hypophysis of a mammal or a mature fish.

Induced breeding of carps in captivity by the use of pituitary hormone injection has been successfully done. Many scientists are involved in doing research to improve induced breeding technique. Several attempts are being made to establish pituitary banks to meet out the need of pituitary hormones throughout the year.

Induced Breeding Technique

Induced breeding technique has following steps:

Collection of Pituitary Extract

Pituitary gland is collected from a mature fish, which is called as a donor fish. Most widely used donor fish is the common carp, *Cyprinus carpi*o. The best time for preparation of pituitary extract is May and June. The Indian major carps like *Catla catla* , *Labeo rohita, Labeo kalbasu, Labeo gonius and Cirrhinus mrigala* do not breed in confined water and have need to be subjected to induced breeding.

To remove pituitary gland, the head of the fish is dissected and brain is exposed. The gland is immediately removed from the brain and stored in a refrigerator. It may be preserved in absolute alcohol at room temperature.

The gland then is homogenized in distilled water. The homogenate is centrifuged and clear supernatant is used as source of hormone to which 0.3% sodium chloride solution is added to it. This extract is ready for the immediate use. If pituitary extract is to be stored for a longer period the glycerin or trichloroacetate acid may be used instead of sodium chloride.

Selection of Breeders

Medium sized fully ripe and healthy fish of around 2 to 4 years of age is preferred for induced breeding. The weight should be 1 to 5 kg. Healthy male and female breeders should be identified and netted out before the breeding season and should be kept in spawning pools. They should be provided with supplementary food.

Injection of Pituitary Extract

To ensure higher success rate of fertilization it is important to coincide time of ovulation with the release of milt of male fish. For this purpose usually ratio of female and male 2:1 is maintained in every set. Dose of pituitary extract to be given is decided according to age, sex, weight size and state of maturity of both donor and recipient. A dose 2 to 3 mg of gland per kg body weight is given to female breeder.

There is no need of injecting dose to the male breeder if it in a state of milt oozing. After 6 hours of the first dose of injection another dose of 5-8 mg of gland per kg of body weight may be given to female if needed. However, a dose of 2-3 mg per kg body weight is recommended for the male breeder. More than 2 injections should not be given.

The injection given may be intra-macular at caudal peduncle or shoulder or intra-peritoneal at the bases of paired fins. The first injection should be given at the early hours of the day, while the second one in the evening. Weather should be rainy or cloudy for easy and early spawning. The fishes should be transferred to the breeding hapa after injecting the pituitary hormone.

Spawning in Breeding Hapa

A pair of breeder is released into the breeding hapa for spawning after injection of pituitary extract. The breeding hapa is a rectangular case of fine netting. For larger fishes its size is 82 x 32 x 32 , but for the smaller fishes it is 5'x 3'x 32 . It is held on four bamboo poles, one at each corner of the rectangular case. The roof of the hapa may be open or closed.

The hapa is made of mosquito net cloth through which laid eggs and milt cannot escape out. Three-fourth part of hapa is summarized in water whereas upper one-fourth part remains in air. After 3 to 6 hours of injection of pituitary extract spawning takes place.

The fertilized eggs are white and opaque whereas unfertilized eggs are transparent and bead-like. A hatching hapa is also rectangular and made of muslin or malmal cloth and is open from above. The mosquito net hapa is present inside the hatching hapa.

Precautions for Induce Breeding

To avoid diseases and parasitic infections, breeders should be properly washed with $KMnO_4$ solution (0.5 g in 100 litres of water) for a few minutes. After this they should be kept in formalin (200 mg/lt of water) for one hour.

POLYPLOID INDUCTION IN SHELLFISH

Polyploid induction in shellfish is different from that in fish as both polar bodies I and II can be present after fertilization. Therefore, tetraploids and aneuploids can be produced, as well as triploids, during triploid induction for Pacific oysters, *Crassostrea gigas*. Temperature and salinity can affect the rate of induction of polyploidy in shellfish and invertebrates, specifically the geoduck clam, *Panope abrupta*. The timing of the meiotic divisions and extrusion of the first and second polar bodies are different for fish and shellfish. The timing of application of cytochalasin B (CB) affects the ploidy induction and the segregation of the chromosomes in Pacific oysters. Fertilized eggs are 2N, 3N or

4N if the treatment affects polar body I, 3N if the treatment affects polar body II and 5N if the treatment affects both polar bodies I and II.

However, researchers produced at least some of the three ploidy levels, 2N, 3N and 4N, when attempting to block either polar body I or II. If polar body I is blocked, 60% of the polyploids are aneuploid, but if polar body II is blocked, the number of aneuploids is no different from that in diploid controls. An unusual configuration of the maternal chromosomes occurred following the blockage of polar body I, which possibly leads to the high proportion of aneuploids and also abnormal tetraploids. The 20 maternal dyads formed three division planes oriented in a tripolar configuration.

Temperature affects the rate of triploid induction in Pacific oysters, and Downing (1988a) recommended that the optimum temperature for producing triploids (88%) was 25°C for 30–45 min after insemination when using CB. However, the concept of τ_0 could probably be applied to shellfish, and the application of the shocks or treatments applied based on developmental time (temperature minutes) rather than absolute time associated with a single temperature. CB also adversely affects survival for the first 48 h after treatment.

Sperm quality affects rate of triploid induction in Pacific and American oysters and their hybrids, and poor sperm quality leads to very low triploid induction rates. The application of cytochalasin B (CB) to induce the polyploidy results in the loss of some embryos, and usually the survival of monospecific embryos is higher than that of their interspecific diploids and triploids. Triploid induction for eastern oysters was 100% at 20 ppt, but was reduced to 84% at 30 ppt, whereas triploid induction for Pacific oysters was 100% at both salinities.

Tetraploid Pacific oysters were produced by using polyethylene glycol (PEG) treatment to fuse blastomeres. This resulted in 1–4% tetraploids. Viable tetraploid *C. gigas* have been produced by inhibiting the first polar body of eggs from triploids that had been fertilized with sperm from diploids, but repeatability for producing high yields of tetraploids is inconsistent. Varying the duration of the treatment to inhibit polar body I of triploid eggs had definite effects for optimizing tetraploid production.

Short treatments 15–35 min after fertilization (approximately half the period of meiosis I in triploid eggs) yielded individuals with tetraploid and heptaploid cells. Longer treatments 7–43 min after fertilization (about three-quarters of the period of meiosis I in triploid eggs) yielded embryos with 100% heptaploid cells. Tetraploid induction was most consistent when treatments were applied to eggs from individual triploid females rather than from multiple pooled females. Eggs from individual triploids were fertilized and 0.5 mg/l CB added after 10 min. A subsample of the fertilized eggs was untreated. When 50% of the untreated eggs exhibited polar body I extrusion, the CB treatment was terminated. Percentage tetraploidy ranged from 13 to 92% after 8 days, with a mean of 55%. Seven of eight replicates went through metamorphosis and settlement, and at settlement the percentage of tetraploids ranged from 7 to 96%, with a mean of 45%, allowing the establishment of tetraploid brood stock for the natural production of triploids.

Application of this technique – the crossing of triploid females with diploid males to produce tetraploids – was also attempted in the pearl oyster, *Pinctada martensii* and eastern oysters. Ploidy in the embryos ranged from 2N to 5N. At 1 year, the majority of the survivors were diploid, triploid and aneuploid, but 2% viable tetraploids remained. The number of ripe female triploid eastern oysters was extremely low, making production of tetraploid individuals from CB-treated 2N × 3N crosses extremely difficult, but it has been accomplished.

INDUCTION OF TRIPLOIDY

Triploidy is induced directly by blocking of second polar body extrusion during second meiotic division shortly after fertilization of fish eggs using various physical shocks and chemical treatments. The mating of normal diploid and tetraploid fish is an alternative method for producing hybrid triploids.

Temperature treatments (both heat and cold shock) of fertilized eggs have widely been used to suppress second polar body extrusion for induction of triploidy in fish. Temperature shock treatments are inexpensive to apply and can be successfully adapted for mass production by fish farms. Cold shock in general has been most successful in warm water fishes. However, heat shock was 100% effective in tilapia (*Oreochromis mossambicus*).

essfully used to block second polar body extrusion in rainbow trout (*Onchorhynchus mykiss*), zebrafish (*Danio rerio*), common carp (*Cyprinus carpio*), Nile tilapia (*Oreochromis niloticus*), yellow perch (*Perca flavescens*), coho salmon (*Oncorhynchus kisutch*), and hybrid channel catfish (*Ictalurus punctatus*), blue catfish (*Ictalurus furcatus*).

Table 1. Methods for inducing triploidy in fish

Method	Species
Cold shock	Oncorhynchus mykiss
Heat shock	Oreochromis mossambicus
Pressure shock	Danio rerio
Nitric oxide	Salmo salar
Freon	Salmo salar
High pH & Ca++	Oncorhynchus mykiss
Cytochalasin B	Salmo salar
Colchicine	Salvelinus fontinalis

Chemicals were also used to block polar body extrusion in fertilized eggs of fish. Researchers produced mosaic polyploid–diploid Atlantic salmon (*Salmo salar*) after exposing fertilized eggs to cytochalasin B. Researchers reported similar production in brook trout (*Salvelinus fontinalis*) after exposing fertilized eggs to colchicine. Among anaesthetics, nitrous oxide induced high triploid yield (79.7%) followed by freon induced lower yield in Atlantic salmon. Either sperm or freshly fertilized eggs of rainbow trout when treated with high pH and calcium resulted in high incidence of triploidy (Table 1).

TRIPLOID FISH PERFORMANCE

Growth performances in triploids

Growth rate is an important characteristic for the genetic improvement in fish for commercial production. Triploids are expected to have a higher growth potential due to sterility and reduced gonadal development. Growth performances between diploids and triploids vary between species. Growth rates in triploids are not significantly different compared to their diploid controls in mud loach (*Misgurnus mizolepis*), Atlantic salmon and two species of tilapia (*Oreochromis niloticus* and *O. aureus*). Triploids in species such as rainbow trout, sunshine bass (a hybrid), coho salmon, Atlantic salmon tend to grow poorly when compared to their diploid counterparts.

However, triploids have a better growth rate than diploids at maturation in catfish and tilapia. Better growth has been reported in triploids of tilapia and European catfish at the age of 14 weeks. No weight gain was observed in female triploid brook trout even after long-term treatment with anabolic steroid, estradiol.

Haematology of triploids

The cellular and nuclear dimensions of erythrocytes are increased in various species of triploid fish. However, this increase in cell size is compensated by a proportional decrease in number of erythrocytes, thereby maintaining a constant haematocrit value for a particular species. The one exception was no decrease in erythrocyte numbers in rainbow trout *O. mykiss*. Increases in the size of erythrocytes in triploid fish were followed by higher haemoglobin content per cell, but total haemoglobin levels in triploid fish showed variable results in different species. Triploids of *Atlantic salmon*, coho salmon, white crappies (*Pomoxis annularis*) and rainbow trout exhibited lower haemoglobin levels than diploids, whereas fish like hybrids of grass carp and bighead carp, ginbuna, ayu (*Plecoglossus altivelis*), brook trout had haemoglobin levels similar to their diploid counterparts.

There was a significant decrease in the haematocrit value and total blood haemoglobin after long term treatment with estradiol-17β. Increased enzyme activity was assayed in triploid ginbuna and spinous loach (*Cobitis biwae*). However, this enzyme activity was balanced by fewer erythrocytes in triploid fish, keeping the activity level similar to diploids in per unit volume. Physiological parameters of blood such as oxygen carrying capacity, pH and nucleotide triphosphate (NTP) levels in the blood of triploid Atlantic salmon and size of the heart in triploid white crappies were not significantly different from diploids.

Plasma osmolality and oxygen consumption recovered earlier in triploid brook trout after exhaustive exercise when compared to diploids.

Food conversion efficiency in triploids

The feed conversion efficiency has been well studied in triploid grass carp. African catfish (*Clarias gariepinus*) and rainbow trout. No significant differences were observed in protein efficiency ratio, net protein utililization, efficiency of energy gain, total energy intake and nitrogen balance in these species. Triploid females, however, showed higher fat deposition in viscera and better muscle pigmentation than maturing diploid females. During development, embryo, larvae and juveniles of triploid rainbow trout do not differ from diploids with respect to ammonia excretion rate, energy expenditure, oxidation of glucose and amino acids. Still, utilization of protein for energy production was significantly lower in juvenile triploids.

Behavior of triploids

The central nervous system (CNS) and hormones regulate the initiation, coordination and execution of all kinds of behavior. The reduction in cell numbers in CNS and reduced levels of hormones in triploid fish suggest lower responsiveness to various environmental stimuli leading to expression of abnormal behavior in triploid fish. Triploids were found to be less aggressive in rainbow trout, ayu, fighting fish (*Beta splendens*), Atlantic salmon than diploids. Triploid ayu were less responsive to sound and light stimuli than diploid fish. Indicators, such as minimum voltage required for eliciting a response and number of days spent in learning to avoid electric shock, in triploid brook trout, and lower jumping efficiency in triploid ayu when crossing a barrier, did not differ from diploids.

Endocrinology of triploids

Gonadotropin releasing homone (GnRH) has central role in vertebrates by regulating the release of gonadal steroids mediated through gonadotropin (GtH). A positive correlation has been observed between brain GnRH and gonadal maturity in teleosts. However, there are only two reports available on the nature of GnRH in triploid fish. The first is reduced level of GnRH in triploid rainbow trout and the second finding is decrease in size and number of immunoreactive GnRH cells in preoptic area (POA) of Indian catfish and low immunoreactivity in pituitary.

The low activity of GnRH in triploid fish may be due to absence of positive feedback stimulation by sex steroids and/or decreased responsiveness of sensory cells to environmental cues required for gonadal maturation in teleosts.

Female triploid fish exhibited low levels of gonadotropin and sex steroids in various species. Triploid females have a low biosynthesis of vitellogenin due to low levels of sex steroids. However, exogenous estradiol-17b (E2) can induce production of vitellogenin in coho salmon and rainbow trout.

Researchers observed similar ovarian steroidogenesis and gonadotropin induced E2 production in both diploid and triploid female brook trout. Either progesterone or E2 was able to upregulate the mRNA levels of estradiol receptor (ER) in the brain of triploid female rainbow trout but they have no co-operative effect. This suggests occurrence of ER in the brain of triploid female as ER-mRNA and immunoreactive ER were also previously detected in triploid rainbow trout. Female triploids showed reduced levels of E2 and T when compared to diploids.

Female triploid brook trout showed low plasma E2 and testosterone levels after long-term treatment with E2 in the diet. However, the plasma levels of sex steroids were positively correlated with advancing developmental stages of ovaries. Plaice (Pleuronectes platessa) and flounder (Platichthys flesus) hybrid females had comparable levels of sex steroids in both diploids and triploids.

In contrast to females, male triploid fish have shown normal levels of sex hormones in most species except Indian catfish. A key enzyme in steroid biosynthesis, steroid dehydrogenase (3b-HSD) and immunoreactive testosterone was localized in steroidogenic cells of male tilapia. Male tilapia also had normal levels of testosterone similar to diploids.

Triploid males of Indian catfish showed reduced levels of maturational gonadotropin (GtH-II) and sex steroids in plasma. 11-ketotestosterone is a prominent male sex steroid in fish. Further study on this male specific hormone in triploid fish may show the impact of triploidy on reproductive endocrinology in male fish.

Disease resistance

Disease resistance has been little studied in triploid fish. Physiological studies indicate that triploids should have inferior disease resistance. However, diploid and triploid Thai walking catfish had similar resistance to *Aeromonas hydrophila*. Triploid ayu, *P. altivelis*, had similar resistance to *Vibrio anguillarum* to that of diploids.

Differences in haematology between triploids and diploids could lead to differences in disease resistance. Researchers found no differences in total or differential leucocyte counts between diploid and triploid 3-year-old tench, in which lymphocytes dominated (> 90%) the total leucocyte count. When the fish were 4 years old, there were no differences in total leucocyte counts, or in differential lymphocyte, monocyte and myelocyte counts but triploids had lower metamyelocyte counts and higher neutrophil granulocyte counts.

The immune system of triploids may be inferior to that of diploids. Sibling diploid and triploid Atlantic salmon injected intraperitoneally with lipopolysaccharide were evaluated for the alternative complement pathway activity (measured by the titre of haemolytic activity against rabbit erythrocytes). Serum iron concentrations decreased to very low levels by day 2 post-injection in the diploid fish and by day 3 in the triploid fish. The longer time taken for the triploids to recover complement activity and the slower onset of the hypoferraemic response suggest that triploids may be at a disadvantage compared with their diploid siblings in their defence against bacterial invasions.

THE EVOLUTIONARY CONSEQUENCES OF POLYPLOIDY

One of the most striking features of genome structure is its lability. From small-scale rearrangements to large-scale changes in size, genome comparisons among species reveal that variation is commonplace. Even over the short time course of laboratory experiments, chromosomal rearrangements, duplications/deletions of chromosome segments, and shifts in ploidy have been observed and have contributed to adaptation. Changes in genome structure typically have immediate effects on the phenotype and fitness of an individual. Beyond these immediate effects, changes in genome structure might allow evolutionary transitions that were previously impossible.

For example, by introducing an additional complement of chromosomes, polyploidization might release gene duplicates from the constraints of having to perform all of the functions of a gene (pleiotropy), providing extra "degrees of freedom" upon which selection can act to favor new functions. Polyploidization can also stimulate further structural changes in the genome, providing polyploid lineages with genomic variation not available to diploid organisms. Indeed, it has been proposed that tetraploidy may be an intermediate stage in some cancers, facilitating a cascade of structural changes that disrupt normal controls to cell growth. Here, researcher discuss the evolutionary impact of polyploidization, beginning with the prevalence of polyploidy, and then he explore the longer-term consequences of polyploidy on the rate and nature of evolutionary transitions.

Incidence of Polyploidization

Polyploidization is the increase in genome size caused by the inheritance of an additional set (or sets) of chromosomes (Fig 1). The duplicated sets of chromosomes may originate from the same or a closely related individual ("autopolyploid") or from the hybridization of two different species ("allopolyploidy"). When polyploidization involves duplicated sets of chromosomes that share homology but are sufficiently distinct due to their separate origins, these pairs of chromosomes are referred to as homeologs (see Fig. 1).

Polyploidy is especially prevalent among hybrid taxa, an association thought to be driven by problems with meiotic pairing in diploid hybrids, which are solved if each homeologous chromosome has its own pairing partner. Additionally, diploid hybrids form unreduced gametes (which have the same number of chromosomes as somatic cells) at unusually high rates, increasing the rate of formation of polyploids from hybrid lineages. By combining traits from two parental species and ensuring fair segregation of these traits, allopolyploids potentially benefit from "hybrid vigor" (where hybrids have characteristics that make them superior to both parental species) and an altered ecological niche without the problems associated with segregation and breakdown at the F_2 generation that occurs among diploid hybrids.

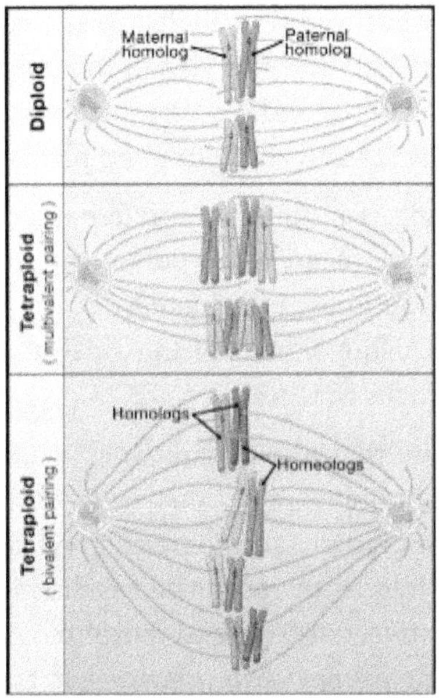

Fig. 1. Two nonhomologous chromosomes are shown (long and short), with each X-shaped chromosome representing a pair of sister chromatids joined at the centromere.

In diploids (A), each chromosome consists of a homologous pair, with one chromosome inherited from the mother and one from the father. In tetraploids (B and C), the chromosomes are further doubled. When the duplicated chromosomes are very similar to one another, they might align randomly in pairs during meiosis (bivalent pairing; not shown) or all align together (multivalent pairing; B). In either case, gametes may inherit any combination of parental chromosomes (multisomic inheritance), and mutations that arise on one chromosome can spread to all other copies, inhibiting their divergence. When polyploidization involves chromosomes that are sufficiently distinct (that is, "homeologs"; differentiated by blue and purple), the more similar pair of chromosomes tend to align together to the exclusion of the other pair (C). With strict bivalent pairing, the homeologs behave as distinct chromosomes and segregate independently (disomic inheritance), allowing their divergence. Newly formed autopolyploids typically exhibit multisomic inheritance, whereas newly formed allopolyploids exhibit a variety of patterns of inheritance, depending on the cross.

Chromosomes that have previously diverged and been brought together by hybridization typically segregate as bivalents (Fig. 1C). Genomic analyses have begun to unravel the genes responsible for bivalent pairing, ensuring that homologs rather than homeologs pair during meiosis (e.g., at the *Ph1* locus in polyploid wheat). In contrast, autopolyploids more often exhibit multivalent pairing than allopolyploids (Fig. 1B), ~3.5 times more so in plants.

In most cases, descendants of a polyploidization event in the distant past ("paleopolyploids") exhibit bivalent pairing of chromosomes and disomic inheritance (as if diploid). This observation has traditionally led to the conclusion that autopolyploids are ephemeral whereas allopolyploids give rise to the majority of long-lasting lineages. Bivalent pairing, however, can occur and is even more prevalent (63.7%) than multivalent pairing (28.8%) among newly formed autotetraploid plants.

Although bivalent pairing is initially nonpreferential in autopolyploids, leading to tetrasomic inheritance (Figure 1), segregating polymorphisms, especially rearrangements and indels (insertions and deletions), may increase pairing fidelity over time ultimately yielding disomic inheritance. Increased pairing fidelity may also stem from genetic changes, such as at the *Ph1* locus in wheat.

Thus, one cannot assume that a paleopolyploid is necessarily allopolyploid solely because it exhibits disomic inheritance; additional evidence is needed, such as phylogenetic evidence that different genomic regions are more closely related to different parental species. Autopolyploids are also less frequently recognized as distinct species than allopolyploids, even when they are reproductively isolated and morphologically differentiated from their diploid parents. Thus, recent papers have argued that autopolyploidy may contribute more to evolution and species diversification than traditionally thought.

Mutations affecting ploidy occur relatively frequently in both plants and animals. Plants produce unreduced gametes, a common route to polyploidization, at an average rate of ~0.5% per gamete. Both unreduced gametes and polyspermy contribute to the production of polyploid animals. Among chicken embryos, 0.9% are triploid or tetraploid, and among spontaneous human abortions, 5.3% are triploid or tetraploid. Placing these rates in context, gene duplication events are much rarer, with a roughly 10^{-8} chance of occurring per gene copy per generation.

Conversely, aneuploidy, the gain or loss of a single copy of a chromosome, can be much more frequent; for example, aneuploids were four times more common than polyploids among the aborted fetuses examined by researchers.

The evolutionary contribution of structural alterations to the genome depends on their ability to persist. Although the probability of a gene duplication is low, the half-life of gene duplicates is very long (over a million generations. Conversely, aneuploids often have low fitness and, in mammals, rarely survive to reproduce. Indeed, it is exceedingly rare for a homologous chromosome pair to be lost or duplicated among all members of a population. In mammals and birds, ploidy changes are also typically fatal, with polyploids dying early during development. Interestingly, polyploidy is lethal regardless of the sexual phenotype of the embryo (e.g., triploid XXX humans, which develop as females, die, as do triploid ZZZ chickens, which develop as males), and polyploidy causes much more severe defects than trisomy involving the sex chromosomes (diploids with an extra X or Y chromosome). Thus, in mammals and birds, evidence suggests that a general disruption of development—not problems restricted to sex determination—is the root cause of the failure of polyploids to persist. Specific developmental problems in human polyploid aborted fetuses have been attributed to abnormal imprinting and placental development.

In many taxa besides mammals and birds, however, polyploidy is surprisingly well tolerated. Polyploid lineages often persist in plants, which is immediately apparent from the excess of even over odd chromosome numbers. Because doubling a number always generates an even number, this excess of even chromosome numbers is a predictable consequence of polyploidization. By contrast, the fusion or fission of chromosomes simply switches whether a cell has an even or odd number of chromosomes. The excess of even chromosome numbers can be used to infer how often changes in chromosome number are due to polyploidization (42% of the time in ferns, 32% in monocots, and 18% in dicots.

In various plant genera, the rate at which polyploids arise and persist is on the order of 0.01 per lineage per million years, roughly $1/10^{th}$ the rate of speciation. With such a high rate of polyploidization per speciation, we would expect a large fraction of plant species to have undergone polyploidization at some point in their evolutionary past. Previous studies had suggested that polyploidy occurred sometime in the past of 57% of flowering plants, based solely on chromosome numbers among extant species.

Recent genomic analyses indicate that an early polyploidization event may predate the radiation of flowering plants, suggesting that 100% of angiosperms are paleopolyploid. Unfortunately, evidence for such ancient polyploidization events is almost always tentative because of the loss of sequence homology and synteny over evolutionary time. Although polyploidization is less prevalent in animals, nearly 200 independent examples of polyploidy have been reported in insects and vertebrates (Table 2), with many more cases known among other invertebrate groups.

Table 2. Polyploidization in Insects and Vertebrates

Reproduction	Insects	Fish	Amphibia	Reptiles	Birds	Mammals	Total
Parthenogenesis	89	9	3	15	0	0	106
Sexual	2	23	26	1	0	1	54
?	0	18	1	0	0	0	19

If polyploidy were an evolutionary dead end, we would expect polyploid taxa to be near the tips of the tree of life and to be relatively species poor. Instead, several polyploidization events are ancient, and several of these events gave rise to species-rich groups.

For example, multiple independent polyploidization events occurred early in the evolution of plants, fish (the diverse group of ray-finned fishes, Catostomidae, Salmonidae, and different groups of Cyprinidae), and amphibia (Syrinidae), with evidence for two additional rounds of genome duplication at the base of the vertebrate tree of life. This data is sufficient to conclude that polyploidy is not an evolutionary dead end, but it does not prove that polyploidization contributes to evolutionary success. Whether polyploidization increases the longevity and species richness of a group relative to taxa that have not undergone polyploidization is a question that has yet to be answered.

Why is it that the duplication of a whole genome often gives rise to lineages that persist over evolutionary time whereas the duplication of a single chromosome virtually never does? One plausible explanation is that polyploidization preserves the balance of gene products. This "balance hypothesis" is an old idea that traces back to the pre-genomics era; for instance, Haldane (1932) pointed out that morphological changes were more marked in trisomic than in triploid plants, arguing that "In the latter case the number of genes of all sorts is increased equally, in the former the balance is upset". Consistent with the balance hypothesis, genes duplicated by polyploidization persist longer, on average, than genes duplicated individually.

Another explanation is that organisms have evolved mechanisms to cope with changes in ploidy because of the natural variation in genome copy number associated with mitotic and meiotic cell cycles (as DNA replicates and cells divide). According to this "evolved-robustness hypothesis," organisms with a regular alternation of generations, with mitoses in both haploid and diploid phases, are predicted to be especially tolerant of shifts in ploidy. In addition, somatic variation in ploidy ("endopolyploidy") is a normal part of development in many animals as well as plants.

Most famously, the chromosomes of the salivary gland are highly replicated in flies, leading to visible polytene chromosomes. In mammals, multinucleate cells are found in hepatocytes and osteoclasts, whereas megakaryocytes, trophoblasts, and hepatocytes display endopolyploidy (that is, have a nucleus with multiple copies of the normal complement of DNA). In summary, the existence of regular mitotic cell cycles, an alternation of generations, as well as endopolyploidy ensures that organisms have experienced, and survived, an evolutionary history at different ploidy levels.

The balance hypothesis and the evolved-robustness hypothesis are not opposing explanations. Instead, they may serve as proximate and ultimate explanations for the same phenomenon—present-day organisms function better with a balanced set of chromosomes because their evolutionary past involved changes in ploidy that preserved the balance, but not the absolute number, of chromosomes.

Immediate Effects of Polyploidization

At a phenotypic level, the effects of polyploidization are often mild and idiosyncratic. Cell volume generally rises with increasing genome size, although the exact relationship between ploidy and cell volume varies among environments and taxa. In yeast, for example, the volume of diploid cells in rich media is 2.4 times that of haploid cells at 30°C but only 1.1 times at 37°C. Larger cells tend to have smaller surface area to volume ratios, a phenomenon thought to lower the growth rate of polyploid cells. Whether or not cell geometry affects growth rate depends on the environment, as expected given that transport across the membrane limits growth only under certain circumstances. Perhaps as a consequence of a slower metabolism, polyploids tend to exhibit slower development, but this pattern is not universally true and can be reversed (e.g., in groups where polyploid eggs or seeds are larger). Interestingly, not all features of a cell scale with ploidy level, which can have important side consequences.

For example, researchers showed that kinetochore size and length of the pre-anaphase spindle do not scale with ploidy level, whereas the spindle pole body does. These authors argue that changes in ploidy thus upset geometric relationships among key components of the machinery used to segregate chromosomes during meiosis, potentially explaining the higher rate of chromosome nondisjunction in tetraploids. To prevent the detrimental effects of genomic instability, animals may have evolved mechanisms limiting the proliferation of tetraploid cells, which would present another barrier to the establishment of polyploid lineages.

Although cell size typically is larger in polyploids, adult size may or may not be altered; as a rough generalization, polyploidization is more likely to increase adult body size in plants and invertebrates than in vertebrates. The poor correlation between cell size and organismal size was even remarked upon by Albert Einstein, who wrote "Most peculiar for me is the fact that in spite of the enlarged single cell the size of the animal is not correspondingly increased". The key to accurately predicting the effects of ploidy on body size must come from developmental biology. In cases where morphogen gradients guide development, ploidy need not affect adult body size because ploidy need not alter the overall density of cellular material, only how it is packaged (i.e., into cells that are twice as large and carry twice as much DNA). By contrast, where growth is determined by cell-cell interactions or where there is a fixed number of cells in the adult, ploidy, by altering cell size, should directly influence adult size.

A surprising feature of many newly formed polyploids is that their genomes are unstable and undergo rapid repatterning. For example, researchers observed extensive genomic rearrangements and fragment loss within five generations in newly created polyploid *Brassica* hybrids, and more recent studies have documented genomic changes soon after formation of wheat and *Arabidopsis* allopolyploids (but not in cotton or *Spartina*. In most examples studied to date, rapid genomic repatterning has been observed in allopolyploids, and there are many reasons to expect that hybridization may be causally responsible.

ransposable elements that are repressed within each parent lineage but activated in hybrids can facilitate the movement of genes and promote unequal crossing over. For example, researchers found that maternally derived siRNAs are not sufficient to repress retrotransposons in the paternal genome of *Arabidopsis thaliana* × *A. arenosa* hybrids. Divergence of centromeres and centromeric histones can lead to segregation distortion and nondisjunction in hybrids. In addition, nonhomologous recombination and nonreciprocal exchanges are particularly likely among homeologous chromosomes that bear structural rearrangements.

Nevertheless, genomic repatterning in polyploids is not entirely driven by hybridization. In autotetraploids of both *Candida albicans*, reduction in genome size through chromosome loss has been observed, largely restoring the diploid complement. The exact mechanism by which this reduction occurs is unknown, but similar reductive divisions (termed "neosis") have been observed in human cell lines that have undergone endopolyploidization in response to carcinogens.

By altering the genomic context of genes, genomic repatterning can increase the genetic variability available to newly formed polyploid populations. This variability might be especially important given the bottleneck in population size associated with the founding of new polyploid lineages.

That said, genetic variability can only fuel the evolution of a polyploid population if individuals can survive the onslaught of genomic mutations. If the genomic mutation rate is too high and the fitness effects of mutations too severe, extinction is the likely outcome. Polyploid lineages that have survived are almost certainly a biased subset of those that have been generated; we only witness those lineages that chanced upon a particularly fit and stable genomic configuration soon after polyploidization.

In addition to structural changes, polyploids often exhibit changes in gene expression. This is especially true of allopolyploids, which exhibit changes in methylation (e.g., 30% of the methylation patterns were altered in allopolyploids of *Spartina*, disruption of heterochromatin (potentially explaining activation of the retrotransposon *ATHILA* in *Arabidopis thaliana* × *A. arenosa* allopolyploids, and biased expression of homeologs (e.g., 43% of genes exhibited biased expression in allopolyploid cotton. These changes in gene expression are strikingly idiosyncratic and tissue specific.

For example, among newly formed cotton allopolyploids, researchers found that alleles from the two hybridized genomes (designated A and D) differ in expression patterns among the tissues examined for 11 out of the 18 genes considered. In the most extreme case, *adhA* was entirely expressed from the A gene in carpels but from the D gene in petals and stamen. In some cases, it has been shown that the majority of changes in gene expression are due to hybridization, rather than ploidy changes *per se*. Even where ploidy matters, disruptions in gene regulation have been associated with divergence between the parental species (e.g., *A. thaliana* contains fewer *ATHILA* elements and these elements differ in sequence from those in *A. arenosa*. Consequently, we would expect much smaller effects on gene expression in autopolyploids. Indeed, microarray analysis has revealed relatively few changes in the relative expression of genes in autotetraploids, for example in maize.

Although genomic repatterning and gene expression changes are more extensive in studies of allopolyploids, we have only limited information about how the extent of change depends on the genetic distance between the parental genomes. It may be that the changes in expression observed are particularly dramatic when the allopolyploids studied involve very divergent genomes (e.g., 7.5 million years [MY] in the case of the cotton studied by researchers; 11 MY between the *Brassica rapa* and *B. nigra* studied.

Indeed, observed that less extensive genomic rearrangements occurred in the allopolyploid formed from the more closely related *B. rapa* and *B. oleracea* (separated by ~6 MY, based. There remains a need to explore how the extent of genomic repatterning and changes in gene expression depend on the divergence between the parental genomes. It is possible that autopolyploids, if formed from distantly related individuals within a species, might also exhibit rapid genomic repatterning and gene expression changes. Alternatively, dramatic changes might be limited to allopolyploids formed from distantly related lineages.

Establishment of Polyploidy

What allows a polyploid lineage to become established once it has arisen? Although many potential evolutionary advantages and disadvantages of polyploidization have been proposed (reviewed below), they pale in significance relative to the immediate ecological differences between a newly formed polyploid and its diploid relatives. The reason has to do with timing: a polyploid lineage must survive long enough for evolution to act, and it will do so only if it is not immediately outcompeted by its diploid relatives. Like most mutations that affect fitness, polyploidization often reduces fertility and/or survival; researchers found, on average, a 20% reduction in pollen viability and a 50% reduction in seed production in their survey of newly formed plant polyploids. However, occasionally, the altered suite of characters displayed by a polyploid—for instance, drought tolerance or pathogen resistance—might better suit the environment or shift a polyploid into a distinct ecological niche. The polyploidy can permit ecological traits from two parental species to be combined together and stabilized as fixed heterozygotes (assuming the hybridizing genomes are sufficiently distinct that disomic inheritance occurs). This advantage may partially account for the unusually high proportion of polyploidy plants in previously glaciated regions, where the hybridization of previously isolated populations may be frequent, where new environments are common, and where competition may be limited.

One hurdle to the establishment of polyploids is reproductive: mating between a newly formed tetraploid and a diploid relative produces triploids with low fitness. Triploid seeds often have lower germination rates. Interestingly, germination success often depends on the source of the haploid versus diploid gamete.

Data reviewed by Researchers indicate that triploid seeds are 19.9% as viable as diploid seeds when the ovule is diploid but only 4.8% as viable when the pollen is diploid. It has been argued that these failures result from an imbalance of maternally and paternally imprinted genomes in the endosperm, a hypothesis that is supported by the effects on seed development observed when imprinted genes are disrupted (e.g., the *MEDEA* mutation in *Arabidopsis*. Even if they reach maturity, triploid individuals also have severely reduced fitness in most sexual taxa because of the fertility problems associated with producing aneuploid gametes; on average, pollen fertility of triploids is only 31.9%.

Several factors enable polyploid plants to overcome the hurdle posed by triploidy, including the ability to self or asexually reproduce, the ability to persist for long periods of time until a suitable mate is found (perenniality), or the tendency to mate assortatively (e.g., according to body size, flowering time, or habitat). Moreover, both genetic and environmental factors are known to alter the frequency at which polyploids are formed, making it more likely that where there is one new polyploid, others will be found. Finally, even the production of triploids is no longer considered to be the death knell of new polyploid lineages: of the viable gametes produced by triploids, a disproportionately high fraction are euploid (having one or more full sets of chromosomes) and can contribute to the production of polyploid offspring.

PRODUCTION AND PERFORMANCE OF TRIPLOID OYSTERS

Introduction

According to the Food and Agricultural Organization (FAO), molluscan shellfish aquaculture, including oysters, clams (including cockles and arks), mussels, scallops, and abalones, accounts for more than half of the world's marine aquaculture production. Among the aquaculture shellfish groups, oysters led the annual production between 2003 and 2007, and since 2008 were second behind clams. Aquaculture information from the FAO lists over 15 oyster species from genera *Ostrea* and *Crassostrea* that are currently being farmed worldwide with a total production of more than 5 million tonnes and sales value exceeding $4 billion.

In the United States, oyster aquaculture includes five species: the Pacific oyster *Crassostrea gigas*, eastern oyster *Crassostrea virginica* species, Kumamoto oyster *Crassostrea sikamea*, Olympia oyster *Ostrea conchaphila* (or *O. lurida*, a native species in the US West Coast), and European flat oyster *Ostrea edulis*. According to the most recent aquaculture census conducted by the U.S. Department of Agriculture (USDA) in 2012, 756 farms in 18 states are involved in the molluscan aquaculture industry with a sales value of $329 million.

Among the cultured shellfish species, the Pacific oyster accounted for 26.4% of the production for an economic value of $87 million on the west coast, while the eastern oyster accounted for 19.7% of the production for an economic value of $68 million on the east and Gulf of Mexico coasts, and other oysters accounted for 7.6% of the total shellfish aquaculture production.

Triploid-tetraploid technology is likely the genetic breeding most widely recognized by the oyster aquaculture industry worldwide and is utilized for commercial oyster seed production. This publication reviews triploid oyster production and performance, and explains methods used to produce and raise triploid oysters.

Triploid oyster aquaculture

Like most animals, normal oysters have two sets of *chromosomes* (which means that they are diploid) with one set inherited from the eggs and the other inherited from the sperm. Triploid oysters have three sets of chromosomes and are valuable to the commercial aquaculture industry because of their potential for fast growth, superior meat quality (especially in summer), year-round harvestability, and low environmental pressure on wild populations (due to their sterility). Tetraploid oysters possess four sets of chromosomes and are valuable in the production of 100% triploid seed when crossing with normal diploids.

Triploid oyster aquaculture has become an important part of the global industry. Currently, triploid Pacific oysters account for about 50% of the production on the west coast of the United States and 100% of the hatchery seed production in France. Triploid eastern oysters account for nearly 100% of the seed production in the Chesapeake Bay. Triploid Sydney rock oysters *Saccostrea glomerata* account for about 15% of the production in Australia. Commercial triploid oysters are typically produced by crossing tetraploids with normal diploids to yield 100% triploid seed. This approach allows for implementation by commercial hatchery operations and has supported oyster farming operations throughout the world.

Methods for triploid oyster production

There are two approaches used for production of triploid oysters. The first approach is to inhibit the release of *polar body I* or *polar body II* after fertilization. This method has been used since 1981 when the first report of triploid production was published. The second approach involves crossing normal diploids with tetraploids to produce 100% triploid offspring. This method is currently used in commercial triploid seed production but relies on the availability of tetraploid breeding stocks.

Triploid production by inhibiting the release of polar body I or II

Oysters, especially *Crassostrea* species, reproduce by releasing their *gametes* (eggs from females and sperm from males) into the surrounding water for fertilization and embryo development. For many bivalve mollusks (oysters, clams, mussels, and scallops), eggs released from females are not completely mature (arrested in the prophase of *meiosis* I). Once fertilization occurs, the eggs resume the maturation process (meiosis I and II) and release polar bodies I and II. The fertilized eggs develop into embryos and then free-swimming larvae after 24 hours. After about 2–3 weeks, swimming larvae go through metamorphosis and transform into sessile juveniles (spat), and continue to grow into adults.

Based on oyster reproductive characteristics, production of triploids can be achieved by inhibiting polar body I or polar body II, which alters the way chromosomes are inherited. This approach has been used to produce triploids in many shellfish species. Polar body I contains two sets of joined matching chromosomes and when inhibited can change the way chromosomes are separated, resulting in a mix of *ploidies*, including tetraploids, triploids, diploids, and *aneuploids* (organisms that do not possess whole sets of chromosomes).

Often, only triploids and diploids survive to spat stage. Inhibition of polar body II usually produces a majority of triploids because the extra set of chromosomes in polar body II is kept. Therefore, inhibition of polar body II has been used as the effective approach for triploid production in shellfish.

Effective methods used to inhibit polar body I or II include: 1) physical treatments, such as high-pressure shock, heat shock, and cold shock; and 2) chemical treatments, such as cytochalasin B (CB), 6-dimethylamnpurine (6-DMAP), caffeine, colchicine, and nocodazole (see summary in Guo et al. 2009). For different shellfish species, the effective treatment conditions, such as concentration, duration, and starting time, are slightly different; details for the species studied so far were summarized in a review. For oysters, CB and 6-DMAP were the most commonly used methods for effective inhibition of polar bodies.

Overall, inhibition of polar body I or II is an effective direct approach to produce triploids. Generally, the inhibition of polar body II produces higher percentages of triploids compared to the inhibition of polar body I, but no significant differences were identified when a systematic comparison was made using the same parents in hard clams.

Although inhibition of polar body I or II is an effective approach to produce triploids, this approach is not commonly used for commercial triploid seed production because: (1) 100% triploid offspring are rarely produced, and they vary greatly depending on gamete quality, egg development, treatment timeframe, duration, and water temperature; (2) triploid larval survival is lower than larval survival in control groups due to the treatment for polar body inhibition; and (3) chemicals used to inhibit polar bodies, such as cytochalasin B, may be a threat to human health.

To overcome the disadvantages of inhibiting polar body I or II for commercial triploid seed production, another method involving tetraploid oysters can be used. Tetraploid oysters offer a solution because they can be crossed with normal diploids to produce 100% triploids. A review of tetraploid oyster production will be covered in a follow-up publication.

Triploid production by crossing diploids with tetraploids

Theoretically, offspring from diploids (2N) crossed with tetraploids (4N) can produce 100% triploids (3N). Applications of this method are recognized in the successful establishment of tetraploid breeding stocks for several oyster species. Currently, triploid seed production in most commercial hatcheries uses eggs from normal diploids that are fertilized with sperm from tetraploids. To ensure predictable spawning success and avoid contamination, gamete collection for triploid seed production is performed using the strip spawning method, which is applicable for most oyster species (but may not be for scallops, mussels, and clams).

The strip spawning procedure includes the following steps: open mature adult oysters (clean the knife between oysters to avoid contamination), determine the sex of each oyster by observing a smear of gonad under microscope (40× magnification), keep females and males separate, strip gonad using a sharp scalpel, and rinse and suspend gametes into individual beakers with fresh seawater. Eggs may be collected on a 20-µm screen by filtering through a 150-µm screen to remove debris and may then be resuspended in fresh seawater. Sperm (from tetraploids) can be collected by filtering through a 15-µm screen and resuspending in fresh seawater. Fertilization is performed by mixing eggs and sperm. Concentrations of ~10–20 sperm/egg are confirmed via microscopy (100-× magnification).

Performance of triploids over diploids

The intention of developing triploid oysters is to meet the industry's need for year-round marketable oysters, especially during the summer months when normal diploids are thin and watery after spawning. Typically, oysters spawn during the spring and fall. Before spawning, the stored energy is used for egg and sperm development within the gonad, which can occupy up to 50% of the soft tissue weight. After spawning, the oyster meat is considered unpalatable because tissue begins to soften and appears watery. Triploids are considered sterile with poor gonad development and are thought to redirect energy typically used for gonadal development to *somatic* growth, allowing them to increase in body size and maintain meat quality. Therefore, triploid oysters are well suited to meet the demands of the industry for a year-round marketable product.

The performance of triploids has been demonstrated at juvenile and adult stages in almost all of the studied shellfish species, most prominently in oysters and scallops after *sexual maturation*. An updated summary of the growth of triploid oysters at adult stage based on a review by Guo, X., Y. Wang, Z. Xu and H.

Fast growth of triploid oysters

At the larval stage, triploids and diploids did not show significant differences in survival. For larval growth, chemically induced triploids grew either faster than or about the same as diploids in the Pacific oysters. Superior growth was confirmed in triploid larvae produced from tetraploids in the Pacific oyster.

Data collected from eyed larvae showed that triploid larvae had grown to 280–300 microns at settlement, while diploids had grown to 250–260 microns.

At the adult stage, superior growth of triploids has been confirmed in almost all oyster species studied. For the eastern oyster, triploids produced by inhibiting polar body I or II typically grew 12–41% faster than diploids. Only one study displayed triploids with 0.7% smaller body size when compared to diploids. Impressively, triploids produced by diploids crossed with tetraploids (2n × 4n) showed 192% faster growth than diploids and in some cases even higher due to selective breeding of tetraploid broodstock. Commercially, triploid oysters cultured in Virginia can reach market size (over 3 inches, or 75 mm) in 1.2 years, while diploids need 1.6 years. In Florida, triploids take 10–14 months to reach market size from spawn while diploids need about 12–16 months. For the Pacific oyster, chemically induced triploids showed 14–150% increased body size compared to diploid controls, and triploids produced from tetraploids (2n × 4n) showed up to 159% faster growth than diploids. For the other four oyster species, triploids produced from chemical inhibition of polar body I or II also showed a significant increase in body size.

Fig. 2. Diploid and triploid oysters harvested after eight months of growout in off-bottom gear at commercial farms along the west coast of Florida (CK=Cedar Key, OB=Oyster Bay, AH=Alligator Harbor). Oysters on the top of each tray are diploids (blue tags) while oysters on the bottom of each tray are triploids (orange tags).

However, it is important to recognize that environmental factors (for example, food availability, temperature, and salinity) can also affect growth and survival of triploid oysters. In general, warm environments that are nutrient-rich produce triploid oysters with substantially higher growth rates. Environments that are nutrient poor may explain the observations where triploid and diploid oyster growth are similar.

Survival of triploid oysters

Despite the benefits of summer meat conditions and higher growth rates, the triploid advantage in terms of survival varies across studies and environmental conditions. Recent experimental work has suggested that triploids appear to be more vulnerable to summer mortality. Triploid eastern oysters have experienced significantly higher mortalities than diploid oysters at sites along the Gulf of Mexico, and the results highlight the need for a better understanding of the triggers of summer mortality in triploids.

Disease resistance of triploid oysters

The effects of triploidy on disease resistance have not been clearly established. Studies on susceptibility to dermo (caused by *Perkinsus marinus*) suggested that triploid and diploid eastern oysters reacted similarly and were equally susceptible to the disease. On the other hand, triploid eastern oysters have shown resistance to the MSX disease (caused by *Haplosporidium nelsoni*) compared to diploids. When exposure to the Roseovarius oyster disease (ROD, caused by *Roseovarius crassostreae*), triploid eastern oysters showed increased survival compared to diploids.

For the Pacific oyster, triploids challenged to *vibriosis* did not show any advantage for susceptibility over diploids. Although disease resistance of triploids was not confirmed in these studies, it is worth noting that increased growth rates in triploid oysters could allow them to reach a "refuge" size to be harvested sooner, and thus reduce the risk of vulnerability to pathogens during the culture cycle.

Ploidy determination

Accurate ploidy determination is absolutely required for the application of triploid-tetraploid technology. Ploidy verification of broodstock before spawning can ensure that hatchery efforts are carried out as planned; ploidy confirmation of larvae at swimming stage can ensure early detection of possible polyploid outcome; and ploidy validation of seed labeled as triploids before shipment can ensure that no mishandling occurred during the larval and seed rearing process. The methods for ploidy determination include the following:

Counting of chromosome number

This is a direct method for ploidy determination. Chromosomes are densely packaged chromatin (DNA and histone protein) within the nucleus of the cells and are visible with a light microscope, particularly during metaphase (a short period of the cell cycle). This method requires a tedious and time-consuming process that includes making cell suspensions, treating samples to assure cells arrest at the stage for visible chromosomes, making chromosome slides, staining chromosomes, and counting chromosomes. This method is usually used for *ploidy* determination at embryo stages or *aneuploidy* detection.

Evaluation of hemocyte nucleus diameters or volume

Blood cells (hemocytes) and nucleus dimensions in triploids are significantly bigger than those in diploids; therefore, these parameters can be used effectively for ploidy evaluation. The measurement of cell size (nuclear volume, cytoplasmic volume, and nucleus surface area) can be performed by microscopy after proper staining.

SEvaluation of dna amount by flow cytometry

Flow cytometry is a technology that measures the characteristics of single cells in a fluid as it passes through a bright-beam laser. The cells can be labeled with specific fluorescent stains that can be excited to emit light at varying wavelengths as they pass through the laser.

For ploidy determination, fluorescent stains for DNA amount (ratio to the chromosome numbers), including 4',6-diamidino-2-phenylindole (DAPI) and propidium iodide (PI), can be used to stain the single cells (sperm, hemocytes, or cells isolated from gill or pooled larvae). The intensity of exit lights is proportional to the DNA amount and allows the different ploidy to be distinguished. Therefore, flow cytometry is a fast, easy, and accurate method for ploidy determination. For flow cytometry data collection, diploid oysters should be used as a reference because flow cytometry measures the relative DNA content.

Molecular genetic markers

Polymorphic genetic markers, such as microsatellite markers, have been used for ploidy determination and parentage assignment in shellfish (Miller et al. 2016) because of their high polymorphism, multi-alleles, co-dominance, and reproducibility. The analysis can be achieved by using a simple multiplex polymerase chain reaction (PCR) containing specific primers of microsatellite markers.

5

GYNOGENESIS, ANDROGENESIS AND NUCLEAR TRANSPLANTATION

The gynogenesis is defined as the process of development of ovum after penetration of sperm, but in the process the sperm nucleus is unable to form syngamy with the nucleus of ovum. So gynogenetic fish inherit genes only from mother.

The penetration of sperm is responsible for initiation of development alone. Gynogenetic haploid embryos contain a haploid chromosome complement representing a single product of a meiotic division, one quarter of a tetrad embryo do not develop entirely normally.

Gynogenetic diploid embryo contains a diploid chromosome complement, also derived from mother only and are obtained by treatments to prevent either the second division in an egg fertilized with UV-irradiated sperm.

In fish, the gynogenesis is achieved both naturally and artificially. In some fishes, gynogenesis is a natural mode of reproduction, these unisex fish may be often triploid. In a few fishes such as silver crucian carp (Prussian crucian carp of European goldfish) and Poeciliapsis, unisex form present in natural population.

Population, which are produced by natural gynogenesis. Artificial gynogenesis can also be produced under suitable experimental conditions. During artificial gynogenesis, manipulations in event of fertilization are made by destroying DNA of the spermatozoa and to restore diploidy to the ovum with either meiosis or cell cleavage.

In this process the male inheritance is excluded, the offspring obtained in this way possesses inheritance of female characters and they are of only one sex, i.e., female. The unisexual population obtained by artificial gynogenesis are used for fisheries management.

Gynogenesis can also be used to produce clonal lines of tilapia. Hybrid among such inbred lines are expected to exhibit significant heterosis and uniform performance with respect to growth rate and body conformation.

The most important breeding goal of tilapia is to improve growth rate and feed conversion efficiency, which are prime factor regulating profitability of the industry. Sex determination mechanisms, genetically male tilapia have higher growth rates than even hormonally reverse female.

Researchers observed that gible carp, *Carassius gibelio* exhibits gynogenetic reproduction by utilizing the sperm of other species which resulted in the increase population in the lake as gible carp females have little opportunity for mating with gibel carp male.

According to them *Ylikiensis roach* is the best available sperm donar option for gynogenetically gibel carp of the lake. These authors made artificial fertilization of gibel carp eggs with sperms of other species such as ylikiensis roach (Cr×r), gold fish (Cr.×G) adb common carp (Cr×C), hatching occurred after three days at 14 °C in all groups.

MICROSATELLITE DNA ANALYSIS OF GYNOGENETIC FAMILIES

Using the here established optimal condition for stickleback diploid gynogenesis induction, additional six gynogenetic families designated Family 1-6 were produced for microsatellite analyses. In families 1-4, all gynogenetic embryos/larvae were collected at day 7 and used to assess level of heterozygosity in the gynogenetic individuals. Families 5 and 6 were used to determine the type of gynogenesis that was induced in this experiment, hence only the embryos/larvae that hatched and survived past 120 days were collected and tested. In general, all the gynogenetic embryos/larvae subjected to microsatellite analysis using 9 loci were found to carry only the maternal alleles and as expected all the normal controls had biparental genotypes. Data from the four gynogenetic families produced to assess genetic homogeneity of the resulting gynogens revealed that families 1 to 4 had statistically significant lower heterozygosity index values compared to normal control fish ($p<0.05$). As shown in Fig. 1, the microsatellite data were analyzed in two ways: 1) by using the whole dataset and 2) by excluding individuals homozygous for allele shared by both parents (scored as "0.5"). In both cases, gynogens were found to be more homozygous than the control fish. The degree of individual homozygosity, though, was further intensified when alleles shared by both parents were excluded in the test (Fig. 1b).

Moreover, in families 5 and 6 (Table 1), 97-100% of the samples were heterozygous for the maternal alleles. This suggests that majority of the gynogens are meiotic diploids produced by retention of the second polar body. The remaining 3% found to be homozygous for the maternal alleles could be mitotic gynogens produced by inhibiting first mitotic division. These embryos can be good candidate for a completely homozygous inbred fish.

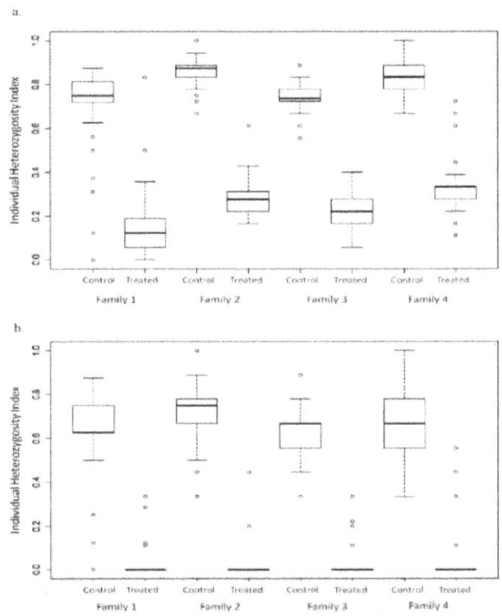

Fig. 1. Heterozygosity of gynogenetic families revealed by microsatellite DNA data.

Ploidy level of gynogenetic fry

A total of 145 gynogenetic and 16 normal control embryos/larvae were subjected to flow cytometry for relative DNA quantification. Of the 145 gynogenetic embryos/larvae, 102 exhibited abnormal features while the remaining 43 were morphologically normal. A histogram of the relative DNA content of a haploid (n) embryo. It had a dominant peak at mean channel 230 which indicates that most of the cells are in the n(G1) phase and a minor peak at mean channel 460 representing cells in the n(G2/M) phase of the cell cycle. The ploidy of the sample is reflected by the dominant peak suggesting that abnormal embryos are haploid (1n, mean channel 230) and the morphologically normal gynogens are diploid (2n, mean channel 460) with twice as much genetic material as the latter.

Table 1 Summary of microsatellite genotyping data of the two gynogenetic families used to assess the mode of gynogenesis occurring in three-spined stickleback

	Number of embryos	**Number (%) of surviving**			
Family	Fertilized	Abnormal	120 days old	homozygote	heterozygote
Family 5	86	20	66	2 (3.03)	64 (96.07)
Family 6	53	13	40	0 (0)	40 (100)

Development of the gynogenetic embryos/larvae

The daily larval development of representative stickleback embryos activated with UV-irradiated sperm and with further HS treatment at 34°C for 4 minutes 5 mpf was monitored for one week (Fig. 2). However the growth of the haploid embryo started to slow down 48 hours after fertilization. It managed to cover only one half of the circumference of the yolk sac and growth from here on was restricted horizontally increasing the volume of the fish but not its length (Fig. 2b).

The diploid embryos hatched at day 6 or 7 post fertilization, whereas haploid embryos did not hatch and remained within their enveloping membrane until they die. Death of the haploid embryo could be attributed to the many deformities observed mainly in the head and tail regions, as well as in the abdomen where the yolk sac is attached (Fig. 3).

Tissue degeneration at the abdominal region prevents absorption of the yolk sac thereby cutting the source of nutrients of the developing embryo leading to the death of the embryo.

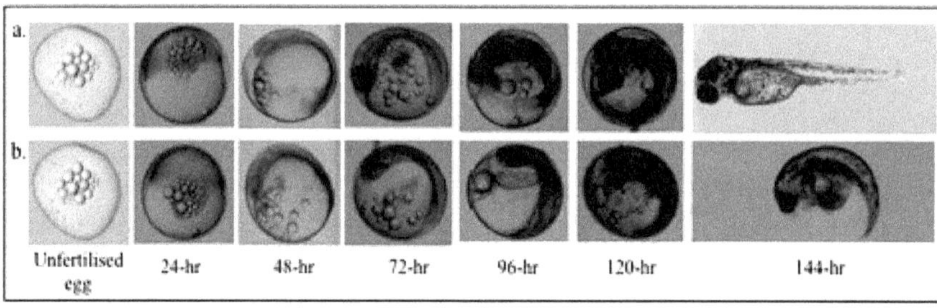

Fig. 2. Embryologic development of representative gynogenetic three-spined stickleback embryos

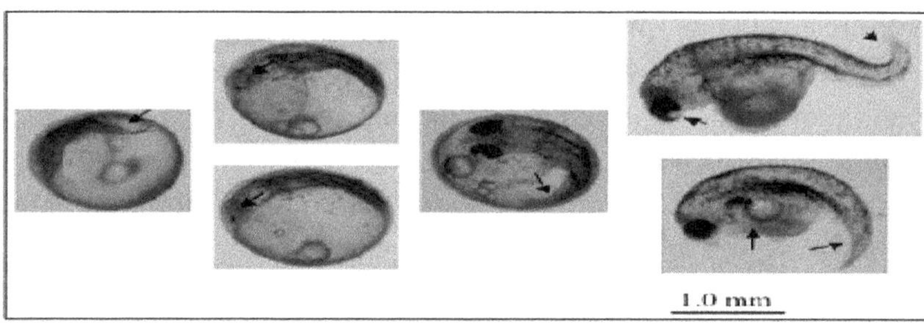

Fig. 3. Commonly observed abnormalities in gynogenetic three-spined stickleback embryos

In this study, UV irradiation of the sperm and heat shock (HS) of the resulting embryos were used to induce gynogenesis. UV irradiation was used because aside from being effective in deactivating the DNA of the sperm, it is cheap and easy to set-up as compared to ionizing irradiations which are more dangerous and require special containment area. Since DNA irradiation is a key step to successful gynogenesis, it is therefore important to determine optimal condition for sperm irradiation. UV irradiation should inactivate the sperm DNA maximally and avoids chromosome fragmentation, because partially inactivated sperm DNA interferes in the embryonic development whereas over-exposure to UV light may already kill the sperm.

In this study, 2 to 4 minutes exposure to UV irradiation can be regarded as optimal with fertilization success of 93-97%. All the resulting embryos were morphological abnormal, confirmed to be haploids by flow cytometry, and had only the maternal genotype. However, UV exposure of more than 4 min in our case was already lethal to the sperms resulting in zero fertilization success. This was also indicated by the observed absence of sperm motility in these treatment groups after UV exposure.

Although both 2 and 4 min UV irradiations were successful in inactivating the genetic material of the sperm, we used the former in the succeeding experiments to minimize possible "Hertwig effect" wherein haploid embryos resulting from the total destruction of the genetic material in the sperm by high doses of radiation would survive longer than those exposed to lower UV dosage.

Application of HS was observed to be effective for the retention of the second polar body in eggs activated with UV irradiated sperms resulting in morphologically normal gynogenetic diploid larvae. For HS to be successful, right combination of the different parameters namely temperature, duration, and initiation time of application is critical. From the review of researchers, it is suggested that optimal temperature for HS is species-specific and could vary from 30°C to up to 42°C. Since HS acts through depolymerization of protein complexes, then the best temperature should be close to the upper limit of tolerance.

Likewise, optimal time for applying heat appears to be slightly different. Timing of HS is very crucial as this decides the kind of gynogens produced. Generally, when HS is applied early, it targets second meiotic division leading to the retention of the second polar body, thereby producing diploid meiotic gynogens. Late application of heat treatment, on the other hand, results in suppression of the first mitotic cleavage, producing an all homozygote female fry (inbred clone). Here, the best yield of diploid gynogens (65%) was observed in eggs activated with UV irradiated sperm and heat shocked at 34°C 5 minutes post fertilization (mpf) for 4 minutes.

Although aiming at different end products, the optimal temperature and duration of HS established here correspond to what was used in his first attempt to produce haploid and triploid three-spined sticklebacks without UV irradiation. Independently tested, this clearly shows that HS at 34°C for 4 minutes can effectively destabilize and disorganize the microtubules of activated stickleback egg, thereby facilitating diploid gynogenesis. The comparatively low diploid gynogen yields at lower (30°C) and higher (38°C) temperatures indicates that 30°C can still be well tolerated by the developing embryo.

The effect of temperature shock at 30°C is very minimal, whereas shock at 38°C could have already interfered in a wider range of developmental mechanisms resulting to further abnormalities that drives the embryos to death. Like in common carp, the present study clearly demonstrated that temperature, duration, and initiation time of HS play an important role in the success of diploid gynogenesis in three-spined sticklebacks.

Success of gynogenesis induction in three-spined stickleback is verified by microsatellite analysis. Microsatellite DNA technology is commonly used because of its

PCR-based, co-dominant Mendelian inheritance, and highly polymorphic characteristics which facilitate molecular typing of the gynogenetic embryos/larvae. Success is measured by the absence of paternal alleles in the resulting gynogenetic individuals. In the six families tested here, it was clear that the paternal alleles are absent and only the maternal alleles are passed on to the gynogenetic embryos/larvae.

The predominance of heterozygotes (97%) for the maternal alleles only suggests that the established optimum HS condition promotes retention of the second polar body, thereby producing mostly heterozygous meiotic diploid gynogens which have the expected high degree of homozygosity (Fig. 1). With the relatively early timing of heat shock (5 mpf) and due to the mode of egg collection (mainly hand stripping), it could be that the eggs used in this study are still undergoing meiosis when the HS was applied. Flow cytometry remains the best method for ploidy determination in gynogenetic stickleback. This method clearly shows that stickleback gynogenetic embryos/larvae that developed normally were diploids (2n) whereas those that showed abnormalities were haploids (n) with peaks at mean channel numbers 460 and 230, respectively. We observed no mosaic or aneuploid, chromosomal anomalies that are also reliably detected through flow cytometry.

With the accuracy of the result obtained here, we find no reason to use more laborious and time consuming classical methods of ploidy determination like chromosome preparation and silver staining of nuclear organizer regions (NORs). Since in the present study, possible effects of doublets on fluorescence were excluded in all samples, the observed predominance of haploid cells at n(G1) phase with a significant 5% of the cells staying at the n(G2/M) is quite striking. This is comparable to what was observed in muskellunge by researchers which suggests that haploid cells have an abnormal cell cycle that probably takes longer in time than that of their diploid counterparts. This could also mean production of lower cell numbers, hence the observed stunted growth of haploids.

ANDROGENESIS

Androgenesis is a method for producing fish in which all the nuclear genetic information originates from the male parent (i.e., from the sperm) while the mitochondrial DNA is still maternally derived. Androgenesis is induced by gamma or ultra-violet (UV) irradiation of the egg to inactivate chromosomal DNA followed by fertilization with normal sperm. This results in a haploid zygote containing a single chromosome set. The normal diploid state is restored by applying pressure or temperature shock at the first cleavage division so as to duplicate the haploid set and restore the diploid condition. This procedure has been accomplished in only a few laboratories to date, e.g., in rainbow trout and amago salmon (*Oncorhynchus rhodurus*).

Recently, researchers used diploid sperm from tetraploid rainbow trout to fertilize gamma-irradiated ova, thus obviating the need for pressure shock treatment to suppress the first cleavage division. Survival of these androgenetic diploids was much better than those produced using haploid sperm. Fused sperm have also been used to produce diploid androgenetic rainbow trout. Potential uses of androgenesis include the regeneration of stocks from cryopreserved sperm, the development of identical groups of fish, and the production of YY males for the production of monosex male salmonid stocks.

Androgenesis is essentially when a male is able to reproduce without a female's genes. Unfortunately, even top-tier exports in the field still don't really understand how it works as a result of how rare the process really is.

Some kind of magic must happen during the introduction of sperm cells and egg cells where only the sperm cell's genetics are used in embryo development, but we've not yet been able to capture exactly how this works and it has scientists scratching their heads.

One theory is that the sperm cell of the *alburnoides* contains two sets of chromosomes and the egg cell involved in the reproduction carried none at all, or the egg cell's chromosomes were possibly destroyed when the sperm cell combined with it. While this is nothing more than speculation at best, the ideas pave the way for how a male sperm cell was somehow able to fertilize an egg cell without any female genetic data.

It's possible that this event was a complete accident and that the mother's egg cell simply had no genetic information from the get-go. If so, this was a one-off mutation that may or may not ever happen again. Nevertheless, for the event to occur just once in a randomized batch of 100 fish, it's safe to say that it is definitely an unexplained phenomenon that could be out of the norm.

Androgenesis is a developmental process, facilitating the inheritance of an exclusively paternal genome. It involves two steps:

- elimination or inactivation of the egg's genome and
- monospermic or dispermic activation of embryonic development by the haploid or diploid gamete;

Haploid monospermic activation requires restoration of diploidy by suppression of the first mitotic cleavage.

- Reliable protocol for inducing androgenesis, requiring the suppression of the first mitotic cleavage, however, is available for less than a dozen species only.
- Due to irradiation damage suffered by eggs,homozygous expression of lethal genes and damage inflicted by thermal shock treatment to suppress the first mitotic cleavage, survival of the androgenotes is very low.
- Androgenesis is induced in fish by fertilizing irradiated eggs (gamma or X ray or U-V ray) with normal spermatozoa.
- U-V (254 nm) irradiation is easy, inexpensive and can be easily set up under laboratory condition.

- Diploid androgenetic individuals can then be produced by various treatments (thermal, pressure or chemical shock) to suppress the first cleavage division yielding a homozygous diploid.
- Theoretically 50% XX and 50% of YY homozygous individuals will be produced.
- The resultant YY males, if viable and fertile, can be crossed with females to produce XY isogenic (genetically uniform) all-male progeny.
- Androgenetic trout was produced using sperm from tetraploid males and found that these males showed better viability than androgenetic diploid.
- Among the aquarists the male fishes are often found more colourful and beautiful thus commanding a greater demand. It would also be advantageous for species in which the male shows better characteristics for culture, like tilapia.
- In androgenesis, polyploidization by inhibition of meioticphase I and II is impossible due to the distruption of the female nucleus.
- Androgenesis results in a rather low rate of development compared to gynogenesis since the egg cytoplasm as well as the egg nucleus is affected by radiation treatment.
- Androgenesis has not been widely investigated in fish as gynogenesis.
- Like gynogenesis,androgenesis can be used to generate clonal lines.
- An advantage of androgenesis is the option of storing and regenerating lines from cryopreserved sperm by androgenesis; eggs have not yet been successfully cryopreserved in fish and this precludes the storage of gynogenetic lines in that way.
- Androgenetic fishes were successfully produced in tilapia, rainbow trout and Siberian sturgeon. Androgenetic tilapia can beuseful for monosex fish culture.

PERFORMANCE OF GYNOGENS AND ANDROGENS

Homozygous gynogens of ayu show increased variation for size and for meristic traits (numerical traits such as anal-fin ray number). Rainbow trout mitogynotes exhibit greater amounts of bilateral asymmetry – unequal counts for meristic traits on the right and left sides of the bodies – than controls, isogenic crossbreeds and normal crossbreeds. Bilateral asymmetry has been previously documented in inbred rainbow trout, and is associated with a reduction in biochemical genetic variation. Isogenic crossbreeds – crossbreeds resulting from mating two clonal lines – had similar bilateral asymmetry as outbred crossbreeds of common carp. The coefficient of variation for size was much larger for the common carp clones compared with outcrossed populations.

Homozygous gynogens of common carp exhibit growth depression compared with crossbreeds. Growth reduction was slightly more when compared with outcrossed crossbreeds than when compared with isogenic gynogenetic crossbreeds.

Reproduction

Researchers compared the gonad development and fertility of common carp produced by either full-sib mating (F = 0.25), gynogenesis by retention of the second polar body (F is approximately 0.25) or gynogenesis by endomitosis (F = 1.0) from a common female. Meiogenetic offspring were all female, but 50% of the mitogynotes were homozygous for a recessive mutant sexdetermining gene, resulting in 50% males and intersexes. Another mutation affecting pigmentation in eggs was also detected in the mitogynote offspring but not in meiogenetic offspring.

The variation in body weight, gonad weight and egg size in each group increased with increased inbreeding. Full-sib inbreds and meiogens (F = 0.25 for both) were comparable in mean gonadal development, but gonads from homozygous gynogenetic carp were often retarded. Ovulation response also decreased with increasing levels of inbreeding.

Gynogenetic and polyploid fish often have abnormal ovarian development. Researchers compared the gonadal morphology of normal and sex-reversed juvenile triploid, gynogenetic diploid and normal diploid coho salmon. Gynogens were 100% females, and 34% had reduced ovaries with areas devoid of oocytes. About 30% of gynogenetic rainbow trout females also had aberrant ovarian development. Gynogenetic fish may be more prone to developmental abnormalities because of either increased homozygosity or incomplete inactivation of the paternal chromatin.

In 90% of the gynogen rainbow trout examined, chromosomal fragments were positively correlated with aberrant ovarian development. The atypical gonadal morphology of the gynogens was similar to the ovarian morphology of triploid rainbow trout. Disruption of the normal diploid chromosomal complement appears to alter germ-cell development in gynogenetic female rainbow trout and probably coho salmon due to the unbalanced nature of the genome. Contrary to what has been observed in cyprinids, male germ cells were not observed in the ovaries of the gynogenetic coho salmon, perhaps because of differing mechanisms of sex determination for salmonids and cyprinids. However, this again could be related to paternal chromatin and the incomplete destruction of the paternal genome, which might be affected by species or temperature.

Monosex Populations

In many species of fish, the female is the homogametic sex. Theoretically, when this is the case, gynogenesis can be used to produce monosex female populations. Gynogenetic, all-female, grass carp, silver carp, *Hypophthalmichthys molitrix*, channel catfish and rainbow trout have been produced.

However, it is not feasible to produce commercial quantities of monosex gynogenetic fish because of the high mortality of the manipulated eggs. Hatching rates of eggs activated and treated to produce gynogens and survival of fry are usually less than 1%. If the male is the homogametic species, androgenesis would result in monosex male populations. Again, the low survival of eggs treated to produce androgens prevents largescale production of androgens.

Cloned Populations

Clonal populations of fish can be generated in a non-specific manner, utilizing gynogenesis and sex reversal. Mitotic gynogens are produced by blocking the first cell division, and these individuals are 100% homozygous. However, they are not clones of their mother as their genotype is different from their dam because of independent assortment and recombination during meiosis and because of the absolute homozygosity induced from the gynogenesis. Additionally, each mitogynote is different from its siblings, and each is variably homozygous for a variety of allelic combinations. Further, the breeder has no control over which genotype survives and is generated by the gynogenesis.

Clonal populations can be produced from these mitogynotes by performing a second generation of mitotic gynogenesis on the firstgeneration mitotic gynogen. Presumably, more than one offspring would be produced.

Then a portion of the fry would be sexreversed, both sexes would exist, although genetically identical, and large numbers of identical fry would be produced via natural mating. Future male brood stock would again be perpetuated via sex reversal. Theoretically, clonal populations could be generated in a similar way from androgens. The primary difficulty in this breeding scheme is that the mitogynotes and androgens are maximally inbred and 100% homozygous, often resulting in reproductive problems.

Phenotypic variability

Cloned populations have been produced in zebra fish, ayu, common carp and rainbow trout. Theoretically, clones – individuals with identical genotypes for their entire genome – should have identical, non-variable performance. However, individuals with extreme homozygosity appear to lose significant ability to respond to the environment in a consistent stable manner, and micro-environmental differences affect performance among individuals. Thus, as the component of genetic variation decreases, actually becoming zero, the component of environmental variation increases and at a more rapid rate than the genetic component decreases, resulting in populations with extreme phenotypic variation.

Homozygous gynogens of ayu show increased variation for size and for meristic traits. Rainbow trout clones exhibit greater amounts of bilateral asymmetry – unequal counts for meristic traits on the right and left sides of the bodies – compared with controls, isogenic crossbreeds and normal crossbreeds. Bilateral asymmetry has been previously documented in inbred rainbow trout and is associated with a reduction in biochemical genetic variation. Isogenic crossbreeds – crossbreeds resulting from mating two clonal lines – had a similar bilateral asymmetry to that of outbred crossbreeds.

Almost identical results have been obtained for clonal rainbow trout. Clonal lines of rainbow trout were produced by androgenesis, followed by another generation of androgenesis or gynogenesis, followed by retention of the second polar body. Clonal populations were more asymmetrical and showed higher variance in total meristic counts than either the isogenic crossbreeds or outbred crossbreeds, indicating reduced fitness in the clonal populations and their ability to react to microenvironmental variation. There was no difference in asymmetry between outbred crossbreeds and isogenic crossbreeds. Fluctuating asymmetry of spotting pattern was also investigated as a potential measure of developmental stability.

Asymmetry of spotting pattern was not different in homozygous clones compared with isogenic crossbreeds, but the spotting pattern had a low coefficient of variation within lines and may be useful for identifying isogenic lines and monitoring genetic divergence between lines. The increased developmental stability and decreased variance of meristic traits in the isogenic crossbred rainbow trout indicate that they were less influenced by environmental conditions and may make better research organisms than the homozygous clones. Researchers used a homozygous clone outcrossed to normal common carp as a powerful internal scaled control for the evaluation of mirror carp strains.

Growth

Homozygous gynogens of common carp exhibit depression in growth compared with crossbreeds. Growth reduction was slightly more when compared with outcrossed crossbreeds than with isogenic gynogenetic crossbreeds. Again, the coefficient of variation for size was much larger for the common carp clones compared with outcrossed populations.

Researchers established six different homozygous clonal lines of Nile tilapia, which was verified by analysis of the allozymes of adenosine deaminase and DNA fingerprints. Up to first feeding, clones had a reduced survival rate (4.0%) compared with normal heterozygous fish (43.9%).

Reproductive traits of the second- generation clones varied between and within clonal lines. The observed body weight of all-female clonal groups (45.5 g) kept under three different density classes was slightly less than that of corresponding allfemale heterozygous control groups (50.2 g). In contrast to what has been observed for clonal populations of other species of fish, Nile tilapia homozygous clones had more uniform growth (coefficient of variation: 23.1%) than in all-female heterozygous controls (coefficient of variation: 34.1%).

Disease resistance

Researchers produced clonal lines from different strains of amago salmon, *Oncorhynchus rhodurus*. As expected, the clones immunologically accepted operculum allografts from inbred sisters. Survival was similar among clones produced by dams from the same strain, but survival was different among clones using maternal parents from different strains. Researchers produced clonal ayu, *Plecoglossus altivelis*, that accepted graft tissue transplanted from intraclonal siblings, but the graft tissue was rejected when transplanted into different clonal hosts. Strain and genetic background affect the viability of clonal lines. Non-specific immune responses, serum lysozyme activity and phagocytosis were different among inbred clones of Nile tilapia. Resistance to *Aeromonas hydrophila* infection was positively correlated to non-specific immune responses. When a clonal line susceptible to *A. hydrophila* was crossed with a resistant clonal line, the resulting progeny demonstrated intermediate disease resistance, indicative of additive genetic variation rather than dominance gene action for this trait.

Four gynogenetic clonal lines of common carp were selected as high or low responders on the basis of their primary serum antibody production to dinitrophenyl-keyhole limpet haemocyanin (DNP-KLH). Number of antibody-secreting cells and not serum antibody titres were the primary difference in progeny of high and low responders. High-responder clonal common-carp lines produced 1.5–twofold higher numbers of specific antibody-secreting cells than lowresponder clonal lines, after both primary and secondary immunization. Individual variation within the inbred lines was reduced compared with that in outbred common carp.

Haematocrit, haemoglobin, red blood-cell count and mean corpuscular volume were measured for stressed and non-stressed clonal lines of ayu and also verified by DNA fingerprinting. Differences among clonal lines were observed for haematocrit and mean corpuscular volume under the no-stress environment. Heritability values were moderate for the no-stress measurements and very low or zero for the stressed groups' traits.

GYNOGENESIS IN SHRIMPS

Over the world, penaeid shrimp has been known by the entrepreneurs and aquaculturist to be the important culture species due to its higher growth rate and tolerance to the varies environment. Basically, penaeid shrimp is an important culture species which carry on external fertilization. Due to its external fertilization, the chromosome manipulation can be induced in the penaeid shrimp as an alternative to sustain production.

In prawn and shrimp, they have been identified with WZ/ZZ sex determination system where female were identified hetero-gamety where female is the one that determines sex and for male sex differentiation is controlled by the adrogenic gland of the shrimp. To meet the higher demand, advantages of increasing sterility are exploited during commercial growing to obtain suitable size for commercialize shrimp. Induction of triploidy, gynogenesis, inter species hybridization can be applied in this type of penaeid shrimp.

From the previous study, there is very scanty research has been conducted on the gynogenesis induction in shrimp. One of its which was induced gynogenesis in Chinese shrimp, Penaeus chinensis and identified that sperm irradiated at 365 nm ultraviolet (UV) can fertilized the egg; however the activated eggs died in the early stages of embryonic development and could not hatch. The method 365 nm of the UV irradiated (UV) for 5–8 s can be applied to get the genetic inactivation of the sperm to carry on the gynogenesis induction. Researchers applied UV irradiated dosage at 254 nm and 365 nm from 20 to 80 s for the initial development of gynogenesis in *Banana shrimp, Fenneropenaeus merguiensis* and identified sperms abnormalities with damage on malformed bodies also with short or missing spike on the irradiated sperms.

However, by applying the 365 nm UV the sperm have a chemical changing which call chromosomal denaturation. Researchers induced gynogenesis in Chinese shrimp, Penaeus orientalis kishinouye to produce all female shrimp populations. The unfertilized eggs were fertilized with irradiated sperm of same species and then were treated with cold shock or cytochalasin B to induce gynogens progeny. This gynogenesis process successfully produced 14.9% and 37.2% of haploid gynogenetic embryos. Through the cytological experiment, study found that the irradiated male nucleus fails to fuse with the female shrimp nucleus to form a zygote. Other study conducted by researchers on *Penaues chinensis* to induce gynogenesis using 30 W of 105 UV irradiation successfully produced 14.0% and 37.2% hatched diploid gynogenetic embryos and identified that the most efficient sperms irradiation methods is the direct irradiation of sperm from the seminal receptacles of the female.

GYNOGENESIS IN FISHES

The higher demands on the animals protein supply focused on developing new method for increasing the fisheries production. The new technique currently known as gynogenesis which identified promotes an opportunity to produce higher production of fish stock. This technique has been successfully applied in many commercial fish species such as in carps, in catfish, in tilapia and in salmonids for the selective breeding purposes. Researchers found out several reports described the techniques to induce polyploidy and uniparental chromosome inheritance which applying gynogenesis and androgenesis technique in fish. According to researchers the technique of polyploidy and gynogenesis were applied in fish culture in order to produce inbred line and also for the production of monosex or sterile populations.

Gynogenesis has been introduced with different purpose in genetic and breeding studies and the previous use of gynogenesis in the production of inbred lines in fishes. According researchers, the early work on the gynogenesis in fish only concentrated on the suppression of the second division of meiosis to induce diploidization of the eggs fertilized with spermatozoa that already inactivated using gamma or UV irradiation. Gynogenesis have been successfully induce in number of fishes such as in grass carp, in sliver carp, in tilapia, *Oreochromis niloticus*, and in zebrafish, *Danio rerio*. Researchers in their study applied pressure shock and cold shocks to produce gynogenesis and triploid in the European sea bass, *Dicentrarchus labrax* species.

They obtained meiogenesis by fertilizing eggs at UV irradiated at 32,000 erg/mm^2. The optimum treatment they obtained for cold shock is at 0–1°C for 15–20 min and 8500 psi of pressure shock for 2 min at 6 mic a.f. Researchers applied cold shock treatment to produce diploid gynogenesis in Stinging catfish, Heteropneustes fossilis and discovered that the optimum cold shock at 2°C for 10–15 min was successfully produced the maximum percentage of gynogen progeny. The finding of the project also expected to be the platform for the production of all diploid gynogens and also the technique to produce of monosex female populations in stinging catfish.

GENETIC MATERIAL PRODUCED BY GYNOGENESIS

In gynogenesis, the male broodstock does not contributed to the genetic material to the progeny and its only contains maternal genome as the sperm only act as the stimulator for the egg to develop. From the previous study conducted identified that the diploid gynogens fish were grow faster and built stronger disease resistance than the haploid progeny. In addition, other study conducted by researchers identified ruptured chromosome fragments from the karyotyping analysis conducted in the haploid gynogenesis embryo of Zebra fish, *Danio rerio*.

Meanwhile, in three-spined stickleback fish, diploid gynogenesis successfully identified by microsatellite DNA analysis that exposed material inheritance in the gynogens progeny. Through the microsatellite DNA analysis also 97%–100% of the samples were identified heterozygous for the maternal alleles which suggest that the majority of the gynogens progeny are meiotic diploids. Usually for the diploid mitotic gynogens, all the gene loci will be homozygous due to the diploidization occurrence by the doubling chromosome in somatic division. Meanwhile, for the meiotic gynogens which produced by inhibiting of second polar body, the gene loci should be recombinant genotypes due to high rate of gene centromere recombination. Thus, the progeny from the female with heterozygous gene loci it should have heterozygous genotypes in the loci. The obvious genetic differences between meiotic and mitotic gynogens are the gene centromere recombination rate, coefficient of inbreeding and variability of quantitative traits.

Clonal hybrids

Salmonids are genetically unusual in many ways, and it may be possible to develop interspecific clonal hybrid lines. Clonal fullsib progeny groups of Atlantic salmon×brown trout hybrids were produced by gynogenesis. One hybrid female had a large portion of the oocytes undergo a premeiotic chromosome doubling, or possibly a complete suppression of meiosis, resulting in ovulation of diploid eggs, each possessing one full set of both Atlantic salmon and brown trout chromosomes identical to those in the maternal somatic cells. This opens the possibility of developing lines of clonal salmonid hybrids for experimentation.

Fig. 5 World's first somatic cell nuclear transplant fish

NUCLEAR TRANSPLANTATION

Introduction

Nuclear transfer is an important technique in animal cloning and also for demonstrating the germline transmission of genetically engineered genomes. Since the birth of Dolly, the first cloned sheep, attempts to produce animals through adult somatic cell nuclear transfer have been successful in a variety of mammalian species. However, despite their importance as an animal resource in fundamental biology, biomedical research, and food production, reports on efficient nuclear transfer in fish are still limited. Indeed, several parameters remain to be determined before nuclear transfer can be widely and efficiently applied to various fish species.

Nuclear transfer in fish has been studied for a half century, starting with the initial work of Tung and colleagues in the 1960s. In the transfer of embryonic cell nuclei to enucleated eggs, transplants have been generated in cyprinid fish, loach, *Misgurnus fossilis*, medaka, *Oryzias latipes* and zebrafish, *Danio rerio.*

However, until our recent work using newly established methods in the small freshwater fish model, medaka, nuclear transfer using adult somatic cell nuclei had not yet been reported. In those studies, we used non-enucleated and diploidized eggs as recipients for the transfer of adult somatic cell nuclei and consistently obtained diploid and fertile transplants that originated from the donor nuclei. Application of this method to other fish species will facilitate new studies involving genetic modification and cloning in both basic and applied fields of fish research. We therefore sought to apply this method to another fish model, the zebrafish.

Since no studies on nuclear transfer using non-enucleated recipient eggs have been reported to date in zebrafish, we first attempted to determine whether non-enucleated eggs were suitable for use as recipient for nuclear transfer in this species by employing a nuclear transfer approach that we successfully implemented in medaka to transfer embryonic cell nuclei to non-enucleated unfertilized eggs. Our results revealed that, as in medaka, non-enucleated recipient eggs could also be used to produce diploid or triploid adult transplants in zebrafish.

In the early 1980s, by means of the serial nuclear transplantation technique, IHB scientists transferred a short-term cultured kidney cell nucleus of triploid crucian carp into the enucleated oocytes of diploid crucian carp and obtained the first sexually matured somatic cell nuclear transplant fish. This is the world's first evidence that the differentiated somatic cells of vertebrates retain developmental totipotency and are able to undergo dedifferentiating and reprogramming. This cloned fish, bred by IHB, was born some 10 years earlier than Dolly, the world's first somatic cell cloned sheep that was born in the UK.

In 1985, IHB researchers conducted the world's first successful transgenic study with agronomic traits and reported the first successful gene transfer in fish. A complete theoretical model of transgenic fish was then built up, leading to a new research era of fish gene-engineered breeding. IHB's relevant studies have been at the forefront of the world. In recent years, IHB researchers generated the champion carp which grows fast and the sterile triploid carp which has a broad potential for breeding promotion. With more experiments conducted and more detailed data collected, IHB's studies on the evaluation of the potential ecological risks posed by the champion fish, and its reproductive containment strategies and aquaculture pattern are more advanced than the United States, and are widely acknowledged by the international peers.

Development of nuclear transplants

The development of NTs is summarized in Table 2. In total, we operated on 184 eggs in six independent experiments. Of the NTs produced, 139 (75.5%) developed to blastulae, 83 (45.1%) survived to the segmentation stage (20-somite stage), and 60 (32.6%) hatched. Fifty-three (28.8%) of the hatchlings exhibited wild-type pigmentation in the skin and RPE and the remainder exhibited the golden phenotype. A total of 31 (16.8%) larvae developed normally and reached the adult stage within two months after hatching. Twenty-eight (15.2%) of these 31 fish had the wild-type phenotype and the remaining three (1.6%) were golden. All of the subsequent analyses were performed using these adult NTs.

Table 2. Development of nuclear transplants produced by transfer of blastula cell nuclei into non-enucleated unfertilized zebrafish eggs

		No. of operated eggs	**No. of developed nuclear transplants**					
			Cleavage	Blastula	Gastrula	Segmen-tation	Hatching	Adults
Total		184	154	139	109	83	60 (32.6)	31 (16.8)
			(83.7)a	-75.5	-59.2	-45.1		
Body color	Wild type Golden					53 (28.8) 07 (03.8)	28 (15.2) 03 (01.6)	

a Numbers shown in parentheses represent the percentages with respect to the total number of operated eggs.

Fertility of nuclear transplants and transmission of the wild-type body color to progeny

The characteristics of the NTs are summarized in Table 3. Of the 28 adult NTs showing the wild-type body color, two (Nos. 1 and 2) (7.1%) were female and the others (92.9%) were male.

Table 3. Characters of adult fish generated by transfer of blastula cell nuclei into non-enucleated unfertilized zebrafish eggs

Fish	Body color	Sex	Fertility	Ploidy	allele of slc24a
No					
1	W	F	-	2N	W
2	W	F	+	2N	W
3	W	M	-	3N	W+V[2]
4	W	M	-	3N	W+V
5	W	M	-	3N	W+V
6	W	M	-	3N	W+V
7	W	M	-	3N	W+V
8	W	M	-	3N	W+V
9	W	M	-	3N	W+V
10	W	M	-	3N	W+V
11	W	M	-	3N	W+V
12	W	M	-	3N	W+V
13	W	M	-	3N	W+V
14	W	M	-	3N	W+V
15	W	M	-	3N	W+V
16	W	M	-	3N	W+V
17	W	M	-	3N	W+V
18	W	M	-	3N	W+V
19	W	M	-	3N	W+V
20	W	M	-	3N	W+V
21	W	M	-	3N	W+V
22	W	M	-	2N	W
23	W	M	-	2N	W
24	W	M	+	2N	W
25	W	M	+	2N	W

26	W	M	+	2N	W
27	W	M	+	2N	W
28	W	M	+	2N+4N[1]	W+V
29	G	ND	-	2N	V
30	G	ND	-	2N	V
31	G	ND	-	2N	V

F, female; G, golden; M, male; ND, not determined; V, variant; W, wild type

1 mosaicism of diploidy and tetraploidy, 2 both of wild-type and variant allele

All of the NTs were individually crossed with specimens of the golden variety to examine their fertility and the transmission of the wild-type body color, one of the donor markers, to the resulting progeny. Twenty-one males (Nos. 3 through 23) exhibited active reproductive behavior and their golden partners spawned 76 to 303 eggs per spawning event per week.

However, the majority of these eggs were unfertilized. In those cases where fertilization was successful, all of the embryos died during embryonic development; most died at the blastula or gastrula stages, and a small number showing varying degrees of abnormal morphology developed until the hatching stage. One female (No. 1) showed active reproductive behavior with golden male partners, but she did not spawn in ten mating sessions over a two-month period. Thus, of the 28 NTs with the wild-type phenotype, 22 (78.6%) were infertile.

On the other hand, the six remaining NTs (21.4%), which consisted of five males (Nos. 24 through 28) and one female (No. 2), were all fertile.

All of F_1 progeny produced by four males (Nos. 24, 25, 27 and 28) and No. 2 female showed the wild-type body color. They developed normally from the embryonic and postembryonic development stages through to the adult stage. In a cross with male No.26, three (1.7%) of the 176 F_1 progeny showed the golden phenotype and died at the hatching stage, whereas all of the other F_1 progeny (98.3%) showed the wild-type phenotype and grew normally to the adult stage.

None of the three NTs with the golden phenotype (Nos. 29 through 31) exhibited active spawning behavior or produced any eggs. In the F_2 generation, produced by crossing wild-type F_1 progeny with the golden variety, body color segregated into wild-type and golden progeny in a 1:1 ratio. Thus, except for male No. 26 with the wild-type phenotype, all of the fertile NTs transmitted the wild-type phenotype to their progeny in a typical Mendelian fashion.

Ploidy status of nuclear transplants and their progeny

Ploidy status was determined by flow cytometry and the data are summarized in Table 3. Of the 22 infertile NTs with the wild-type phenotype, 19 (67.9%) males (Nos. 3 through

21) were triploid, and two males and one female (No. 1) were diploid. The embryos showing abnormal morphology that were obtained from some of the infertile NTs all exhibited aneuploidy.

Of the six fertile NTs with the wild-type phenotype, four males (Nos. 24 through 27) and one female (No. 2) were diploid, and their F1 progeny were also diploid (Fig. 2C). The last male NT was a mosaic of diploid and tetraploid cells, but produced only diploid F_1 progeny. Thus, a total of eight (28.6%) of the NTs with the wild-type phenotype were diploid, and only one (3.5%) was a mosaic of diploid and tetraploid cells. All of the F_2 progeny were diploid, as were the three NTs with the golden phenotype and the control fish from the donor strain.

Molecular genotyping of *slc24a5* in nuclear transplants and their progeny

All of the NTs and the six F_1 and eight F_2 progeny from each of the six fertile NTs were assayed for the other donor marker, *slc24a5*, by PCR. The wild-type allele of the gene was confirmed in all of the NTs with the wild-type phenotype, but not in the NTs with the golden phenotype. A variant allele of the gene was detected in NTs with the golden phenotype, as well as in all triploid NTs and the mosaic NT which exhibited both diploidy and tetraploidy (No. 28). However, the variant allele was not observed in the diploid NTs with the wild-type phenotype.

In addition to the variant allele, the sequence of the wild-type allele was detected in all 36 of the F_1 progeny examined (Fig. 3B). In the F_2 generation, the wild-type and variant alleles were detected in individuals with the wild-type phenotype, but the variant allele was only detected in individuals with the golden phenotype.

In this study, we transferred blastula cell nuclei from a wild-type donor strain to non-enucleated unfertilized eggs of a recipient golden variety of zebrafish to determine whether non-enucleated eggs were suitable for use as nuclear transfer recipients in this species. As a result, 15.2% of the operated eggs reached the adult stage exhibiting the wild-type phenotype, a donor marker, with most being either triploid or diploid. PCR analysis of the *slc24a5* gene suggested that the triploid NTs arose due to fusion of the diploid-donor and haploid-recipient genomes, whereas the diploid NTs originated solely from the diploid donor nuclei.

Half of the diploid NTs were fertile and transferred the donor genetic marker to their F_1 and F_2 progeny in a Mendelian fashion, while one of the diploid NTs was also fertile and produced a small number of F_1 progeny with the golden phenotype and numerous progeny with the wild-type phenotype. The origin of these golden F_1 progeny and the mechanism by which they are produced is currently unknown and will need to be elucidated in future research. However, the remainder of the diploid NTs and all of the triploid NTs were infertile.

One NT with wild-type body coloration was a fertile diploid and tetraploid mosaic. Results of genetic crosses and PCR analysis of the *slc24a5* gene in this individual suggest that the diploid cells originated from the diploid donor nucleus, but they did not undergo fusion with the recipient nucleus. Conversely, the tetraploid cells appear to have formed as a result of the fusion of the diploid donor nucleus and a diploidized recipient egg nucleus. Interestingly, the F_1 progeny originated from the diploid nuclei but not from the tetraploid nuclei.

Three individuals with the golden phenotype were also obtained in the present study. These fish were diploid and carried only the variant allele of *slc24a5*, not the wild-type allele. While these findings suggest that the golden individuals originated from recipient nuclei, the diploidization mechanism of the recipient haploid nuclei has not yet been clarified.

The results of the present study corroborate those of our previous studies in medaka, and the characteristics of the NTs described here are considered to be closer to those reported. For example, as in the report on medaka by Bubenshchikova et al., both diploid and triploid NTs with donor markers were obtained in the zebrafish of the present study, whereas only triploid NTs with donor markers were obtained in the studies. Several genetically distinct wild populations of medaka are known to exist. Bubenshchikova et al. used both donor and recipient strains that originated from a population in southern Japan, while researchers used strains from populations in both northern and southern Japan for the donor and the recipient strains, respectively.

In zebrafish, such differences in the genetic characteristics of different populations are not known. It has been reported that the cytoplasmic environment of the recipient affects the development of NTs in cases where donors and recipients are derived from different families or orders. The results of this and previous studies suggest that the likelihood of successfully developing diploid NTs with donor markers increases when the donor and recipient have similar genetic backgrounds.

Although males accounted for most of the wild-type NTs, the reason for this marked sex bias in NTs is not known. Widely varying sex ratios in the offspring of different pairs of parental fish have been reported in zebrafish, but the mechanism underlying sex determination in this species has not yet been clarified.

The successful generation of NTs with donor markers was approximately ten times greater in the present study on zebrafish (15.2%) than it was in a previous study on medaka (1.5%). While the reasons for this disparity are unclear, there were several differences in the experimental systems between these two studies. For example, in the zebrafish experiments, donor nuclei were transferred to activated eggs, whereas in medaka, donor nuclei were transferred to non-activated eggs. In addition, in the medaka experiments, egg activation occurred at the same time as nuclear transfer by stimulating the eggs through pricking with transfer needles.

Over the past 25 years of mammalian cloning studies, enucleated metaphase II oocytes have been used as recipients in nuclear transfer experiments. This is because using zygotes as recipients is not well suited to the development of NT embryos. Similarly, nuclear transfer using recipient eggs arrested at metaphase II by incubation in fish body fluids has also been reported in fish. On the other hand, researchers reported successful cloning in mice using zygotes at the first mitotic phase as recipients. Zygote nuclei have been reported to contain factors that are necessary for the reprogramming of the introduced genome and for embryonic development. The high nuclear transfer success rates achieved in the present study using activated eggs with haploid maternal nuclei as recipients may be indicative of similar conditions in the recipient eggs of fish and mammalian zygotes.

6

SEX REVERSAL AND BREEDING

INTRODUCTION

Sex reversal is a redirection of sexual phenotype during embryonic development. Most aspects of sexual phenotype, including sex ducts and genitalia, depend on the presence of steroid hormones produced by the testis or ovary. The testis or ovary fate of the gonad is decided through a process called sex determination. During sex determination, the testis and ovary pathways are mutually antagonistic and compete for control of gonad fate.

Sex reversal occurs when the sexual trajectory of the gonad is changed to the opposing pathway, switching the sexual phenotype of the organism to the opposite sex. These fish are sequential hermaphrodites, meaning that, as adults, they transition from one mature sex to the other through a process of degeneration and regeneration of their testes or ovaries. Depending on the species, sex change can be size dependent and/or socially mediated. In clownfish, a dominant female and dominant male make up a breeding pair while all juvenile fish are male. Loss of the dominant female fish triggers the most dominant male to transition into a female. A juvenile male then takes the place of the former dominant male in the breeding pair.

In vertebrates, sex reversal has been defined as a mismatch between the phenotypic and the genetic sex. In fish, as well as in other poikilotherm vertebrates, sex reversal can be easily induced by hormonal and sometimes by environmental treatments. Because sex-reversed individuals can be used to produce genetic all-male or all-female populations, they have long been applied in aquaculture and for research on sex determination and differentiation. Following the identification of sex-linked markers, sex-reversed individuals have been identified more recently in wild populations of various fish species.

Although their prevalence has still to be better characterized for most of these species, modelling studies have recently underlined the possible ecological and evolutionary consequences of sex reversal on wild fish populations. Following a brief overview of sex determination and differentiation, the present review will therefore analyze fish sex reversal in both domesticated strains and wild populations, the associated mechanisms, and its possible consequences on the dynamics of wild populations.

As reported by many authors, sex determination systems are conserved and stable within birds and mammals with a few exceptions, whereas they appear to be highly variable among poikilotherms, including fish species. Based upon their sexuality traits, fish species have been categorized into:

1. **hermaphroditic:** individuals in a population produce both male and female gametes simultaneously or successively through a natural sex reversal process;
2. **gonochoristic:** each individual is either a male or a female, and this fixed sex will be kept permanently; and more rarely
3. **unisexual:** populations are composed exclusively of gynogenetic females.

Gonochoristic fish display an amazing diversity of sex determining systems, ranging from a strict genetic control (genetic sex determination, GSD) to an environmental control (environmental sex determination, ESD) or the joint action of both factors. GSD was originally reported in medaka but has been shown to be widespread within gonochoristic teleost species. In most fishes, sex is determined at fertilization by sex chromosomes with a monofactorial system often based on a XX/XY male heterogamety or less frequently a ZZ/ZW male homogamety.

Sex in some fishes as well as in a few other animal species relies upon a polyfactorial system where the sum of various independent loci determines the sexual phenotype. Contrary to mammals and birds, where the sex chromosome pair can be identified by its heteromorphy with a decayed 'heterogametic sex chromosome', in most of the poikilotherm vertebrates including fish, sex chromosomes are homomorphic and poorly differentiated. Because they cannot be directly identified in fish through karyotype analysis, indirect approaches have been developed to search for the sex chromosomes of various species in order to analyze or control sex determination. More recently, molecular sex-linked markers have been identified in some model fish and aquaculture species. All-male or all-female populations (also known as monosex) have been produced for various aquaculture species to benefit from the best sex-linked dimorphisms (e.g., growth, age at puberty).

The sexual phenotype in fish is a plastic trait that can be easily manipulated by hormonal treatments (induction of sex reversal) during critical stages. Complete and functional sex reversal was produced by oral administration of androgens to masculinize or estrogens to feminize medaka fry. The use of Y-linked color genes brought evidence that part of the phenotypic males from the masculinization groups were neomales (genetic females masculinized by the treatment = mXX). Similarly, in the feminization process, part of the phenotypic females were neofemales (genetic males feminized by the estrogens = fXY). By mating mXX with classic fXX females, all-female progenies have been obtained (mXX × fXX → fXX). Conversely, by mating fXY with mXY, viable and fertile YY males (mYY) were obtained (mXY × fXY → 1 fXX + 2 mXY + 1 mYY), and these YY males when mated with classic fXX breeders sired only male progenies (mYY × fXX → mXY).

Following this pioneer work, numerous studies have challenged the hypothesis that estrogens and androgens could be the natural 'gynoinducers' and 'androinducers', respectively. These studies mainly relied upon in vivo treatments using steroids, steroid enzyme inhibitors, or steroid receptor antagonists, *in vitro* steroid metabolism, steroid enzyme immunodetection, and gene expressions, concluding that aromatase and estrogens are the ones which play a key role in fish sex differentiation.

All the hormonal sex reversal treatments that have since been applied to numerous fish species are based upon these pioneer studies. As reported by researchers, this plasticity of sex differentiation can also lead to xenobiotic-induced sex reversal in fish due to pollution. Since these endocrine disrupting effects have been recently reviewed, they will not be discussed further in the present review.

Identification of an ESD system is more recent in gonochoristic fish species. ESD implies that the environment experienced by the offspring during early ontogeny determines the sex of a progeny. Temperature was the first environmental factor demonstrated to be able to determine the sex of an Atlantic silverside *Menidia menidia* progeny through a temperature-dependent sex determination (TSD). Fifteen years later, a temperature-induced sex reversal (TISR) was first evidenced in the Nile tilapia *Oreochromis niloticus*, with high temperatures being able to masculinize sensitive progenies despite the fact that sex in this species is usually determined by major and/or minor genetic factors. TSD was then found in the pejerrey *Odontesthes bonariensis*.

Together, these studies strongly stimulated research on TISR/TSD in various other fish species. EISR/ESR (environmentally induced sex reversal/environmental sex reversal) and especially TISR have been obtained in an increasing number of fish species under controlled conditions. This included species with -

i. a male heterogametic GSD system, such as the Japanese flounder Paralichthys olivaceus, the medaka Oryzias latipes, and the Southern flounder P. lethostigma;
ii. a female heterogametic GSD system, such as the blue tilapia O. aureus, the turbot Scophthalmus maximus, or the half-smooth tongue sole Cynoglossus semilaevis; or
iii. a polygenic GSD system, such as the seabass Dicentrarchus labrax or domestic strains of the zebrafish Danio rerio.

THE USE OF SEX REVERSAL TO UNDERSTAND AND CONTROL FISH SEX DETERMINATION OR DIFFERENTIATION

Sex control is often a good answer to various aquaculture problems mainly related to sex-linked dimorphisms or for reproduction management. Depending upon the species, one sex grows faster than the other, which can be the case for males (tilapia) or for females (e.g., flatfish, sea bass, turbot). The sex-linked dimorphism can also concern the color or the body shape of ornamental fish species, such as the guppies, the Siamese fighting fish, or the platyfish. Besides, skewed sex ratios can be a means to optimize reproduction like in tilapia where a 1:3 (M:F) ratio is usually recommended for fry production. Furthermore, in salmonids flesh quality is affected by the endocrine changes associated with sexual maturity, and consequently, it is more interesting to rely upon female production because they mature later than males. Hence, studies on sex determination and sex differentiation have been strongly stimulated in various aquaculture species due to the benefits generated by controlling sex (i.e., better management and profitability). In some laboratory models and aquaculture species, sex reversal has been used to investigate the process and mechanisms of sex differentiation and/or to develop sex control approaches.

YY Male Production

The pioneer studies of Yamamoto demonstrated that viable and fertile YY males can be produced in the medaka *O. latipes* through estrogen feminization and breeding fXY × mXY. Since then, YY 'super-males' have been developed in various species either through this hormonal approach coupled with progeny testing or through alternative procedures such as androgenesis of XY males or gynogenesis of fXY (sex-reversed females by estrogen treatment) in numerous fish species. In some cases a large scale production of YY individuals exists. Good survival rates of YY males have been reported in rainbow trout, tilapia, and yellow catfish, whereas low survival rates or non-viability of the YY individuals have been reported in the medaka and the guppy versus *Betta splendens* and *Amatitlania nigrofasciata*. In 2 important aquaculture species, the Nile tilapia and the rainbow trout, XX and YY males have been generated in order to produce female (XX) and male (XY) monosex progenies, respectively, through a genetic approach.

WW Female Production

Because of important sex-linked growth dimorphism in turbot, the use of all-female populations could enhance farming profitability. As reported in our previous results on turbot, the breeding of androgen-induced neomales (mZW) with classic females (fZW) provided some 1:3 (M:F) progenies, suggesting the viability of WW females (mZW × fZW → 1 mZZ + 2 f ZW + 1 fWW).

However, due to the late sexual maturity of turbot, this 3-generation process requires at least 6 years to produce fWW (and more to also obtain mWW for WW mass production) in the absence of reliable sex chromosome markers. Moreover, turbot sex also relies upon minor genetic and thermal factors. Therefore, although this approach is technically feasible, it has not been yet developed at a commercial level neither in turbot nor in any other ZZ/ZW commercial species displaying a better female growth rate.

FISH SENSITIVITY TO SEX REVERSAL TREATMENTS

Until recently, based upon the results of Yamamoto, it was generally assumed that effective sex reversal treatments had to be applied during a critical neutral time window which was species specific and generally different for feminization and masculinization, since ovarian differentiation happens prior to that of testis. This window precedes the onset and covers the first steps of gonadal sex differentiation.

Once the natural sex differentiation process had started, a complete and functional sex reversal was considered as no longer possible. Species-specific characteristics of efficient treatments have been extensively described previously.

In cichlid and cyprinid species, where histological sex differentiation usually begins after first feeding, treatments mainly relied upon dietary administration of steroids or inhibitors of steroidogenic enzymes leading to androgens or estrogens.

Interestingly, based upon the expression analysis of several candidate gene profiles, the inhibition of endogenous estrogen synthesis could even result in the production of neomales that are physiologically more similar to genetic males than phenotypic males produced by androgen treatments.

In the Nile tilapia, the standard and very efficient treatment used by most farmers to produce male monosex populations is based on a 3-4 weeks dietary administration of 17α–methyltestosterone (17MT) from the first feeding (around 10 dpf). A precocious sensitive period for sex reversal might also exist in tilapia monosex embryos, since the proportion of females or males increased (20-25%) with 17MT or EE2 immersion treatments of <12 h on XX or XY eggs. Although similar androgen 11KT (11 ketotestosterone) treatment or treatment with an aromatase inhibitor resulted in only 10% masculinization.

Uptake and clearance rates showed that EE2 levels in the eggs dropped at 4 dpf, but the whole body levels were still elevated at 10 dpf. Accumulation of hormones in the egg or vitellus could be delaying the effect on the embryo. However, this hypothesis does not explain the increase in male percentage when precocious high temperatures were applied.

Moreover, similar EE2 bio-concentrations were obtained with a single immersion treatment at 10 dpf resulting in lower female proportions. Surprisingly, no YY embryos were affected even at superior doses of 6,500 μg/l, although both XY and YY progenies were sex reversed (83.7-100%) at later stages of 10 dpf. Perhaps a Y-linked repressor or an X-linked activator is acting at this embryonic stage, which may still be partially active later since female proportions increase but are very rarely 100%. Together, these studies suggest a precocious sensitivity to external factors in tilapia embryos just before the beginning of the morphological differentiation of the brain (31 hours post-fertilization, hpf) and also before the possible identification of the primordial germ cells (46 hpf) in the Nile tilapia.

An important challenge is to compare the mechanisms of the precocious versus classic treatments, considering that the precocious doses were very elevated. We assume that both treatments probably act at different levels of the cascade leading to sex differentiation. The use of species sensitive to environmental treatments will allow bypassing the question of steroid accumulation associated with precocious hormonal treatments.

Until now, identification of the critical inversion window mainly relied upon the empiric analysis of the sensitivity of sex differentiation to external factors (hormones, xenobiotics, and environmental factors) rather than upon specific and precise developmental, histological, or molecular traits. We believe that this imprecise period has to be revisited based on the present knowledge on sex determination/differentiation and using up-to-date genomic tools available for some model species. For example, histological sex differentiation in rainbow trout has been reported to occur around 5 weeks after hatching at 10°C (around 67 dpf) in genetic females, but the physiological or molecular sex differentiation of the gonads characterized by sexually dimorphic expression patterns begins much earlier around hatching time (around 32 dpf).

Fish, like other vertebrates were supposed to have a neutral stage during which the gonads are bipotential before they differentiate towards ovaries or testis. As reported before, sex differentiation towards a specific sexual phenotype can be easily induced in teleost fish with exogenous treatments applied at the 'critical' sensitive window covering at least the beginning of the sex-differentiating process. Gonadal bipotentiality was supposed to be lost after sexual differentiation. However, since the evidence that oocytes can be generated from frozen type A spermatogonia of rainbow trout transplanted into female recipients, it is clear that bipotentiality still persists in adult teleost gonads.

Similarly in adult mice gonads, Sertoli cells and granulosa cells can also be reprogrammed in the absence of their respective transcription factors, DMRT1 or FOXL2. Based upon the hypothesis of the possible persistence of gonadal bipotentiality, functional masculinizations have been obtained in adults of Nile tilapia and medaka through long E2 depletion treatments using an aromatase inhibitor resulting in sex-reversed males that had functional testis and sperm as well as male specific sexual behaviors. Female-to-male sex reversal was also possible in adult zebrafish with a 5-month aromatase inhibitor treatment causing retraction of the ovaries after which testes-like organs appeared, and cyst structures filled with spermatozoa-like cells were observed in sections of these tissues. Therefore, sexual plasticity still exists in adult gonads of teleosts, but functional sex reversal in adults needs longer treatments (3-6 months in tilapia and zebrafish; 2 months in medaka) than in fry (2 weeks 1 month in tilapia).

Based on these results, we assume that several steps of the gonadal development are therefore probably sensitive to external influences but also that sex reversal can be obtained either through a control of sex differentiation or through the inhibition of the maintenance of the phenotypic sex (especially femaleness).

In many species of cultured fish, females grow more than males and mature later. Sex-related differences in growth and sexual maturation can be detrimental to modern aquaculture practices because they can limit or decrease production.

Controlling the proportions of sexes in production stocks to overcome these problems can be easily achieved because the process of gonadal differentiation is generally very labile in fish. This process is naturally controlled by steroid hormones, androgens, and estrogens, which the developing gonads themselves produce. It is thus possible to make a batch of fish develop only into the desired sex, usually females. In aquaculture, this is known as sex control.

HORMONAL TREATMENT

To masculinize fish in those few cases where males are the faster-growing sex – in tilapia, for example – sexually undifferentiated alevins can be treated with androgens known as "male hormones." However, to feminize fish – which is the case in most aquacultured species for which controlling sex is advantageous (eels, carp, salmonids, catfish, sea bass, flatfish) – estrogens or "female hormones" can be used. There are two methods to produce all-female stocks: indirect feminization with androgen treatment and direct feminization with estrogens.

Indirect feminization

In the indirect method of feminization, sexually undifferentiated fish are first masculinized with an androgen treatment. Androgens do not affect genetic males but sex-reverse genetic females. Thus, a genetic female treated with androgens will develop as a male and have testes capable of eventually producing sperm.

However, this treatment leaves the sex chromosomes unaffected. Genetic females will remain so in that their sex chromosomes will be XX, regardless of the androgen treatment. If, in the next generation, eggs obtained from normal females (which always carry the X chromosome) are fertilized with sperm from the genetic females turned into males (which also carry the X chromosome), all-female offspring will result. This is the basis of the indirect method of feminization.

The advantage of this method is that fish that reach market have never been treated with hormones – only their parents were, and for a short period of time. The major inconvenience is that setting up the methodology – masculinizing a sexually undifferentiated batch and identifying the males that are, in fact, genetic females to be used as sperm donors to produce the second generation – is costly and time-consuming. In addition, in order to work, this methodology requires the target fish to have a female XX and male XY system of sex determination, which is not the case in many species.

Direct feminization

Direct feminization involves treating a batch of sexually undifferentiated fish with estrogens. If performed correctly, which is not as straightforward as it may sound, this results in the production of an all-female stock in a single generation. This is the most commonly used method of feminization for its simplicity. In many species, it results in increased production.

However, the fish thus produced have been treated with sex hormones, which can make producers and consumers reluctant to commercialize and consume the fish. Nevertheless, as explained below, there is scientific evidence that proves treating fish with hormones for purposes of sex control does not present any risk for the fish, the operators who administer the treatment, the environment, or the final consumers.

Natural estrogen

The hormone used to feminize fish is the sex steroid estradiol (E_2), the major estrogen produced by all vertebrates from fish to humans. This naturally occurring estrogen is sufficiently potent to feminize many species of fish if applied correctly. Therefore, there is no need to use potent synthetic estrogens such as diethylstilbestrol (DES) or ethynylestradiol (EE_2). Their use should be discouraged or even prohibited.

Short treatments

Effective E_2 treatments for direct feminization can be as short as hours or, at the most, three months. Taking into account that these treatments must be administered when fish are very small and still sexually undifferentiated, this means the feminization treatments are finished many months – and often years in many species – before the fish are marketed.

Therefore, not only are fish treated with a hormone they naturally produce, but the treatment is terminated so far in advance of marketing that there are no residues present. In fact, several studies have demonstrated that essentially all the administered hormone is cleared in two to three weeks after the end of the treatment.

Low doses

Fish have a labile period during which the sexually undifferentiated gonads are most responsive to the effects of administered E_2. If this period is scientifically determined, then effective dosages for feminization can be very low.

To compare, the intake of EE_2 for women who take the contraceptive pill is in the range of 30 to 50 ìg per day. To ingest the same amount of the natural E_2, one would have to consume tens or hundreds of kilograms of treated fish every day.

SEX-REVERSAL IN ADULT FISH

Adult sex-reversal – the change of primary sex (gonadal sex) and secondary sex characteristics and to another sex during adulthood, occurs in many fish species and is triggered by social or environmental conditions. This is an extreme example of phenotypic plasticity – the ability of animal to change its form due to a cue from the environment. Sex reversal requires considerable changes to both the reproductive system (testis or ovary) and changes to secondary sex characteristics (such as pigment, body shape).

One example of sex-reversal occurring in nature is observed in some species of goby fish. These fish get around as a group of females (harem) with a single dominant male. With the loss of the male from group, one of the adult female fish (usually the largest) undergoes sex reversal to become the male of the group .

The prized fish model for studying development and modeling of human disease is zebrafish. Adult zebrafish can also undergo sex-reversal induced by certain environmental conditions such as water temperature or the presence of aromatase inhibitors. Despite being used for decades as a developmental and genetic model, the primary sex determination mechanism is unknown. Recent research indicates that the primitive germ cells have a key role. The developing gonad is made up of somatic cells that form either the supporting cells or hormone producing cells of the gonad and the primitive germs cells which form the future sperm or oocytes.

In zebrafish gonad development goes through a transient stage where – no matter what the ultimate sex will be – both future male and female junvenile fish gonads contain primary oocytes. In the male these early oocytes die off through programmed cell death and the gonad (and the final morphological phenotype) develops as a male. In females, the primary oocytes are maintained and the gonads differentiate into ovaries and the female secondary characteristics develop. For more information.

One method was to make use of a mutant animal that undergoes germ cell loss later in adulthood – nanos 3 null mutants. Nanos proteins are a evolutionary conserved proteins required for *survival* of germ cells thus mutation or deletion of *nanos* results loss of germ cells and infertility.

Zebrafish nanos3 is required to *maintain* the germ line – therefore *nanos3* mutants are initially fertile, but by 5 months old females are now infertile due to the loss of oocytes during maturation.

Researchers examined the sex characteristics of *nanos3* mutant adult fish at 2.5 and 5 months. Initially in *nanos3* homozygotes (ie. have no functional copy of the *nanos3* gene) the females had typical female phenotypic characteristics (at 2.5 months). However by 5 months their overall body shape and pigmentation was now similar to that of a adult male zebrafish. Indicating that the later life loss of oocytes (due to lack of *nanos3*) in the females adult mutants, resulted in female to male sex-reversal.

The second method they used was a transgenic approach to induce germ cell loss after chemical treatment. They used transgenics to produce a line of zebrafish that expresses an enzyme in specifically in germ cells which induces cells to undergo cell death when exposed to a certain drug, metronidazole (Mtz).

One week following Mtz treatment, the loss of oocytes from transgenic gonads was already evident and by two months the females were showing increased yellow pigmentation and body shape changes – ie. they were now more similar to a male phenotype.

They were even able get 4 sex-reversed fish to induce females to spawn and produce viable embryos. This indicates that sex-reversal is complete – initially adult females, following the loss of oocytes, now become sperm producing males. This may give some clues as to the mechanism of natural forms of adult sex-reversal in other fish species.

What about mammals? Evidence using mouse models suggest that somatic sex needs to be maintained/reinforced throughout adulthood. Loss of *FOXL2* (a gene required for ovarian development) expression in the adult ovary results in reprogramming of ovarian cell types (eg grandulosa cell) into cells types typically found in the testis, expressing male marker genes such as *Sox9*. Like-wise deletion a gene required for male development in adult testes, *DMRT1*, results in loss of Sertoli cells (testis cell type) and new expression of female marker genes (i.e. genes that are normally only expressed in the ovary).

THE IMPORTANCE OF SEX REVERSAL

The precocious breeding of tilapia, especially in ponds, with fish maturing as small as 8cm in length, bedevilled their culture for many years. It was soon realised that male fish grew significantly faster than females as the latter turned most of their energy to the production of eggs and fry. Production of all-male fish therefore became essential.

Initial attempts used closely related species such as *Oreochromis hornorum, O. niloticus, O. aureus and O. mossambicus* crosses. Some of these hybrids produced nearly 98% male fish, but hatchery production of large numbers of fry was inconsistent and unreliable. Mixed-sex culture was tried in many tropical countries, but even good quality stock that matured late in the wild started reproducing at an early age in aquaculture ponds, and commercial culture failed time and again.

It was then discovered that the sex of Oreochromis juveniles was only fixed during the first three weeks of life. In the wild, various factors would normally result in a balanced ratio of 50:50 males and females. It was further found that if artificial male hormones (androgens) were fed to the fry for a fixed period of time during this stage, the sex of the fish could be 'determined' and high yields of males achieved.

For example, 17 alpha-methyl testosterone (MT)incorporated into the feed at precise levels produced about 99,8% males.

This is now the established method for producing quality tilapia fingerlings – but does the use of a hormone mean the fish are safe to eat?

No residue

Remember, the fry are only 9mm long when they first feed, and after the short 18-day to 21-day treatment can be from 15mm to 20mm – still a minute fish, weighing less than 0,1g. The amount of MT absorbed is infinitesimal, and scientific tests have proven that no 'residue' is detectable even 10 days after treatment, never mind when the fish is at harvest-size.

All that's happened is that the MT has replaced those slight natural 'cues' that would have turned half the fish male, and half females. Thus they are safe to eat. Unfortunately, though, it has to be noted that misinformation has, in some cases, resulted in customer resistance to the use of sex-reversed tilapia, irrespective of scientific proof of their safety.

Demanding

The methods of feeding the fry are quite demanding. If any other type of food is present in the culture water, you'll have problems. It only takes a few females in the pond to flood the system with fry and juveniles, and the whole sex reversal exercise is wasted. Newly hatched ova or swim-up fry are best collected from mouth-brooding females for incubation while still carrying the yolk sac.

If then reared under the correct water chemistry and cleanliness conditions, and fed on the MT feed for 18 to 21 days, then a high success rate can be achieved. If, on the other hand, the fry are free-swimming and more than a few days old, or green water (containing algae or zooplankton, on which the fry will feed) is present in the culture tank, then the sex reversal fails.

Super males

One other sex reversal method is the production of 'super male' fish. These are fish that have been bred to have YY-chromosomes, as opposed to the normal XY males (females are XX). It is only used in a few hatcheries worldwide, but the advantage is that the actual fish eaten are never hormone-treated. For those who want to farm tilapia commercially and successfully, obtaining good quality all-male fingerlings from a competent hatchery is an essential first step.

COEXISTENCE OF GENETIC AND ENVIRONMENTAL INFLUENCES

Reviews on the environmental effects on fish sex determination/differentiation have generated debates about TSD in gonochoristic fish as well as in other vertebrates. Mainly based upon reptile sex determination, criteria have been proposed for identifying TSD in vertebrates. Amongst them, 2 main criteria have been used by researchers to reclassify 19 out of the 27 fish species previously reported to be TSD species as GSD + environmental effects. This was based on (a) the existence of sex chromosomes which was considered a signature of an exclusive GSD; and (b) for some of the fish species the efficient temperature treatments which skewed the sex ratios were considered to be outside the natural range specifically encountered by the fry or juveniles of these species during their critical period.

Concerning these 2 criteria, it is important to enrich the debate with recent data concerning GSD in vertebrates and TSD in poikilotherm vertebrates which have led to the deconstruction of some common myths about the evolution of sex determination in the animal kingdom. Mainly based upon few mammal models and to a lesser extent upon bird models, sex in animals has long been considered to be stable and determined by sex chromosomes.

This predominant misconception that sex chromosomes are necessarily associated only to GSD has for long led to underestimate that at least in poikilotherm vertebrates, genetic systems of sex determination without sex chromosomes or with the presence of both major and minor genetic factors can also exist, especially in fish.

Indeed, in amphibians, reptiles, and teleosts 'sex-determining mechanisms are diverse and can evolve rapidly', with multiple evolutionary transitions between TSD and GSD having occurred in reptiles and amphibians and therefore, the line between a sex determination system relying on pure genetics or on pure environmental factor(s) is faint. Following the theoretical model of researchers, ESD and GSD have for decades been considered as 2 mutually exclusive systems. However, based upon extensive studies in 2 lizard species, *Pogona vitticeps* and *Bassiana duperreyi*, unambiguous evidences have shown that GSD with sex chromosomes (respectively ZZ/ZW and XX/XY) and TSD can perfectly coexist. Furthermore, *B. duperreyi* has even a heteromorphic pair of sex chromosomes.

GSD systems can become sensitive to the environment and then present genotype-by-environment interactions or evolve towards TSD, allowing the coexistence of environmental and genetic influences. As hypothesized previously, the major role of sex reversal to induce this type of transition has been recently confirmed in the Australian lizard *(P. vitticeps)* by an experimentally temperature-induced rapid shift between a predominantly GSD (with heteromorphic sex chromosomes ZZ/ZW) and a TSD system using sex-reversed ZZ females identified in the wild. Although the W chromosome was eliminated in one generation under laboratory conditions, a similar process should be much more gradual in wild populations, and therefore the transition between GSD and TSD will probably need multiple generations. Altogether, these recent data confirmed that at least in these lineages TSD and GSD are not mutually exclusive systems but the 2 extreme endpoints of a continuum, underlying the coexistence of both genetic and environmental influences in the middle of this spectrum: 'many species have differentiated sex chromosomes, but also show a temperature override, where genes and environment interact to determine sex'. In fact just recently, a three-ended continuum that can be visualized as a triangle has been proposed in which genes, environment, and random factors (i.e., any stochastic fluctuation in the expression of key genes of the cascade) could interact to determine sex.

Regarding fish, such a continuum with possible coexistence of genetic and environmental factors has also been proposed by some authors. In the Atlantic silverside, whose TSD system reached a consensus, thermal sensitivity of the wild populations varies from high to low levels within a natural thermal gradient. Because large females of *M. menidia* have a better reproductive fitness than large males, long growing seasons are more beneficial to females than to males. Therefore, at low latitudes when growing seasons are longer, early spawning (low temperatures) favors female differentiation. Temperature is a reliable cue of the future length of the growing season.

Sex determination in Southern (high thermal sensitivity with an adaptive role of TSD) or in Northern (low sensitivity) populations, respectively, relies upon a polygenic control, with both temperature and genetic factors interacting, or upon major genetic factors. Between these 2 extreme conditions, a mixture of TSD and GSD determines the sex of individuals. Further studies analyzing 31 wild populations have demonstrated that the pattern of sex determination in the Atlantic silverside is actually non-linear, with none of the studied populations displaying a complete TSD or pure GSD, since in the first case some genetic component always exists, while in all the GSD populations some thermal sensitivity still persists.

Hence, fish TSD relies upon genetic × environmental interactions with gradual shifts in sex ratio, unlike the abrupt pattern found in many reptiles. Furthermore, the identification of a major sex determinant (the *amhy* gene) in another species, the pejerrey *O.bonariensis* with a high thermal sensitivity again underlines that in sensitive fish species, both genetic and environmental factors can determine the sexual phenotype. Finally, it is also important to remember that differences exist in TSD traits between reptiles and fish. In reptile TSD species, especially in turtles and crocodiles, specific temperatures usually determine the sex of all individuals, producing therefore all-male or all-female progenies.

Conversely in fish with TSD, such as *M. menidia,* extreme temperatures usually do not induce 100% male or female broods. For some individuals in a progeny, sex is therefore not determined (or not only) by temperature. Similarly, apart from the pejerrey, temperature-induced monosex populations are very rarely reported in the Nile tilapia, the European sea bass, as well as in other thermosensitive species. Besides, high between-family variance in sex ratios have been reported in tilapia, with contributions from both male and female breeders. Because skewed sex ratios are strongly heritable, they can be selected.

Moreover, in the same species, these 3 characteristics can also be applied to its thermosensitivity. Skewed sex ratios have also been shown in the rainbow trout, *Oncorhynchus mykiss*, following high temperature exposure (18°C treatments applied from 42 days post-fertilization [dpf] onwards). In this species there is a strong genetic XX/XY sex determination where the major sex determining gene *sdY* has been identified, but the presence of a minor loci (a recessive mutation), which was called *mal,* associated or not to other loci and/or to environmental variations can lead to spontaneous masculinization. The study of this transmission was performed through several generations showing that it was a heritable trait. High temperatures of 18°C were able to masculinize XX females that carried *mal* in gynogenetic doubled haploid individuals.

In tilapia (but probably also in other thermosensitive species) thermosensitivity has been demonstrated to be highly heritable (h = 0.69), and therefore, it can be selected. In the turbot, *S. maximus*, sex determination mainly relies on a female heterogametic system (ZZ/ZW), but again, thermosensitivity and family temperature interactions have been reported at least in some progenies. These types of sex systems resemble polyfactorial systems. Until now, very few cases of 'true' polyfactorial sex determination have been reported in fish. However, a polygenic sex determination is characterized by a large between-family variance in sex ratio, paternal and maternal effects on sex ratios, and a sex ratio response to selection. Interestingly, species such as rainbow trout and tilapia that are usually classified as GSD can present some of these characteristics.

It is also interesting to draw a parallel between the hypothesis of an ancestral polygenic sex determination in fish and the recent demonstration that domesticated strains of zebrafish have lost the natural sex determinant present in wild zebrafish populations that rely upon a classic monofactorial GSD of ZZ/ZW type.

Interestingly, besides the ZZ/ZW system, other factors might be at play, because sex-reversed females have been reported. As suggested by the authors, domestication or/and selection or/and inbreeding of zebrafish laboratory lines have resulted in the evolution of the sex determination system possibly by using cryptic minor genetic factors. Such minor factors could coexist with the sex determinant in the wild population and partially explain why sex-reversed zebrafish can be found in both wild populations and domestic strains.

These sex reversals could also be associated with zebrafish sensitivity to various environmental conditions (oxygen, density, temperature, thermocycle, poor nutrition), and/or to the unveiling of minor factors. In haplochromine cichlids from Lake Malawi in East Africa, where multiple interacting loci control sex determination, sex-determining loci or sex determination systems can hide other loci or systems through epistatic dominance mechanisms or sex determination hierarchies. Altogether, this could explain why in tilapia but also in species considered to have a strong GSD (e.g., medaka, rainbow trout, common carp), minor factors have been detected at least under extreme genetic (i.e., clonal lines in the rainbow trout) or environmental (including those that are 'out of the normal incubation range') conditions.

Vertebrate sex-determining systems are therefore probably more complex than previously thought, especially in poikilotherms, and the hypothesis of simultaneous involvement of multiple interactive factors should not be ignored even in species where major factors have been identified. Often considered as a transient state, polygenic sex determination is nevertheless stable over time at least in the house fly, *Musca domestica*

Regarding the second criteria that sex ratio deviation should occur within a range of temperatures naturally occurring during decisive developmental steps, recent data also bring some new perspectives. In the Californian grunion fish, *Leuresthes tenuis*, both temperature and photoperiod can influence the sex ratio. However, because the seasonal thermal variation is very low (4°C) along the North American Pacific coast, photoperiod has become a more informative cue of season length than temperature. Longer day lengths and lower temperatures induced higher proportions of females. This is the first case of a photoperiod-dependent sex determination system in fish.

Temperature and photoperiod seem to act in synergy in this fish to determine the sexual phenotype. This suggests that -

i. temperature in fish is not the only environmental factor involved in ESD systems;
ii. in some environments, photoperiod might be a more informative factor of the future environment of the fry or juveniles than temperature; and
iii. in species for which at least 2 environmental factors might have synergetic effects on sex ratios, application of a treatment of only one environmental factor probably requires higher levels of this factor to obtain a similar effect than when the 2 factors are applied.

Therefore, in sensitive species the effects of additional environmental factors on sex ratios have to be searched for. If other factors (at least a second one) are also efficient to influence the sex ratio, then synergic effects will have to be searched by applying decreasing values of temperatures.

Because minor and/or environmental factors can both induce sex reversal in gonochoristic fish, we consider that their sex reversal can be defined as a mismatch between the phenotypic sex and the sex theoretically governed by the major genetic factor(s).

MECHANISMS OF SEX REVERSAL

Sex reversal can be considered as a change in the sex threshold trait. In fish this threshold appears to be extremely susceptible to environmental modifiers (e.g., temperature, pH, density) shown to act at cellular and/or gene levels. Modifiers that affect the onset of ovarian meiosis, germ cell proliferation, or the number of germ cells lead to sex reversal. As previously questioned, the number of germ cells could be an important factor or even the threshold modifying the gonad sex fate at least in some species. The presence of germ cells is indeed necessary for ovarian formation and differentiation in some fish species. Loss of germ cells induced by high temperatures or by other means in zebrafish, medaka, and fugu induced sex reversal of females into males, with somatic cells seemingly predisposed to develop into male cells. However, this was not the case in the goldfish. In the TSD pejerrey, germ cell apoptosis by high temperatures alone or coupled to busulfan did not affect the sex ratio of the sterile gonad. Likewise, our group have shown that germ cell depletion by high temperatures does not apparently cause sex reversal in the Nile tilapia.

NEW ACTORS OF SEX DETERMINATION AND DIFFERENTIATION

The first step at which environmental factors like temperature may be overriding the genetic sex is by somehow affecting the primary sex-determining gene. We are in a fascinating era with the discovery of several different master genes in teleosts that will all be valuable and reliable sources to establish the genotypic sex of wild populations of these species in order to find sex reversals. Most of these genes have emerged from duplicated copies of downstream sex-differentiating genes that have been recruited to become master sex determinants. *Dmy/dmrt1bY,* which arose from the testis *dmrt1* gene, was first identified in the medaka *O. latipes*.

Spontaneous XY female medaka are attributed to mutations of the male-specific region, the *dmy* coding region, or a single nucleotide mutation in exon 3 resulting in a truncated *dmy*. In another medaka, *O. luzonensis*, sex is directed by the male-specific expression of *gsdfY,* a Y allelic form of the gonadal soma-derived growth factor, a member of the TGF-β superfamily. *Gsdf* may also be the sex determinant in the sablefish *Anoplopoma fimbia*.

In the Patagonian pejerrey *O. hatcheri*, maleness is directed by a Y copy of the anti-Müllerian hormone *amh* gene also belonging to the TGF-β superfamily. Despite having a strong TSD, another pejerrey, *O. bonariensis,* also expressed *amhy* at early stages of male promoting temperatures followed by the expression of the autosomal *amha* gene, showing the coexistence of both temperature and genetic sex determination in this species.

Two tandem Y copies of the *amh* gene have been identified in the Nile tilapia with a 233-bp deleted region in exon VII and a 5-bp insertion in exon VI that leads to a truncated *amh* gene lacking the TGF-β domain, defined as *amhΔY,* and another *amhy* having a missense SNP which may be determining sex in some strains, notably the Japanese strain. In several pufferfish *(Takifugu)* species an allelic variation in the *amhr2* gene, the *amh* receptor, has been shown to be responsible for maleness. Surprisingly, in the rainbow trout it is a copy of an immune gene, the interferon regulatory factor 9, which has become the master *sdY* gene.

Sex-determining genes are being searched for in several other fish species with many advances achieved in identifying linkage groups (LG) with quantitative trait loci and SNPs associated to sex using genome-wide sequencing or restriction associated DNA sequencing (RADseq). These techniques have, for instance in the Nile tilapia, narrowed down an inverted region on LG1 to 8,8 Mb that harbors a sex-determinant but not the *amh/amhy* and contains 8 genes amongst which is the Wilm's tumor protein homolog *wt1b* in the Manzala and Ghana strains.

In tilapia, sex-linked loci seem to vary in function of the strain and population. Environmental and parental factors are additional factors that influence the phenotypic sex of tilapia. Genotyping of functional SNPs evidenced that 2 were heterozygous for males, while females were homozygous, with misassignments corresponding to XX males. Associations were confirmed in a temperature-masculinized family which had a 66% male sex ratio.

Study of LGs linked to temperature-induced phenotypic male tilapias was performed by comparing a thermosensitive line versus a low or non-thermosensitive line showing a strong association of LG1, LG3, and LG23. A strong SNP association was also shown in LG20 in another family sensitive to temperature-masculinization using RADseq. In zebrafish, sex-determining studies have followed a long windy road: RADseq revealed a polyfactorial system in domestic strains, but it found a ZZ/ZW system in the wild populations with the sex determinant on chromosome 4R].

Some female-to-male sex reversals (female genotype *f*/*m*) were found but not the reverse, suggesting that either there is a dominant *f* allele which is required but not sufficient for a female phenotype, or alternatively there is a 2-dose male 1-dose female hypothesis. The authors suggest that environmental factors might be behind these female-to-male reversals affecting the meiotic oocytes by inhibiting primordial germ cell proliferation, meiosis entry, or stimulating oocyte apoptosis.

Gsdf is a novel gene, specific to fish and possibly an endogenous inducer of testis development. In medaka, *gsdf* seems to be the target of *dmy* since they co-localize in somatic cells (pre-Sertoli cells). It is detected just after *dmy*, with a downregulation seen in XY females treated with estradiol. *Dmy* knockdown with RNAi resulted in male-to-female sex reversal induced by suppressing the male pathway genes *(gsdf* and *sox9a)* and upregulating the female pathway (R-spondin1, *rspo1*). Complete female-to-male sex reversal of the medaka *O. sakaizumii* but not of *O. latipes* stimulated *gsdf* and subsequently *dmrt1*. In tilapia, upregulation of *gsdf* transcripts were initially suggested to be at 8 dpf prior to *dmrt1* and localized also in the somatic cells surrounding germ cells.

However, CRISPR/Cas9 knockouts revealed that *gsdf* is in fact expressed after *dmrt1* and an aromatase inhibitor treatment rescued the testicular phenotype of XY gsdf$^{-/-}$ gonads [Jiang et al., 2016]. In vitro analysis showed that *dmrt1* co-transfected with *sf1* was able to activate *gsdf* in a dose-dependent manner. G*sdf* action in XY tilapia is thought to be related to the suppression of estrogen production possibly via the inhibition of ovarian-differentiating genes.

Ontogeny expressions show that both the sex determining gene *amhy* and the autosomal *amha* gene are implicated in testis differentiation in the TSD pejerrey even in males lacking *amhy,* while in the GSD pejerrey *amha* is not implicated. In the Nile tilapia, *amhy* and *amh*Δ*Y* are expressed specifically in XY larvae from 9 dpf, but only the *amhy* knockdown resulted in male-to-female sex reversal. Overexpression of *amhy* in XX fish but not of *amh*Δ*Y* resulted in female-to-male sex reversal, suggesting that *amhy* is the sex determinant at least in the Japanese strain.

A non-synonymous SNP in exon VI was found in one of the tilapia *amh* genes (but which of the 3 genes it corresponds to is unclear) associated to temperature-induced males, that could be useful to increase the proportion of males in the selective breeding of temperature-induced lines. *Amh* and *amhr2* together with *gsdf* appear to play important roles in primordial germ cell proliferation which might explain why they have been recruited to become primary sex determinants. In the medaka XY *hotei,* a mutation in the *amhr2* gene causes hyperproliferation of the germ cells and male-to-female sex reversal due to lack of Amh signaling on the mitotic self-renewing germ cells. In tilapia *amhr2* knockout caused 100% male-to-female sex reversal but in contrast, only 60% were obtained with *amhy* due to F_0 generation being mosaic with the mutation varying between individuals. Therefore, both *amhy* and *amhr2* are critical to determine sex in tilapia, but it remains to be known how the signaling occurs. Knockdown of both *amhy* and *amhr2* in XY tilapia causes upregulation of the aromatase *cyp19a1a* gene as well as higher E2 levels, but no increase in *cyp19a1a* expression was seen in *amhy/amhr2* overexpressed XX fish.

REVISITING OLD PLAYERS OF SEX DIFFERENTIATION

The classical view of hierarchy and a sex-differentiating pathway is being replaced by a scenario of a gene network with a complex interplay of signals acting in a modular way. The underlying molecular mechanisms modifying testis or ovarian thresholds leading to sex reversal in fish are slowly being better deciphered with transcriptome microarrays studies or using next-generation sequencing techniques (NGS) coupled to overexpression and knockdown studies. The hormonal estrogen status has been known for a long time to be an important factor affecting the sex threshold towards ovarian differentiation via changes of aromatase, the estrogen-producing enzyme or its gene *cyp19a1a.* The critical role of estrogens with the earlier upregulation of the female pathway in fish sex differentiation but not androgens has been reconfirmed.

Depletion of estradiol levels was the prerequisite for adult sex reversal in both Nile tilapia and medaka. In both species the aromatase inhibitors first caused female germ cell degeneration, depletion of E2 levels followed by the development of testicular portions where interstitial cells became steroidogenic-active Leydig cells and only later the rise in 11KT levels, effects that were rescued with E2 supplement. In this transdifferentiation, granulosa became Sertoli-*dmrt1* positive cells with suppression of female-specific genes followed by upregulation of male-specific genes.

In medaka some cyst germ cells were germline stem cells with bipotential activity that differentiate into oogonia but in the absence of E2 are capable to differentiate into spermatogonia. The mode of action of feminizing EE2 (ethinylestradiol) treatment was studied in rainbow trout fry/juveniles with a meta-analysis of microarray data showing a rapid and strong decrease of key testis genes, notably those involved in androgen synthesis.

It highlighted new pathways such as the progesterone-oocyte maturation pathway with several genes of the germline, cell division and oocyte meiosis represented and the peroxisome proliferator-activated receptor (PPAR) signaling pathway containing genes from the nuclear hormone (steroid/thyroid/retinoid) receptor family that perhaps regulate aromatase levels.

Some studies suggest that induced masculinizations might not follow the same pathways as natural 'genetic' testis differentiation. Different fish masculinizing compounds, consisting of 17MT (an aromatizable androgen), dihydrotestosterone (DHT, a non-aromatizable androgen), and fadrozole (the aromatase inhibitor) were evaluated for their androgen activities, inhibition of aromatase activity, and efficiencies in female-to-male sex reversals in tilapia and using a species for which the male blue coloring is associated with high androgen levels. The masculinizing actions of 17MT and DHT are most likely due to binding to androgen receptors seen by the inhibitory effects and sex reversal capabilities when combined with the androgen receptor agonist flutamide.

The fadrozole action (the most potent in inhibiting aromatase activity) might be an excess of un-aromatized androgens accumulated in the differentiating gonad. Masculinization treatments with androgens (11βOHΔ4) of XX rainbow trout or long aromatase-inhibitor treatments of XX tilapia caused a 'transdifferentiation' with a rapid inhibition of the female-specific genes followed by upregulation of male genes but with disrupted expression profiles.

Temperature-masculinizing treatments in tilapia caused a very quick and strong overexpression of the main male-differentiating genes *amh* and *dmrt1* to levels above those found in differentiating genetic XY males, followed only after by the repression of the principal female-differentiating genes *cyp19a1a* and *foxl2*. High temperature accelerates the molecular mechanisms leading to testis development, since other male genes also showed earlier upregulations such as *sox9a*, *sox9b*, *dax1*, and particularly *cyp11dh* which encodes 11-β hydroxylase that participates in 11KT production, suggesting that an increase of androgens might be required for temperature-induced masculinization. This suggests that temperature-induced masculinization requires the male-differentiating pathway to be turned on and only after that, the ovarian-differentiating pathway is suppressed. In the TSD pejerrey, a heterologous microarray study revealed that in the masculinization at male-promoting temperature (MPT) there is an interplay of pro-apoptotic and anti-apoptotic genes. In the sea bass, high temperatures (HT) also promote males, and microarray analysis showed elevated *dmrt1* expressions in sex-differentiating HT fish together with increases in cholesterol transporting genes and stress response genes as well as upregulation of epigenetic regulatory genes, while low temperature fish showed high *cyp19a1a* levels.

Cyp19a1a expressions were compared during sex differentiation between a GSD Atlantic silverside population from Nova Scotia and another TSD population from South Carolina which were both treated at feminizing (15°C) and masculinizing (28°C) temperatures. In the GSD population, *cyp19a1a* expression was earlier, stronger, and clearly dimorphic and was not affected by temperature. In contrast, in the TSD population only the feminizing temperature treated fish had elevated *cyp19a1a* levels but they were considerably lower than the presumptive females of the GSD population. This suggests that *cyp19a1a* is perhaps predisposed to be suppressed easily in this population (lower threshold?) and perhaps under different regulating mechanisms. It is possible that a difference in DNA methylation exists in the *cyp19a1a* promoter region (Fig. 1) as shown in the sea bass HT males.

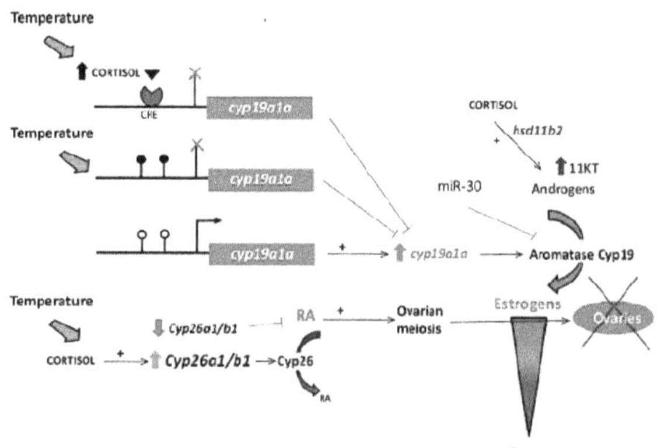

Fig. 1. Schematic representation of different regulating mechanisms that have been described amongst fish species in temperature-induced males, of cortisol action, cortisol binding, and DNA methylation on the aromatase cyp19a1a gene promoter suppressing the gene expression, which affect aromatase activity and estrogen levels leading to testis differentiation. Cortisol also leads to lower retinoic acid which delays ovarian meiosis inducing testes differentiation.

BRAIN IMPLICATION IN THE SEXUAL PHENOTYPE

Fish possess a large plasticity for sex reversal which is also evident in the brain with a reversibility existing even in adulthood. In contrast, brain sex differentiation in birds and mammals occurs only during embryo development and is shown to be mediated not only by gonadal steroids but also by genes on the sex chromosomes as well as epigenetic regulations. Several neurosteroids are produced in the teleost brain, notably estrogen with an extremely elevated aromatase activity that increases with age, probably for the continuous neurogenesis.

Most teleosts possess 2 aromatase genes *(cyp19a1a* and *cyp19a1b)* generated during teleost genome duplication, with *cyp19a1b* being the predominant brain gene. The *cyp19a1b* gene is thought to be implicated in a feedback loop in the brain regulating teleost steroid production through the kisspeptin and the hypothalamo-pituitary-gonadal axis. In mammals, testosterone exerts many physiological actions in the brain, affecting male sexual behaviors that are mediated by the aromatase neural conversion of androgens into estrogens and therefore, the brain aromatase is a critical factor in the masculinization of the brain.

Differentiating rainbow trout males had overall higher *cyp19a1b* expression levels in developing brains. During the onset of gonadal sex differentiation, female tilapia brains had higher aromatase activity than males, but no sex differences were found for *cyp19a1b* expressions. The precocious sex reversal of tilapia XY and YY into females caused an increase in brain *cyp19a1b* expression that was dependent on the EE2 dose and was also associated with an increase in both testosterone and estrogen concentrations. Upregulation of *cyp19a1b* expression was also seen when we masculinized XX females with high temperature treatments, together with higher *erα* and *er* β expressions.

We saw different *cyp19a1b* expression profiles for the different phenotypic males with levels remaining low till adulthood in the 17MT males compared to fadrozole-treated males and temperature-induced males, all lower than genetic males. This study suggests that the gene endocrine environment is probably different between these phenotypic males, and it would be interesting to investigate if they display differences in their sexual behavior. In the medaka brain, *esr1* has a male-biased expression suggesting that it has a role in the masculinization or defeminization in medaka while *esr2*β and *ar*β were implicated in the feminization or demasculinization. In this study the ventral telencephalic and preoptic regions showed strong female-specific expressions of ERs and ARs which were activated by estrogens and inhibited by androgens. In the TSD pejerrey, *cyp19a1b* is expressed in the brain at MPT before the onset of gonadal sex differentiation.

We have shown in tilapia a very precocious sexual dimorphism of *amh* expression in male brains occurring between 10 and 15 dpf which is the stage when the gonad aromatase *cyp19a1a* gene only starts to be expressed in differentiating females. This suggests that masculinization of some brain regions might not be mediated by gonad steroids. It could be the result of *amh* present on the presumptive Y chromosome of tilapia, the *amhy* gene, although no brain expression was detected by researchers. We have developed a precocious sexing procedure for tilapia using brain *amh* expression that reflects the male phenotype of individuals at 14 dpf correlating accurately with the sex ratio of a progeny. This procedure could be an efficient way to accelerate the selection of YY individuals or for generating a thermosensitive line.

The adult treatments of medaka and tilapia with aromatase inhibitors (AI) clearly showed the brain sexual plasticity since the XX treated fish displayed male-specific territorial and sex behaviors. AI treatment of the African cichlid *Astatotilapia burtoni* adult females did not cause sex reversal of the ovaries retaining female gene profiles; however, partial masculinization was observed with changes in coloration, body shape and form, steroid levels, and notably modifications in the brain gene expressions which affected sexual behavior.

Cyp19a1a present in the brain in this species was upregulated in AI females while both *arα* and *cyp19a1b* were downregulated, resembling the profiles of males which could be to compensate E2 levels or due to higher testosterone levels.

EPIGENETICS: A LINK BETWEEN THE ENVIRONMENT, GENE EXPRESSION, AND SEX REVERSAL

One of the underlying mechanisms leading to sex reversal in some fish species might be epigenetics. It is now evident that the environment can induce epigenetic modifications of histones affecting the compactness of the chromatin structure and DNA accessibility without causing actual nucleotide changes that play essential roles in gene regulation and ultimately modify the phenotype. One of the most studied epigenetic mechanisms is DNA methylation of cytosines (C5 position) at CpG dinucleotides because of its implication in a large number of cellular functions such as development and cell differentiation, tissue-specific gene expression, genome imprinting, X chromosome inactivation, and in disease susceptibility.

Hypermethylation is associated with gene silencing, but methylation also acts at the level of transcript splicing. Higher DNA methylation levels have been found in sea bass gonads of differentiating males in the aromatase *cyp19a1a* promoter that were inversely correlated with the *cyp19a1a* gene expression. Exposure during early life to high temperatures (>17°C) induced a male bias and also increased the methylation levels in both males and females.

Interestingly, sex-specific differences were seen for the 7 CGs sites, while only 2 showed temperature-induced differences. The *in vitro* study suggests that the hypermethylation blocked the stimulatory effect of SF1 and FOXL2.

Opposite methylation patterns were evidenced in the half-smooth tongue sole, with males having a low methylation rate at the *dmrt1* promoter (located on the Z chromosome) associated to an elevated *dmrt1* expression. Tongue sole is a ZW/ZZ species but temperatures of 28°C lead to ~73% masculinized ZW females, and these phenotypic males show a demethylation of the *dmrt1* promoter. Genome-wide DNA methylation and the transcriptome were studied in this species, permitting a much larger analysis of the methylation status of gene regulatory elements and gene expressions.

Differences in the methylation rates were found between ovaries and testes notably for the sex-differentiating pathway genes. Interestingly, higher levels of methylation were found in a 2-Mb region of the Z chromosome which differed from the W chromosome that clustered for testes, which might lead to some dosage compensation in ZW males. In addition, the tongue sole showed splicing alternatives but not silencing, for a female W gene *fig1a* (factor in the germ-line alpha) associated to differential expression between ZW females and ZW males. Very recently, the global DNA methylation status was analyzed in the Nile tilapia between mature (>4 months) females, mature males, and from a matured mixed-sex group treated at high temperatures.

Overall higher CpG methylated domains were found in the gene bodies of females compared to males. High temperature treatments increased the methylation levels in both males and females. The orphan nuclear receptor *dax1 (nr0b1a)* gene had higher methylation levels in induced males when compared to control females, while the steroidogenic enzyme *hsd17b8* gene as well as the *erb2* and *gsdf* genes had lower levels.

Cytosine methylation is catalyzed by the DNA methyltransferase enzymes (DNMT). In mammals, DNMT1 is involved in the maintenance of the CpG methylation pattern on the new DNA strand after replication, while DNMT3A and DNMT3B are considered de novo methyltransferases implicated in the initial CpG methylation pattern. DNMT3L is an additional member with no enzymatic activity that acts as a cofactor of DNMT3A and DNMT3B activity. In teleosts, due to the genome duplication, up to 6 *dnmt3* genes have been identified in zebrafish which complicates the analysis of the Dnmt enzyme contribution. We have started to analyze the expression levels of Nile tilapia *dnmt* genes during sex differentiation.

Sexual dimorphism and temperature differences were observed for some of the de novo DNA methyltransferases, *dnmt3* while *dnmt1* remained unchanged which suggests that the *dnmt3a* and *dnmt3b* paralogues have specific roles during tilapia genetic sex differentiation and temperature-dependent masculinization. It might indicate differences in gene methylations occurring between genetic males and females but can also be induced by a temperature- induced masculinization. This stage coincides with repression of the *cyp19a1a* gene in temperature-induced males. Whether its promoter is hypermethylated in tilapia males similarly to what has been seen in sea bass remains to be investigated.

DNA methylation associated with sex-determining and -differentiating genes has begun to be investigated in other fish species, but a whole avenue is now possible with genome-wide studies of the methylome associated with RNAseq which will enlighten how the methylation marks are activated or erased, implicated in transcript silencing or dosage compensation, between genetic sexes and sex-reversed fish. Epigenetic marks other than methylation are only starting to be analyzed in fish during sex determination and differentiation. For instance microRNA, small non-coding RNAs that inhibit at post-transcriptional level target-gene expressions, have begun to be profiled between differentiating ovaries and testes in the Nile tilapia.

Integration of miRNA and mRNA was compared for early differentiating XX and XY Nile tilapia identifying 130 novel miRNA of which 49 were unique for XX and 45 solely found in XY gonads. Sex-biased miRNA were negatively correlated with genes from the steroid hormone biosynthesis pathway *(Cyp11a1, Hsd3b, Cyp19a1a, Hsd11b)* and to the ovarian or testes differentiating network genes such as *Foxl2, Amh, Star1, Sf1, Dmrt1,* and *Gsdf.*

Epigenetic mechanisms can be reversible and heritable. The study of trans-generational epigenetic inheritance is easily accessible in fish, notably those with short life stages, and a better understanding of these mechanisms will be possible by studying successive generations of sex-reversed individuals. It would be very intriguing to see how sex-reversed fish have adapted to natural high temperature regimes and how populations have been potentially affected in the wild, depending upon the presence or absence of such epigenetic marks.

Sex Reversal in the Wild

Many fish species have been manipulated with exogenous factors to reverse the primary sex under controlled conditions, but several questions are raised on whether this can occur under wild conditions, at the individual level, at what frequency, does it impact the population, and finally does it have an ecological significance. Apart from *M. menidia* where the existence of TSD has been associated with a better reproductive fitness of one sex following in this way, possible advantages for the existence of environmental sex reversal in fish species are not known.

It is first essential to study how frequent environmental sex reversals are in wild populations to then study their fitness and the effects of biased sex ratios on population dynamics. Because the majority of fish species with GSD have homomorphic sex chromosomes, karyotype analysis does not allow the study of mismatches between the genotypic and phenotypic sex. To study these discordances in wild fish populations requires sex-specific markers.

The role of *dmy/dmrt1bY* as the sex determinant in the medaka *O. latipes* was partially established by researchers due to the finding of 2 wild-derived XY female mutants. The finding of spontaneous sex-reversed medaka led to the *Dmy/dmrt1bY* genotyping in 8 medaka strains to try to find discordance between the genotypic and phenotypic sexes. Both PCR analyses and FISH using a Y-specific BAC probe identified 40 males out of a total of 764 fish (5.2%) lacking the *dmy/dmrt1bY* marker which were further confirmed to be XX males when some were used in progeny testing giving female-biased offspring. Frequency of these XX males ranged between 3.7-15.4% of males in 6 strains analyzed, but no XY females were found amongst them.

Existence of the XX males indicated that *dmy/dmrt1bY* and the Y chromosome were not indispensable for testis development, and this was explained by putative autosomal modifier genes overriding the sex chromosomes by either suppressing a female-determining locus on the X chromosome or by acting as male inducers. The absence of XX males or their presence in low frequency in some strains led the authors to suggest that the modifiers were polygenic and strain specific.

A larger *Dmy* genotyping was performed in a survey of 3004 wild-caught and wild stock medaka from 109 sites in Japan, Korea, China and Taiwan. In the majority of fish, the phenotypic sex coincided with the genotypic sex, but in 8 localities 15 *dmy*$^-$ males, presumably XX males (1.03% of males), were found while 26 *dmy*$^+$ XY females were located in 13 sites (1.68% of all females). A progeny testing by crossing these XY females with inbred XY males gave functional XY females in the F_1 generation that all carried the maternal Y chromosome. This study showed not only that *dmy* was commonly found in wild populations of medaka but also the existence of wild XY sex-reversed females that either carried *dmy* or a mutated *dmy* gene which apparently was not lethal. The study to identify the genetic basis leading to the spontaneous XX female-to-male sex reversal showed that it was a recessive mutation, controlled by the *sda-1* locus located on the autosomal LG8.

The absence of the Y chromosome in successive generations of medaka from the Shirone wild population suggests that LG8 has taken over the role of the sex chromosomes. The genetic basis of XY lacking *dmy* resulting in male-to-female sex reversal was found to be of 2 types, one due to mutations in the amino acid coding sequence of *dmy* and the other due to low expressions of *dmy* at 0 days after hatching at the critical sex-determining period while the coding sequence was normal. This suggested that *dmy* expression below a threshold did not cause pre-Sertoli (*dmy*-positive cells) to develop into Sertoli cells, disrupting primordial germ cell proliferation and testis development leading to female phenotypes. Regulatory mutations in the *dmy* flanking region might be the cause of the mutants having reduced *dmy* expression. In addition to the reduced *dmy* expression levels there might be autosomal loci associated with sex reversal.

The sex-determining gene *sdY* identified in rainbow trout was searched in several salmonids species with the finding of a single sdY^{+} female in Chinook salmon *O. tshawytscha* and a single sdY^{-} male in lake trout *Salvelinus namaycush*. *SdY* was used further to search for mismatches between the genetic and phenotypic sex in wild populations of Chinook salmon from Alaska, Idaho, and Washington. While all males were sdY^{+} (45/45 males), 13 (8.3%) out of 157 phenotypic females were found to be also sdY^{+}. Seven of these putative XY females contained all 4 exons of *sdY* and were also positive for the Y chromosome male markers *OtY1* and *GHc*, suggesting possible unknown environmental factors causing sex reversal.

In the Nile tilapia the 233-bp deleted *amh* gene, which was initially called *amhy* and later *amhΔY*, was identified in males and considered to be specific for the Y chromosome and was always associated to the male sex determinant *amhy* in this species. We analyzed the existence of the *amhΔY* in different sexual genotypes of the domesticated Manzala strain using primers that covered both the deleted and undeleted regions of the *amh* genes producing a fragment of 800 bp and 1,000 bp. respectively. XX females had only the 1,000-bp fragment, while XY and YY individuals had both (Fig. 2A).

Similarly, only the undeleted 1000-bp fragment was observed in XX males whereas both were seen in XY females. The presence of the extra 1000-bp fragment (the putative X-*amh*) in YY individuals was later explained when researchers showed that 2 tandem Y *amh* genes existed, one corresponding to the 233-bp deleted *amhΔY* gene and the other, the sex determinant *amhy* gene with a missense SNP (C/T) and with a 5608-bp deletion in the promoter. Using *amhΔY*, we genotyped wild-caught Nile tilapia from Lake Turkana and Lake Baringo (Kenya) living in 27-28°C waters and fish caught in the Loboi Hot Springs (Kenya) that are constantly at 35-36°C. The large majority of fish showed a good agreement between the phenotypic sex and the genotypic *amh*Δ*Y* sexing (Fig. 2B).

However, we found 10 out of 91 individuals (~11%) that were *amh*Δ*Y*⁻ males (putative XX males) and 5 females (5.5%) that carried the *amh*Δ*Y*⁺ gene (putative XY females), in both constant temperature regimes and hot springs. We do not know if individuals in the wild that carry the *amh*Δ*Y* also have the *amhy* gene, but both were linked to the presence of the Y chromosome in the Japanese strain. We have nevertheless found wild African populations that lack both the *amhy* SNP and the *amh*Δ*Y*. The presence of *amh*Δ*Y* in these wild Kenyan populations indicates that it is an old truncated form that did not rise from Nile tilapia domestication. We are currently analyzing the discordance between the phenotypic and genotypic sexes by genotyping both *amhy* and *amh*Δ*Y* to confirm these surprisingly high proportions of putative sex-reversed tilapia. We are also performing progeny testings using wild individuals considered to be sex reversed.

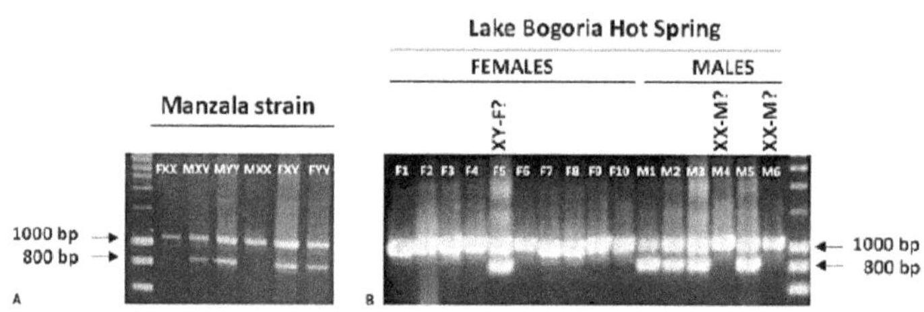

Fig. 2. PCR genotyping of Nile tilapia males (M) and females (F) using the amhDY marker.

A Presence of a 800-bp band corresponding to the putative 233 bp deleted amhDY gene associated to the Y chromosome in XY and YY males and in XY and YY females of the Manzala strain. The 1000-bp band was observed in all sexual genotypes and corresponds to 2 putative amh genes: the X chromosome amh, and the sex-determining Y chromosome gene amhy just downstream of the amhÄY gene B Mismatch between the phenotypic sex and the sexual genotype when using the amhDY marker in some wild caught tilapia from Lake Bogoria Hot Spring (35-36°C) (Kenya) suggesting possible XX males and XY females.

The pejerrey *O. bonariensis* has been considered until very recently to be exclusively a TSD species, where 100% males are produced at 29°C (MPT) and 100% females produced at 17°C (FPT). However, at intermediate temperatures of 25°C, mixed sex ratios ranging from 20 to 80% males have been found between progenies from different parents. The sex determining *amhy* gene identified in the GSD Patagonian pejerrey was searched and cloned in the pejerrey *O. bonariensis* and first tested on a laboratory broodstock to identify mismatches between the genotypic and phenotypic sex.

Out of 24 broodstock fish, 14 carried the *amhy*⁺ while 10 did not (*amhy*⁻), with males being all heterozygous for *amhy* (represented here as *amhy*⁺). Furthermore, an *amhy*⁺

female was also identified and shown by progeny testing to be heterozygous (*amhy*$^{+}$). The *amhy* was then used to analyze possible TSD and GSD interactions of *O. bonariensis* progenies treated at intermediate temperatures. One of these progenies gave a male-biased sex ratio of 70% with 33 phenotypic males found to be *amhy*$^{+}$ while 34 were *amhy*$^{-}$, out of which 13 (38.2%) were female-to-male sex-reversed males indicating the coexistence of TSD and GSD in this species. This coexistence was also evidenced in wild populations, since *amhy* could be used to genotype pejerrey *O. bonariensis* from Lake Kasumigaura in Japan with the finding of 3 *amhy*$^{-}$ males (7.5% of males) and 1 *amhy*$^{+}$ female (2% of females). This study together with our work on tilapia raises the question on the impact of high temperatures due to climate change on wild populations and how they will adapt.

CONSEQUENCES OF SEX REVERSAL IN WILD POPULATIONS

In some fish species, such as the Nile tilapia, the pejerrey *O. bonariensis,* and the Patagonian pejerrey *O. hatcheri,* genetic factors determine the development of a phenotypic sex within a specific range of environmental conditions, but under extreme conditions, temperatures can override the genetic factors at least in some individuals. Due to TSD/ESR, some individuals will have opposite sexual phenotypes and genotypes. Both pejerrey and Nile tilapia have a XX/XY heterogametic system, and both sex-reversed XX males and XY females have been already described in some wild populations. Modelling studies have analyzed the possible effects of their reproduction on sex ratio, dynamics, and resilience of wild populations. The prerequisite for a sex-reversed breeder to impact a population is its relative reproductive capacity and fitness when compared to genetic males or females. As already reported in this review, numerous sex-reversed individuals have been generated for sex control or research purposes: neomales (mXX or mZW), neofemales (fXY or fZZ), but also YY males, YY females, and WW females have been demonstrated to be viable and functional in most species analyzed.

Moreover, recent meta-analysis concluded that: (i) induced sex reversals by exogenous hormones/endocrine disrupting chemicals (EDCs) can negatively affect both the size and the gonadosomatic index (GSI) of the fish during treatment. However, at least size was rapidly restored after the end of the treatment. As suggested by the authors, these negative effects are associated to EDCs rather than to sex reversal; size and gonads of EDC-treated genetic males and EDC sex-reversed males should be equally affected; (ii) following the functional masculinization of genetic females by EDCs/androgens, sperm traits (ejaculate volume, sperm motility, duration or linearity) are similar between genetic males and sex-reversed males. Altogether, these studies suggest that sex-reversed males could efficiently compete with EDC non-reversed individuals. Similar results could probably be expected from temperature-induced males.

Nile tilapia has been found to adapt to extreme thermal conditions in the wild, including warm temperatures in hydrothermal springs with temperatures up to 40°C.

Most of the other wild populations are found in tropical/subtropical regimes where the temperature fluctuates between 28 and 34°C. In this mouthbrooding species, after a first guarding phase, the fry swim out of the mother's mouth during the day to search for food and stay mostly in shallow waters where temperatures are frequently higher than 32-34°C.

As demonstrated earlier in the Nile tilapia, XX fry can be strongly masculinized by high temperature treatments (>32°C) applied during at least 7 days from 10 dpf onwards. Interestingly, fluctuating temperatures that mimic natural thermal regimes (35°C during the day and 27°C during the night) can also masculinize the sensitive progenies of the blue tilapia. Working on different wild populations including one (Metahara) living under high temperature regime (between 32 and 39°C), researchers ound male breeders that sired all-female or highly skewed female (98%) offspring under controlled conditions. This strongly suggested that these individuals were XX males.

Moreover, wild fry shoals were also collected in shallow waters, their genetic structure was characterized, and the phenotypic sex determined. Amongst them, some contained 98% sibling female fry. This indicates that the father had to be an XX male. More recently, analysis of sex-linked molecular markers also suggested the presence of sex-reversed XX males in other wild populations. Taken together, these results strongly suggest that XX sex-reversed breeders exist in the wild where they reproduce and sire all or almost all-female offspring. Similarly, under controlled conditions androgen-induced XX males and XY or YY males of Nile tilapia were found to have quite similar male behavior and sperm quality. Besides, our previous unpublished data revealed that under controlled conditions, progeny testing of 19 single hormonally-induced sex-reversed XX males sired progenies strongly skewed toward females but in some of them, unexpected XX males (1-25.5%) were identified. Interestingly, these unexpected F_1 males when crossed with genetic XX females also sired progenies with strong female bias but again containing some unexpected males (2.5-9.2%). Similar results have been reported in the medaka by researchers. This suggests that in the Nile tilapia and the medaka (i) the sex determining factor is not obligatory for producing a male phenotype; and (ii) the possible existence of minor/cryptic sex-related genes unveiled under extreme genetic or environmental conditions. Finally, in the Nile tilapia thermosensitivity has been demonstrated to be a heritable trait.

Based upon the modelling studies and taking into account the biologic specificities of the Nile tilapia, what could happen when a Nile tilapia population is confronted to a rapid increase of the water temperature due to climate change? The first effect will be a masculinization because of its thermosensitivity, with sensitive progenies becoming partially masculinized. Based upon our previous studies, important variance of thermosensitivity (proportion of males following a 36°C treatment during the critical period) exists within (61.4-94% males) and between (61.4-80.6% males) wild populations. This will increase the male proportion but also favor the production of XX sex-reversed males.

Around 6 months later, sexual maturity will occur and sex-reversed XX males will be able to sire only XX offspring. Because reproduction in tilapia is a continuous activity (as long as temperature is higher than 23-24°C), a dominant male can reproduce daily for 8-9 months per year. For each reproduction, on average a female will produce between a few hundred to 3,000 eggs. A single male can fertilize several females per day (3-5). Therefore, a single XX male can potentially sire more than 1,000 individuals/day. Based on the brooding behavior, survival rates of these fry will be very high. Part of these XX offspring will then be masculinized, and part will stay females.

As suggested by the models, this could progressively increase the proportion of the XX genotype and decrease the proportion of XY individuals. In an extreme scenario, the XY genotype could even disappear from the population. In tilapia, population thermosensitivity can be characterized by 2 parameters: the percentage of thermosensitive families (Tf) and the proportion of males obtained following temperature-induced masculinization (M%). Our previous results reveal that thermosensitivity was also present in the Metahara wild population living in a hydrothermal spring (32-39°C) but, however, XY males were still identified. This can first result from a rapid adaptation of the thermosensitivity.

Indeed, the Metahara population had an important Tf but the M% was lower than in the other wild populations living in more standard thermal regimes. Moreover, in other wild populations, we also identified through progeny testing a XY female probably resulting from the presence of a minor factor rather than from an environmental effect. The reproduction of such females with classic XY males leads to an excess of males (75% males) composed of 2/3 XY and 1/3 YY males. Probably because of these 2 mechanisms, the Nile tilapia might be able to adapt to important temperature changes. Current studies will further verify these hypotheses.

Very recently researchers experimentally demonstrated that climate change can strongly and quickly impact the sex determining system (GSD-TSD transition and loss of the W chromosome) of the Australian dragon lizard: under high temperature regimes, a TSD system can be fixed in small populations where the heterogametic sex chromosome (W) could disappear through the reproduction of the sex-reversed homogametic individuals with non-reversed homogametic breeders (ZZf × ZZm) leading to the decline and progressive extinction of the heterogametic genotype (ZW). This confirms that the resilience of some wild populations facing a rapid climate change is of importance. However, a rapid TSD adaptation could lead to stable balanced sex ratios. Therefore, it will be of major importance to understand and predict the evolution of thermosensitivity and subsequent sex ratios at the population level after the establishment of TSD. Finally, a strong plasticity of sex determination systems could favor an adaptation to a rapid climate change.

HORMONAL INDUCTION OF SEX REVERSAL

Sexuality in fish has a great role to play in aquaculture as there are differences in growth rate, body form and external characters between males and females in several species. The sex determining mechanism of some of the ornamental fishes is given below:

I. Hermaphrodites - Guppy, sword-tail, fighter fish
II. Polygenic determination of sex - sword-tail
III. Sex chromosomal sex determination
  a. XX female - XY male - Goldfish
  b. ZZ male - WZ female - mosquito fish
  c. WXY - XY, YY - males

XX, WX – females – platy

The process of sex differentiation in teleost is protracted and labile rendering the hormonal induction of sex reversal possible in gonochoristic and hermaphroditic species. The induction involves administration of optimum dose of a sex steroid (e.g. 17□-methyl testo-sterone) during the labile period, which reverses the phenotypic expression of a genetic female into a male but the genetic male remains a male. The process of endocrine sex reversal comprises the entire sequence of differentiation, gametogenesis, ovulation/ spermiation, courtship behaviour, secondary sexual characters and sex pheromone production. In most teleosts males grow faster than female (e.g. cichlidae); males of most ornamental fish are more colourful than females and thus have a higher commercial value. Ferminization has been attempted in 30 species to understand the process of sex differentiation and for the development of broodstock for all male or all female population. For example, hormonal sex reversal was induced to produce XY females in *Poecilia reticulata* which when crossed with normal males (XY), sired 25% homogametic (YY) males.

As many as 31 natural and synthetic steroids have been tested for sex reversal in economically important species. In general, the optimum dose required to induce complete sex reversal is species specific and in some species, strain specific. Among androgens, 17 α–methyltestosterone and estrogens, estradiol 17 αB was widely used for brooding (viviparous) poecilids higher dose of 300-500 mg per kg diet is required.

In the direct hormone sex reversal the hormones are administered to the fish in their feed. Almost any feed can be used as the hormone carrier provided it has a sufficient nutritional value for fry and is readily consumed. Simple pellet feeds made locally are suitable, provided that they have a high enough protein content (30-40%). It is not the concentration in the feed that is the determining factor, but rather the actual quantity of hormone consumed; the later is dependent on the feeding rate and how well the fry consume the hormone feed mixture. The treatment is easy to manage and carry out in farms.

Generally, the hormones are administered for only a 40-60 day period, rarely 100 days. The hormones are reduced to unmeasurable quantities within 5 days after their withdrawal from the feed. These hormones are also very similar or identical to the naturally occurring hormones. If used at the correct levels they will cause no adverse changes in the fish under treatment.

In recent years, biotechnological research has led to the development of three additional breeding programmes that can be used to increase yields. One of the most common breeding programmes in aquaculture is the production of sex-reversed brood stock to produce mono sex populations for grow-out. This is done either because one sex is superior or more desirable or to prevent reproduction during grow out.

The production of sex-reversed broodstock is usually accomplished by feeding either estrogens or androgens (sex hormones) to sexually undifferentiated fry to sex-reversed them. Sex-reversed fish are individuals that are one sex phenotypically but the other genetically (XX O) is as example phenotypically male genitically female.

If sex reversal is done properly, sex-reversed fish are capable of producing monosex populations for grow-out. The type of hormone used estrogens to produce sex reverse females or androgens to produce sex-reversed males depends on the sex determining system of the species and whether you want to produce an all male or an all female population.

GENETICS OF SEX DETERMINATION

A sex-determination system is a biological system that determines the development of sexual characteristics in an organism. Most sexual organisms have two sexes. In many cases, sex determination is genetic: males and females have different alleles or even different genes that specify their sexual morphology. In animals, this is often accompanied by chromosomal differences. In other cases, sex is determined by environmental variables (such as temperature) or social variables (the size of an organism relative to other members of its population). The details of some sex-determination systems are not yet fully understood

The XX/XY sex-determination system is the most familiar sex-determination systems. In XY/XO system, females have two copies of the sex chromosome (XX) but males have only one (X0). The ZW sex-determination system is found in birds and some insects and other organisms many other sex-determination systems exist. In some species of reptiles, including alligators, some turtles, and the tuatara, sex is determined by the temperature at which the egg is incubated. Some species, such as some snails, practice sex change: adults start out male, and then become female. In tropical clown fish, the dominant individual in a group becomes female while the other ones are male, and blue wrasse fish are the reverse. In *Bonellia viridis*, larvae become males if they make physical contact with the female and females if they end up on soil.

Some species, however, have no sex-determination system. Hermaphrodites include the common earthworm and certain species of snails. A few species of fish, reptiles, and insects reproduce by parthenogenesis and are female altogether.

In some arthropods, sex is determined by infection, as when Bacteria of the genus *Wolbachia* alter their sexuality; some species consist entirely of ZZ individuals, with sex determined by the presence of *Wolbachia*.

Temperature-dependent sex determination

Temperature-dependent sex determination (TSD) is a system in which the temperature the eggs experience determines the sexes of the organisms that hatch. It is most prevalent and common among amniote vertebrates that are classified under the reptile class, but is also used among some birds, such as the Australian Brush-turkey.

It also happens in some fishes such as salmon, gold fish, tilapia etc. It differs from the chromosomal sex-determination systems common among vertebrates. It is a type of environmental sex determination (ESD); in other ESD systems, some factors such as population determine the sex of organisms (see Polyphenism).

The eggs are affected by the temperature at which they are incubated during the middle one-third of embryonic development. This critical period of incubation is known as the thermosensitive period (TSP). The specific time of sex-commitment is known due to several authors resolving histological chronology of sex differentiation in the gonads of turtles with TSD

A variety of sex determination systems exists in reptiles and fishes, including temperature-dependent sex determination (TSD). While the precise physiological basis of TSD is unknown, numerous past studies have provided a general characterization of this phenomenon. The period of thermosensitivity encompasses the middle one third to one half of embryonic development. It has been shown that in several species of fishes this period starts prior to (histologically defined) gonadal differentiation and extends to a time when sex specific changes are becoming evident in the gonads

Studies on the mechanism of sex determination of the commercially important tilapia, *Oreochromis niloticus*, have demonstrated that this species exhibits a predominantly monofactorial genotypic system with male heterogamety and female homogamety. The effects of environmental conditions during the period of sex differentiation on the sex ratio of the Nile tilapia, *Oreochromis niloticus*. Different sex genotypes were exposed to varying temperatures putative all-female, all-male and all-YY males.and salinities putative all-female progeny only.

For a minimum period of 21 days after first feeding and were on grown prior to sexing by gonad squash. The majority of the putative all-female progeny exposed to high temperature 36.54°C produced significantly higher percentages of males compared to controls reared at ambient temperature 27.87 °C.

Similarly, at high temperature, some of the all-male and YY male progenies had significantly lower percentage of males compared to controls. Sex differentiation in YY males appears to be more labile than in normal XY males although this could possibly be attributable to different levels of inbreeding.

Low temperature 25.78°C and varying levels of salinity 11.30 to 26.65 ppt did not significantly affect sex ratios. The apparent sensitivity of sex differentiation to some environmental factors is considered in the context of a predominantly monofactorial genetic sex determining.

Evidence of the effect of environment on sex ratio has been observed in a number of fish species. A number of studies have reported on the varying degrees of temperature dependent sex determination and its adaptive significance in the Atlantic Silverside, *Menidia menidia*. Other environmental extremes such as low pH have also been observed to affect sex ratio in Cichlids and Poecilids Rubin.

A number of studies have also demonstrated effects of temperature on sex ratio in tilapia. In an early study, researchers observed that at a cold temperature 19°C the sex ratio of *O. mossambicus* in one experiment had a significant excess of males 89% compared to the control reared at 28°C mechanism.

The effect of temperature on sex differentiation is variable among species. Fish species that do not possess a clear distinct chromosomal sex, show strong temperature dependent sex determination/differentiation (TSD).

In gold fish sex differentiation is affected during a specific period of larval development, and that the timing of the sensitive period is itself dependent on temperature. Sex differentiation depended on the initial temperature until Day 17 in the group reared at 30 °C and Day 22 in the group reared at 23 °C. The sex ratio was greatly affected by a high temperature regime (7.7% females at 30 °C). At temperatures between 17 and 30 °C, the percentage of males increased with temperature. In conclusion, water temperature can play an important role in the process of sex differentiation in goldfish.

A temperature sensitive period (TSP) has also been demonstrated in several species of fish they show that Temperature is responsible for sex determination in the European sea bass, *Dicentrarchus labrax*.

Thermolability in phenotypic expression of sex was examined in hime salmon, a land-locked type of sockeye salmon *Oncorhynchus nerka*, thus produce all-female populations in practice to develop effective salmon aquaculture in a safe, simple, environmental friendly. Researchers observed female-biased sex ratios in fish reared at high temperature during embryonic development in hatcheries, suggesting that incubation temperature is probably the factor responsible for a shift in sex ratio in sockeye salmon, *O. nerka*. Although researchers suggested further studies were necessary, there were no rigorous tests on thermosensitivity of gonadal sex determination in salmoids thereafter.

Conclusion

Temperature fluctuation experiments showed that effects of male producing temperature on sex differentiation were greater than the effect of temperature that allowed the expression of the genotypic sex. Although we have seen consistent effects of temperature on sex differentiation of goldfish in laboratory experiments, such effects may not be seen in the fish from the wild. Even though there is great temperature fluctuation at goldfish spawning grounds, sex may be determined by genetic factors, provided that the basal temperature is below male-producing temperature. In conclusion, water temperature can play an important role in the process of sex differentiation in goldfish.

7

POPULATION GENETICS AND INTERACTIONS OF HATCHERY AND WILD FISH

INTRODUCTION

Genetic variation in natural populations reflects population history and the evolutionary potential of a species, an important consideration for aquatic conservation. The presence of population subdivision within a fish species is the result of an interplay between restricted gene flow and independent genetic changes within isolated populations. In the absence of gene flow, conspecific populations generally will diverge from one another as a result of genetic drift, natural selection and mutations. In addition, fish dispersal strategies and life-history traits determine the magnitude of landscape effects in shaping patterns of genetic variation.

GENETIC VARIATION, POPULATION STRUCTURE AND BIODIVERSITY

Genetics is in reality a relatively new field. For centuries, humans moved fish among countries and stocked conspecifics in new watersheds without any knowledge or concern regarding genetic principles, impact or consequences. During the initial 70 years of the twentieth century, fish development and loading were uncontrolled. Starting during the 1970s, preservation hereditary qualities have become perceived and are a thriving issue, as are biodiversity and genetic biodiversity.

As a general rule, individual nations and normal asset organizations presently take a much more conservative approach to stocking programmers, genetic conservation and biodiversity. However, many decisions and policies are made and implemented without data on population genetics and the genetic interactions of fish populations. There is a requirement for substantially more examination around here.

At the point when information is free, absence of precise stocking chronicles muddles information understanding. Another void is an absence of information showing the connection between execution and biochemical and molecular markers, which was introduced in the previous chapter. These are extremely challenging information to create, as it isn't not difficult to duplicate the normal environment, rivers, reservoirs, lakes and oceans in a realistic manner. Geneticists at times offer regular asset supervisors clashing guidance in regards to the desirability of increased or decreased genetic variation and the policies and mechanisms to achieve various goals.

The primary inquiry that should be considered is the significance of hereditary variety in natural populations. Is it better to have more hereditary variety or less hereditary variety? These are troublesome inquiries to respond to, and the appropriate response might be diverse relying upon the individual circumstances. Do populace structures direct the requirement for the amount and sort of hereditary variety In the circumstances leading to different population structures, have the selective pressures led to the ideal genotypes in a specific climate or have limits on quality stream in that environment limited the development of the optimum genetic structure of a population.

Theoretically, genetic variation is beneficial and important. Hereditary variety is significant for the drawn out endurance of an animal types. Genetic variation can ensure the fit-ness of a species or population by giving the species or population the ability to adapt to changing environments.

Clearly, an absence of hereditary variety or a lot of homozygosity can be inconvenient to an individual's or a population's survival traits and fitness. The cheetah is a perfect representation of the expected adverse impacts of abundance homozygosity. This profoundly homozygous species has serious regenerative issues. Homozygosity has additionally been connected with two-sided deviation (fluctuating unevenness) unbalanced meristic counts on the right and left halves of the body in fish. Furthermore, profoundly or absolutely homozygous people and populaces really show greater phenotypic variation than outbred controls because they are more greatly affected by environmental or micro-environmental change and have decreased homeostatic capacity contrasted and more heterozygous people and populaces. Inbreeding in little, normal populaces builds elimination rate.

Inbreeding depression resulting from increased homozygosity is well documented in fish. Field crops have been endangered when they did not have the genetic variation to respond to new pathogens or plagues. Unmistakably, the presence of hereditary variety is critical to the drawn out endurance and wellness of a species. Numerous regular populaces react to various types of choice, for example, directional, bidirectional, cyclical and stabilizing selection, which help to ensure the maintenance of the genetic variability and or fitness of population. Levels of homozygosity and inbreeding can be significant not just in homegrown or aquaculture populations, but in wild populations as well. Inbreeding does unfavorably influence regenerative accomplishment in wild deer. Microsatellite heterozygosity was used as a mark of individual inbreeding coefficients among unmanaged deer on the island of Rhum. Scotland. Heterozygosity was related with lifetime reproducing achievement (all out off-spring) in the two guys and females.

The majority of inbreeding experiments on fish and other organisms have been done in aqua-culture and laboratory-type environments. Some have conjectured that inbreeding melancholy would be more extreme and influence fitness more adversely in the harsher natural environment compared with the laboratory environment or aquaculture environment where animals are well taken care of. Nonetheless, the wellness of mosquito populaces declined similarly in normal tree holes as under favorable laboratory conditions for mosquitoes.

EFFECTS OF GEOGRAPHY AND ENVIRONMENT ON POPULATION VARIATION

The ecosystems that support fisheries, together with other economic activities, are subject to a number of alterations of significant relevance to their functioning and resilience and to the goods and services they can provide. Because of our imperfect understanding of ecosystem structure and functioning, as well as the inherent difficulty of distinguishing between natural and human-induced changes, the latter are not always perfectly predictable and/or reversible.

Overall Impacts

Impacts from fisheries on the environment have been abundantly described and reviewed. More specifically, capture fisheries impact target resources. They reduce their abundance, spawning potential and, possibly, population parameters (growth, maturation, etc.). They modify age and size structure, sex ratio, genetics and species composition of the target resources, as well as of their associated and dependent species. When poorly controlled, fisheries develop excessive fishing capacity, leading to overfishing, with major ecosystem, social and economic consequences.

Fishing may also affect ecological processes at very large scale. The overall impact has been described as comparable, in aquatic systems, to that of agriculture on land in terms of the proportion of the system's primary productivity harvested by humans. Overfishing transforms an originally stable, mature and efficient ecosystem into one that is immature and stressed. This happens in various ways. By targeting and reducing the abundance of high-value predators, fisheries deeply modify the trophic chain and the flows of biomass (and energy) across the ecosystem. They can also alter habitats, most notably by destroying and disturbing bottom topography and the associated habitats (e.g. seagrass and algal beds, coral reefs) and benthic communities.

The alteration of the habitat by various human activities may be physical (e.g. by adding artificial structures like artificial reefs, oil rigs, aquaculture installations), mechanical (e.g.

through the "ploughing" effect of dredges and trawls), or chemical (e.g. through injection of nutrients, pesticides, heavy metals, drugs, hormones). Fishing may result in changes in productivity of resources (some positive and some negative) and affects associated species. Some aspects of fisheries can have significant and long-lasting effects, e.g. destructive fishing techniques using dynamite or cyanides or inadequate fishing practices (e.g. trawling in the wrong habitat); pollution from fish processing plants; use of ozone-depleting refrigerants; dumping at sea of plastic debris that can entangle marine animals or be swallowed by turtles; loss of fishing gear, possibly leading to ghost fishing; lack of selectivity, affecting associated and dependent species, resulting in wasteful discarding practices, juvenile mortality, added threat to endangered species, etc.

Poorly-managed large-scale mariculture can damage coastal wetlands and nearshore ecosystems, often used as nurseries by key capture fishery resources, and contribute to ecosystem contamination with food residues, waste, antibiotics, hormones, diseases and alien species.

Impact on Associated and Dependent Species

The Law of the Sea provides that fisheries management must take care also of associated and dependent species. The impact of fishing on these species has been documented in some areas but is still frequently unknown or only partly understood. The decline of primary productivity consumers low in the food chain removes important forage species needed higher in the food web, with cascading effects for the ecosystem.

Conversely, the removal of top predators such as mammals, tuna or sharks, may release an unusually large abundance of preys at lower levels with cascading and feedback effects on the food chain and species composition. For example, as most sharks and some batoid fishes (angel fishes) are predators located at or near the top of marine food webs, their depletion modifies the intricate trophic interactions of their ecosystems.

The removal of predators through fishing in Kenyan reefs resulted in the expansion of sea urchin population, which apparently led to a decrease in live coral and to loss of topographic complexity, species diversity and fish biomass. Researchers reports that the hunting of sea otters (*Enhydra lutris*) in the Northeast Pacific caused a large-scale expansion of sea urchins, the increased grazing of which caused the decline of the important kelp forest. She also reports that, in the Bering Sea, the expansion of the fisheries on pollock (*Theragra chalcogramma*) during the 1970s has been considered as a probable cause of the decline of several populations of marine mammals, e.g. sea lions (*Eumetopias jubatus*) by 76%, seals (*Callorhinus ursinus*) by 60% and (*Phoca vitulina*) by 85%, as well as the decline of several seabird populations (*Urea algae, U. lomvia, Rissa brevirostris, R. tridactyla*). All these non-fish species compete directly with the Pollock fisheries since the target species represent 21-90% of their diet. Impacts can be particularly serious on cartilaginous fish populations, which have a lower productivity and resilience than bony fishes. As a consequence, fisheries targeting shark have a low record of sustainability and some species of skates, sawfish and deep-water dogfish have been virtually extirpated from large regions.

A well documented example of direct impact on benthic species is that of modern towed gear (trawls and dredges) which caused, *inter alia*, long-term changes in abundance and species composition in the Wadden Sea and Australia. The mortality of benthic species associated with or preyed upon by target bottom fish resources resulting from the use of trawls can vary greatly, depending on how the gear is built or rigged.

For example, the addition of a tickler chain on a commercial beam trawl will allow it to catch more of its bottom-fish target but, at the same time, will detach and uproot more benthic species (as bycatch). Likewise, some gear modification can reduce fishing mortality. For example, operating a semi-pelagic trawl 15 cm above the sea bottom has no measurable effect on the benthos community, while the standard demersal trawling dragging on the bottom reduces benthos density by 15.5%. Finally, fishing can have significant impact on the genetic diversity of resources and can permanently change populations characteristics.

Impact on the Environment

Fishing gear can change the living and non-living environment within which the target and other related resources live. Environmental damage may come from the very nature of the fishing technology (e.g. in the use of dynamite or poison) or from the inappropriate use of an otherwise acceptable gear (e.g. using trawls in coral reefs or seagrass beds).

The use of dynamite and other explosives for "blast fishing" is still common in parts of Asia, Africa, Caribbean and South Pacific. A relatively small explosive (beer-bottle size) is capable of destroying a three-metre circular area of stony corals. These practices are generally officially banned by fisheries regulations and laws but often persist because the people involved have little, if any, alternative livelihood.

The impact on the habitat depends on the gear and sediment type. Highly dynamic, soft bottoms (e.g. coarse sand, hydraulic dunes) may suffer limited damage even when exploited by heavy (including hydraulic) dredges. On the contrary, stable, hard, and highly structured habitats (such as coral reefs, seagrass beds, sponge beds) will be easily damaged. One well-documented example is the use of modern towed gear (trawls and dredges) which caused, *inter alia*, destruction of seagrass beds (*Posidonia oceanica*) in the Mediterranean and destruction of the oyster (*Cassostrea virginica*) habitat in Chesapeake Bay. Damage is also related to fishing frequency, gear weight and rigging. Addition of heavy tickler chains to the trawl ground rope increases bottom abrasion and turbidity while adding rollers reduces it. The use of sodium cyanide in the Philippines to catch marine tropical fishes for the aquarium trade has led to the destruction of the coral reef habitat and decline of aquarium and food fish.

Poor Selectivity, Bycatch and Discards

Fishing generates bycatch and discards. A first attempt to address the issue at global level was made in the late 1990s by FAO and the first estimate of the global extent of the problem (about 27 million tonnes of resources dumped per year) was published by this Organization. Most fishing activities are not selective enough to remove from the ocean only the desired targets and will probably never be. This leads to accidentally catching other species (bycatch), part of which has little or no use (at least in the local context) and will be dumped overboard (as discards) together with the offal from fish processing at sea.

The effect is to increase availability of food to scavenger species (including sea birds) and, when concentrated over time, may cause local anoxia of the seabed environment. The resulting amount of organic material may not be negligible. On the North Pacific shelf and upper slope, the amount of offal generated by at-sea production of surimi (a protein extract of fish) is very significant, since the technology used extracts less than 50% of the wet weight from the total catch, the rest being dumped. In the North Sea, 6.5 to 12.5% of the groundfish caught is dumped at sea. Some of this is consumed by sea birds, but a certain amount of offal becomes available to benthic scavengers.

The increase in abundance of dogfish (*Scyliorhinus canicula*) in northern Spain fisheries and that of *Raja radiata* in Greenland shrimp fisheries has been associated with increased discards. Oxygen depletion due to excess organic loading from discards has been recorded in the New Zealand fisheries for hoki (*Macruronus novaezelandie*) as well as in the North Eastern Atlantic.

Bycatch mortality also affects many non-fish species which are relevant to the functioning of the overall ecosystem. For example, surface and sub-surface driftnet and long-line fisheries have serious negative effects on populations of sea birds, e.g. albatrosses and petrels in long-line fisheries in the North Pacific and in the Southern Ocean. High seas drift nets have had a considerable impact on sea birds in the northern Pacific as have gillnets in southwest Greenland, eastern Canada and elsewhere.

Gear Loss and Ghost Fishing

Voluntary dumping or loss of fishing gear may lead to ghost fishing. The scale of the impacts of ghost fishing is basically unknown but there are indications that the effects are not negligible. Species affected by discarded gear include not only teleost fish but sea birds, marine mammals and turtles. For example, the incidence of entanglement of marine mammals in floating synthetic debris in the Bering Sea has been related to growing fishing effort and increased use of plastic. Researchers concludes that entanglement is the principal cause of the current decline in the fur seal population of the Pribilof Islands, accounting for 15% of the mortality of youngsters. As an average, a northern fur seal (*Callorhinus ursinus*) is expected to encounter 3-25 pieces of net debris along the 800-km yearly migration in the Northeast Pacific. Fish traps, unless made of biodegradable material, contribute to the problem. To illustrate this problem, 31 600 pots were lost in the Bristol Bay crab fishery in a period of two years.

OVERVIEW OF THE IMPACT OF CLIMATE CHANGE ON FISHERIES

Natural climatic fluctuations, particularly those at medium (decadal) scale, have always affected fisheries as well as their management performance. The atmosphere and the ocean will continue to warm over the next 50–100 years, sea level will rise due to thermal expansion of water and melting of glaciers, ocean pH will decline (become acidic) as more carbon dioxide is absorbed, and circulation patterns could change at local, regional and global scales.

The major aquatic habitats in SSA include the Great Rift Lakes such as Lakes Malawi and Victoria; man-made reservoirs such as Lake Kariba; large river and floodplain systems such as the Nile and Zambezi Rivers; and coastal habitats including estuaries, mangrove swamps, and deltas. The diversity of the habitats and the species they support respond differently to different impacts of climate change.

Even though it is not possible to generalize the impacts of climate change on fisheries in SSA, they share something in common – climate change is very likely going to lead to fluctuations in fish stocks. Fluctuations in fish stocks will have major economic consequences for many vulnerable communities and national economies that heavily depend on fisheries.

The impacts of climate change on fish stocks in SSA can be classified as physical and biological changes. Physical changes include sea surface temperature rise, sea level rise, changes in salinity and ocean acidification. Biological changes include changes in primary production, and fish stock distribution. These factors when combined together will have adverse impacts on the already strained resource.

Physical Changes

Water Surface Temperature Rise

The oceans play a significant role in regulating global climate. Their heat capacity (and thus net heat uptake) is about 1000 times larger than that of the atmosphere and they therefore absorb significant amount of heat emitted globally. Such changes in ocean temperatures can change the dynamics of aquatic environments of the region. Changes in ocean dynamics could lead to changes in migration patterns of fish and possibly reduce fish landings, especially in coastal fisheries.

For example, increased ocean temperature may affect upwelling along the Gulf of Guinea, which can make the ocean waters become unsuitable for fisheries, causing a reduction in and possible collapse of fishing activities. Inland waters are also equally vulnerable and could be impacted strongly by climate change. The international Dialogue on Water and Climate noted that water stress will increase significantly in those regions that are already relatively dry, such as SSA. One of the stresses that inland waters of SSA are facing is increasing water surface temperature.

Rising water temperature also threatens biodiversity. Generally, fish have a thermal preference that optimizes physiological processes. If water temperature rises above the maximum tolerable threshold of a species, then its existence is threatened. Researchers provide an example from the Lebialem Highlands in Cameroon where women have started to hunt for tadpoles and frogs because there are no fish in most of Bangwa Rivers. However, even the number of tadpoles and frogs have significantly declined (partly) due to the warming rivers that have increased the amount of predator fish in an area they have never inhabited before.

Sea Level Rise

Globally, sea level has already risen by 10 to 20 cm during the 20th century, largely due to thermal expansion, and by 2100 a global rise in sea level of between 9 cm and 88 cm has been predicted, based on the Intergovernmental Panel on Climate Change's full range of 35 climate projection scenarios. In coastal areas, sea level rise may alter the salinity of estuarine habitats, inundate wetlands, and reduce or eliminate the abundance of submerged vegetation, adversely affecting those species which rely on these coastal habitats for reproduction and recruitment. In addition, with high sea levels, sea ports, existing fishing facilities like jetties and fish storage centers built on the coastal fringes slightly above the mean high tide line will be subjected to more frequent tidal and storm inundation. Therefore, sea level rise is very likely going to have a negative impact on fishery production (due to salt stresses on the fish stock and its habitat) and fish landing, processing and marketing facilities.

There are very limited studies done to assess the impacts of sea level rise in the coastal zones of SSA. There is, however, some information on the impact of sea level rise in some of the most populous coastal cities in West and East Africa. For example, Nigeria's 800 kilometer low-lying coastline from Lagos to Calabar makes the region prone to seawater intrusion into coastal freshwater resources. This will have a negative impact on inland fisheries and aquaculture. The people in the coastal areas who used to depend on fishing have seen their livelihoods destroyed by the rising waters. In another report by the UN-HABITAT, a sea level rise in Abidjan is likely to swamp 562 square kilometers along the coastline of the region, as lowland marshes and lagoons dominate the coastal zone. The same report also predicts that Mombasa, in the East African coastal zone, could be submerged by a sea level rise of 0.3 meters, with a large area becoming uninhabitable and less productive due to salt stresses.

Increasing Water Salinity

Climate change can cause an increase or decrease in water salinity in multiple ways. While tropical oceans are increasingly becoming saltier, oceans closer to the poles have become fresher. This highlights that tropical oceans are very likely to suffer more from the potential impacts of increasing water salinity relative to waters in higher latitudes.

Changes in water salinity have different effects depending on the tolerance level of the organisms and the nature of their ecosystem – whether freshwater, marine or estuarine. The salinity of some freshwater ecosystems are predicted to increase as a result of anthropogenic climate change. Such physical changes will negatively impact the population of both plankton and bigger prey fish species by affecting the organisms' ability to osmoregulate.

Some empirical studies illustrate that change in salinity has a negative impact on zooplankton population, particularly in freshwater ecosystems. Researchers depict that zooplankton communities of low-lying, coastal, tidal lake and wetlands are adversely affected by small increases in salinity levels. They further warn that such changes in zooplankton abundance may further disturb the ecological functioning of these valuable but vulnerable ecosystems. As is discussed in the subsequent section, changes in zooplankton populations or other planktonic primary and secondary producers disrupt the food chain, thus having a considerable negative impact on fishery.

Salinity is also considered one of the most important variables determining the survival of organisms in estuarine ecosystems; either by having a direct impact on the organisms or indirectly by destroying their habitat, including their breeding and nursery grounds. Researchers states that all estuarine fish are euryhaline (able to cope with salinity fluctuations), but their ability to do so varies from species to species and hence changes in salinity may influence their distribution. Even though salinity changes may not have a direct negative impact on estuarine fish species per se, it can have a negative impact on their habitat. For example, increase in water salinity has contributed to destruction of 60 per cent of mangrove areas in Senegal). According to researchers, each acre of mangrove forest destroyed leads to an estimated 300 kg loss in marine harvest. Therefore, change in water salinity are going to have a tremendous negative impact on fishery in the region, threatening the livelihoods of many impoverished coastal communities.

Ocean Acidification

Oceans are believed to have the capacity to absorb most of the anthropogenic CO_2 emissions. CO_2 is soluble in water and reversibly converts to carbonic acid. As a result of this chemical reaction, the world's oceans are acidifying at an alarming rate. While this has a positive impact in slowing down global warming, increased acidity as a result of dissolved CO_2 in seawater has negative a impact on ocean ecosystems. The impact to the ecosystem is difficult to estimate as different species at different stages of life history respond differently to different pH changes. According to researchers, ocean acidification research is in its infancy and although the field is moving forward rapidly, good data is still scarce. Although the science base is still limited, there is a clear need to provide timely advice and a balanced perspective on the possible impacts on marine fisheries.

Quantifying the effects of ocean acidification on human communities requires assessing the direct and indirect chemical impacts on valuable marine ecosystem services such as fisheries. The direct effects include changes in physiological processes such as reduced growth of calcified structures, otolith development, and fertilization success. These may ultimately lead to direct impacts at the whole-organism level, including reduced growth and reproductive output, increased predation and mortality, alteration in feeding rates and behavior, reduction in immunocompetence and reduced thermal tolerance. Indirect effects include alteration in predator or pray abundance, effects on biogenic habitats such as coral reefs, and changes in nutrient recycling. While adult fish seem well-equipped to deal with low pH waters, or higher levels of CO_2 in seawater, their egg and larval life stages may not be so fortunate.

For example, increased CO_2 level in oceans can potentially narcotize male gametes indicating that acidification may impair fertilization, exacerbating problems of sperm limitation, with dire implications for marine life. Ocean acidification could potentially slow the growth of plankton and invertebrates that are at the bottom of the food chain. Thus acidification can alter the productivity at certain trophic levels, thereby disrupting the complex food chain of aquatic ecosystems with effects on the productivity of fisheries. One of the very likely socio-economic impacts of ocean acidification in SSA is a decrease in populations of calcifying organisms such as mollusks. This may have tremendous socio-economic effects either by:

1. lowering export earnings of net mollusk exporting nations;
2. reducing jobs for many fishers involved in mollusk farming and harvest; or
3. increasing mollusk prices which may exclude marginal consumers – further widening protein and wealth gaps between the rich and poor.

A study done by researchers to assess the vulnerability of nations to ocean acidification-driven decreases in mollusc harvests, suggests that countries with low adaptability, high nutritional or economic dependence on mollusks, and rapidly growing populations will be most vulnerable. These are common characteristics of most SSA countries. Thus, the region is very likely to be more, if not the most, vulnerable to ocean acidification.

Biological Changes

Climate change is already affecting the trends of some important biological processes, resulting in changes in primary production and changes in fish distribution). Climate induced changes in primary production and fishstock distribution have negative implication on food security in many tropical coastal states in general and SSA in particular.

Changes in Primary Production

The relationship between climate change and future ocean primary production is likely to be a key constraint on fish and fisheries production. Survival of fish larvae during the planktonic stage is thought to depend strongly on the availability of sufficient and suitable food. Therefore, in addition to effects of changes in production, climate induced changes in distribution and phenology of fish larvae and their prey can also affect recruitment and production of fish stocks.

Even though there are some studies done to assess the impact of climate change on primary productivity of aquatic environments in high latitude waters, there is very limited study done in the tropics. One of the very few studies available is the case of Lake Tanganyika.

Lake Tanganyika has historically supported one of the world's most productive pelagic fisheries, which provided an annual harvest in recent years of between 165,000 and 200,000 metric tons, representing 25–40 per cent of animal protein supply for the populations of the surrounding countries.

A rise in surface-water temperature coupled with a regional decrease in wind velocity has increased the stability of the water column restricting mixing and deep-water nutrient upwelling and entertainment into surface waters. This has led to a decrease in primary production by about 20 per cent over the past 80 years.

According to researchers, a 20 per cent reduction in primary productivity can be translated as a roughly 30 per cent decrease in fish yields. The decline in catch was accompanied by breakdown of the previously strong seasonal patterns in catch, suggesting decoupling from ecosystem processes driven by the weakening of hydrodynamic patterns. These changes in the pelagic fishery are consistent with a lake-wide shift in ecosystem functioning.

Even though there are multiple factors that affect primary production in aquatic environment, one of the main factors is surface temperature rise. Researchers suggest that if air temperature increases by about 1.7 degrees, as predicted for the next 80 years, there will be further increases in thermal stability and reductions in productivity in these large lakes. They further warn that the human implications of such faint but progressive environmental changes are potentially dire in this region of the world, where large lakes are essential natural resources for regional economies.

Changes in Fish Distribution

Change in fish distribution is among the most commonly reported ecological responses of marine species. Fish species are believed to respond to environmental changes such as warming water temperatures by shifting their latitudinal and depth ranges. Changes in ocean dynamics could lead to changes in migration patterns of fish and possibly reduce fish landings, especially in coastal fisheries of many African countries. Marine fisheries are an important food source, and therefore, changes in the total amount or geographic distribution of fish available for catch could potentially affect food security.

The effects of changes in fish-stock distribution vary across latitudes. Some fish species will migrate due north in search of habitats with optimal water temperature and thus potentially increasing fish harvest in higher latitudes. On the other hand, counties in lower latitudes such as SSA are very likely going to lose some fish species and stocks. Researchers depict that changes in fish-stock distribution could range from a 30–70 per cent increase in high latitude regions to a drop of up to 40 per cent in the tropics. This poses a tremendous challenge to the predominantly artisanal fishers in SSA mainly because it is economically prohibitive to follow the fish-stock, i.e. restricted to economic exclusive zones. Therefore, according to OECD, such changes in fish stock distribution will change the distribution of benefits and costs of fisheries with some winning and some (including fishers from Sub-Saharan African countries) losing.

INTER-SPECIFIC HYBRIDIZATION

For production of quality seed in aquaculture, many methods were tried in the past, including intergeneric and interspecific hybridization. Hybridization technique was used by aquaculturists in the hope of producing aquatic organisms with specific desirable traits or general improvement in performance. Intergeneric and interspecific hybridization programs have been applied in fish farms with the purpose of producing animals that perform better than the parental species (hybrid vigor).

Quality seed is a fundamental prerequisite for sustainable and successful aquaculture, be it small-scale or commercial farming. Inadequate supply of quality seed is often suggested as a major constraint for aquaculture in many parts of the world. The issue of quality comes to the attention of producers only after a certain period of time when performance indicators (e.g., growth, production, survival, and disease) consistently point a finger towards seed quality. Factors which contribute to production of poor quality seed would have become established as a normal practice in the system.

The mating or crossing of two different species is a process called hybridization with the offspring known as hybrids. Hybrids can have some characteristics of both parental species. A hybrid with selected or favored characteristics of each parent is one of the goals of animal husbandry. When a hybrid has characteristics superior to both parents, it is said to have hybrid vigor or positive heterosis, which of course, the ultimate breeding goal. Hybridization occurs widely in fishes under natural conditions and is observed in fish more commonly than in other vertebrate animal groups.

Several factors have been suggested as contributing to the high incidence of natural hybridization among closely related fish species, including external fertilization, weak behavioral isolating mechanisms, unequal abundance of the two parental species, competition for limited spawning habitat and decreasing habitat complexity. Hybrid zones are defined as areas of narrow regions where genetically distinct populations or species meet, mate and produce hybrids. Intra specific hybridization (crossbreeding) is a classical approach for the genetic improvement of livestock animals.

This method also has shown its potential in aquaculture. Increment of 55 and 22% in growth rate of channel catfish and rainbow trout crossbreeds, respectively, were achieved using this technique. However no increase in growth rates in Chum salmon crossbreeds when compared with parental strains. Cross-breeds of different strains of European catfish, *Silurus glanis*, exhibited higher adaptability under warm water conditions and mixed diet feeding regimes. investigated the heterosis effect between crosses of five Norwegian strains of Atlantic salmon. They did not find a significant heterosis effect for either growth rate or survival rate. Similarly, found no heterosis effect for growth rate of Atlantic salmon fry.

Inter-specific hybrids thus, have attracted attention because they can improve productivity through hybrid vigor, transfer desirable traits or produce sterile animals. Hybridization may also be used to combine other valuable traits such as better growth and flesh quality, disease resistance and increase environmental tolerances. In recent years, hybrids of major carps are being successfully produced in public and private hatcheries and are available for farming due to high resistance against unfavorable ecological conditions. Many molecular biologists and fish geneticists realized that the use of inter-specific hybrids in global fisheries production is not well-reported and examined properly.

On the other hand, there have still been some controversies in global acceptance for using inter-specific hybrid organisms that have been genetically modified. This paper focuses on the crossing among different genetically distinct species and rearing of hybrids to understand the potentiality of hybrids in the world's aquaculture production.

Use of Inter-specific Hybrids in Aquaculture Production

Inter-specific hybridization has long been practiced in various species of fishes to increase growth rate, improve flesh quality, produce sterile animals, increase disease resistance and environmental tolerance and to improve other quality traits to make fish more profitable. Majority of the earlier works on hybridization was conducted for salmon fishes, but these species did not produce hybrids of commercial importance. For this reason, hybrids in these fishes do not draw attention to fish culturists. Due to the increased expansion of fish farming throughout the world, hybrids produced from inter-specific crosses play substantial role for global aquaculture production. The increased use of artificial breeding and fertilization techniques and increased knowledge of reproductive biology encourage the aquaculturists to produce hybrids in a view to improve the quality traits over their pure parental siblings.

Improved Growth Performances

Increased growth rate is the most desirable trait for stock improvement in aquaculture. Growth increase may result from dominant variance or from increasing the number of polymorphic loci in an individual. Increased heterozygosity has been implicated in improved growth and other desirable characters in a variety of species such as developmental compatibility food conversion efficiency and oxygen metabolism.

A hybrid between white bass (*Morone chrysops*) and the stripped bass (*M. saxatilis*) is called sunshine bass, exhibits faster growth and has many good culture characteristics than either parents under captive culture system. Crosses of the black crappiexwhite crappie (*Pomoxis nigromaculatus* x*P. annularis*) stocked in small ponds and impoundments, silver carpxbighead carp (*Hypophthalmichthys molitrix*x*Aristichthys nobilis* in polyculture systems and catfish hybrids between the African catfish and the Vundu (*Heteropneustes longifilis* or *H. bidorsalis*) in intensive concrete tanks were reported to grow faster (positive heterosis) than conspecific parents. Improved growth performances were also obtained from crosses of mrigal (*Cirrhinus mrigala*) and catla (*Catla catla*) and common carp (*Cyprinus carpio*) with rohu (*Labeo rohita*) in pond culture system in India.

Intergeneric hybrids between catla (*Catla catla*) and fimbriatus (*Labeo fimbriatus*) were observed to be combine desirable qualities such as the small head of the fimbriatus and the deep body of the catla and exhibited heterosis in terms of meat yield with higher flesh content than either of the parents. Hybrids between tambaqui (*Colossoma macropomum*) and pacu (*Piaractus brachypomus*) in Brazil and Venezuela raceways and ponds grew faster than either parent. Crosses of the green sunfish (*Lepomis cyanellus*) with bluegill (*L. macrochirus*) and crosses of the gilthead sea bream (*Sparus aurata*) with red sea bream (*Pagrus major*) also had positive heterosis in growth and other culture characteristics. Several hybrids have been produced in the Mediterranean with the cross between red sea bream and common dentex (*Dentex dentex*), being especially fast growing in cage culture management.

Manipulation of sex-ratio

Production of mono-sex populations in fish is often preferable for aquaculture development. This preference may be due to growth differences between sexes, e.g., male tilapia grow faster than females, whereas female salmonids and sparids grow better than males. A specific sex chromosome may produce a valuable product and monosex populations help reduce unwanted reproduction. Hybridization between some species of tilapias such as and the blue tilapia (*O. aureus*) results in the production of predominantly male offspring and reduces unwanted reproduction in grow-out pond culture management. This cross produces predominantly males because of different sex-determining mechanisms in the two species and the hybrid males had superior growth over pure parental species. has the XX, XY system with the male being heterogametic, whereas blue tilapia has the ZZ, ZW with the heterogametic genotype being female.

Similarly, crosses between (*O. niloticus*) and Wami tilapia (*O. honorum*), and long-finned tilapia (*O. macrochir*) and Mozambique tilapia (*O. mossambicus*) and Wami tilapia produce hybrid offspring that are predominantly male with excellent growth and production. The hybrid between striped bass (*Morone saxatilis*) and yellow bass (*M. mississippiensis*) produced 100% females with excellent survival and growth in culture system.

Overall Improvement

The principal aim of hybridization is to combine desirable traits from different species to increase the overall production or marketability of a cultured species. The major hybrid catfish cultured in Thailand is a cross between African and Thai (*C. macrocephalus*) catfish, which combines fast growth rate of the African catfish with the desirable flesh characters of the Thai catfish. The overall product is improved and the flesh is still acceptable to Thai consumers, although it does not grow as fast as the pure African catfish.

The rohuxcatla hybrid grows almost as fast as pure catla, but has the small head of the rohu and is therefore useful in Indian aquaculture. Catlaxfringed-lipped peninsula carp (*Labeo fimbriatus*) were reported to have small heads of the fringed-lipped peninsula carp and deep body and nearly equal growth rate to the catla; dressing percentage also improved in this hybrid. The sunshine bass hybrid (white bassxstripped bass) has a suite of advantageous traits including good osmoregulation, high thermal tolerance, resistance to stress and disease, high survival in culture and modified water-bodies and ability to utilize soy beans as a protein source. The overall growth performances of hybrids (*C. catla*x*L. rohita*) fed on wheat bran was consistently higher followed by rice broken and blood meal.

Among the cultivatable hybrids, red tilapia is more desirable than darker skinned tilapia in Cuba, Venezuela, Thailand, Europe and the United States of America. Most red tilapia are descended from the Nilexblue tilapia cross, but red tilapia also result from the cross of Wami tilapia (*O. urolepis hornorum*) with Mozambique tilapia. It has been reported that red tilapia from Mozambique tilapia x Wami tilapia are being farmed in central Thailand to Lao PDR for aquaculture purposes. The latter cross is also salt tolerant and used for coastal aquaculture in parts of South East Asia. Stability of the skin coloration is often a problem in successive generations and studies have been undertaken to understand the genetic mechanisms of color inheritance.

Hybrids between different species of North American catfish have been researched for more than 30 years. Among the interspecific catfish hybrids, crosses between channel catfish (*Ictalurus punctatus*) and blue catfish (*I. furcatus*) exhibits good culture characters of the channel catfish with the ease of harvesting characters of the blue catfish such as better angling and increased seinability. Once breeding problems are worked out, these hybrids may be useful in culture as they show heterosis for growth rate and are superior to channel catfish in low oxygen tolerance, disease resistance uniformity in body shape, angling vulnerability, seinability and dress-out percentage. Hybrid produced from the crosses between the muskellunge (*Esox masquinongy*) and the pike (*E. luscious*) is sterile and well-adapted to intensive culture systems. However, the hybrid has the similar sport fish characteristics of the pure parental muskellunge, but higher protein requirements than both parental species.

Disease Resistance and Environmental Tolerances

Hybridization may be used to improve disease resistance by breeding a higher resistant species with a less resistant one. reported that crosses of coho salmon (*Oncorhynchus kisutch*) with other species, e.g., rainbow trout had increased disease resistance to a variety of salmonid viruses, but other culture characteristics were poor. Viability was increased when hybridization was followed with triploidization and stated that the rainbow trout (*O. mykiss*)xchar (*Salvelinus* spp.) triploid hybrids had increased resistance to several pathogenic salmonid viruses and early sea water tolerance.

Hybrids may have increased environmental tolerances when one parental species has a wide range of tolerance (e.g., euryhaline species), a specific tolerance (cold tolerance species), or because of increased heterozygosity sometimes being associated with a broad niche. Mozambique tilapia and Wami tilapia can reproduce in saline waters; however has improved culture performance in many aquaculture systems.

Hybridization between Mozambique and yields a red tilapia with salinity tolerance. Hybrids between Mozambique and Wami tilapia, called the Florida red strains, have high growth rates and can reproduce in salinities of 19 ppt. Crosses between and blue tilapia also resulted in progeny with good salinity tolerances. Hybrids also may be used to exploit degraded aquatic environments. Lakes affected by acid rain may not be suitable for native salmonids, but splake, a hybrid between lake trout (*Salvelinus namaycush*) and brook trout (*S. fontinalis*) can tolerate reduced pH levels of 4.9-5.4 of acid lakes of Ontario; lake trout reproduce successfully only in waters with pH values above 5.5. The splake also was shown to have higher survival and growth than both brook and lake trout in lakes with pH in the range of 5.5-7.2.

Hybrid Polyploidization

Hybridization combined with chromosome manipulation may increase the viability and developmental stability of hybrid fishes during early life history stages. Polyploid hybrid salmon appear to be better suited for a variety of culture situations than either polyploid or hybrid salmon are on their own. Although many diploid salmonid hybrids are not used for culture, triploidization of the hybrids may confer increased viability, growth and survival. Triploidization of Atlantic salmon (*Salmo salar*)xbrown trout (*S. trutta*) hybrids increased survival and growth rate to a level comparable to Atlantic salmon. General disease resistance was improved by triploidizing the cross between rainbow trout and char; rainbow trout and coho salmon triploid hybrids had increased resistance to infectious disease, but the latter hybrids grew more slowly. Triploid Pacific salmon hybrids between chum salmon (*Oncorhynchus keta*) and chinook salmon (*O. tshawytscha*) have earlier seawater acclimatization times.

Experimental Hybridization

Laboratory hybridization experiments have been utilized extensively to confirm the probable hybrid nature of certain individuals by demonstrating that two taxa will interbreed when provided with the opportunity to do so, or that gametes from two taxa can be artificially cross-fertilized. Hybrids produced from appropriate cross-fertilization techniques among commercially important fish species have been tested for their growth performance, viability and fertility. Hybrid recently produced experimentally between Sheim (*Acanthopagrus latus*) and sobiaty (*Sparidentex hasta*) in Kuwait appears to have good growth, flesh quality and is fertile.

Hybrids resulting from crossing several sunfish species have been used for the past three decades to improve farm pond fishing. The most desirable hybrids result from crossing the female green sunfish (*Lepomis cyanellus*) with males from one of three other species. These include the bluegill (*L. macrochirus*), the redear, or shellcracker (*L. microlophus*) and the warmouth, or goggleye (*L. gulosus*). The most commonly used hybrid in the Southeast is the male bluegill (BG) x female green sunfish (GS) cross. This BG x GS hybrid has the most desirable set of characteristics which means that the hybrids can outperform their parental species in one or more ways. Rapid and superior growth is one way hybrid sunfish exhibit hybrid vigor.

Experimental hybrids between dusky grouper (*E. marginatus*) and the white grouper (*E. aeneus*) were evaluated, but all the hybrids died within 10-day post-hatching. The camouflage grouper (*Epinephelus polyphekadion*) is more resistant to environmental stress and disease than the marbled grouper (*E. fuscoguttatus*). Experimental hybrids (marbled grouperxcamouflage grouper) exhibited faster growth performances and increased conversion efficiency. Hybrid between the Beluga (*Huso huso*) and Russian sturgeons (*Acipenser gueldenstaedtii*) was evaluated and appeared to have a wide salinity tolerance to both fresh and seawater as well as good growth rate and these hybrids now are being considered for culture in Russia and Iran.

Two loach (*Misgurnus* spp.) are cultured both for food and for ceremonial purposes by Buddhists in Korea: the mud loach (*M. mizolepis*) and the cyprinid loach (*M. anguillicaudatus*). The mud loach grows to a larger size, has a faster growth rate and is more resistant to diseases, while the cyprinid loach has a more desirable body color. These two species of loach was hybridized to combine the fast growth and large size of the mud loach (*Misgurnus mizolepis*) with the desirable body color of the cyprinid loach (*M. anguillicaudatus*). Fertilization, hatching, survival and karyology of the hybrids were very similar to the parents. These hybrids are now being culturing commercially and continued studies are planned to combine other desirable characteristics of the hybrids and their fertility.

Hybrids produced using the eggs of Asian catfish (*Clarias batrachus*) and African catfish (*C. gariepinus*) performs as well as either parental control during the alevin stage and better in fry and advanced fry stage, while the reciprocal hybrids are inferior in all performance traits. During the different experiments, this hybrid group showed highest survival from post-larval stage to market size fish. Growth performance was always better than maternal control and in some cases better than or close to paternal control. Preliminary observations of organoleptic testing revealed that the hybrid showed superior taste performance compared to parental groups. Further researches are needed to examine other desirable traits of the hybrids and their sterility.

Hybridization between giant catfish (*Pangasiodon gigas*) and giant pangas (*Pangasius sanitwongsei*) are now been practicing in Thailand. Both of these catfishes are extraordinarily large, reaching 3 m and 300 kg and the giant catfish is considered as an endangered species whose trade is restricted under the Convention on International Trade in Endangered Species of Wild Flora and Fauna. Hybrids between these two catfish species shows good growth performances and should be used to reduce pressure on the giant catfish so as not to endanger it through excessive catch of brood fish from the wild or through genetic introgression of the two parental species.

Due to the wide geographical distribution of yellow bass (*Morone mississippiensis*), hybridization tests with stripped bass and comparisons with the sunshine bass have been conducted. The yellow bass hybrid exhibited 65% survival to harvest as compared to 45% for the sunshine bass, but poorer growth rate and condition factor when raised in tanks continuously supplied with pond water. Further research have been undertaken to explore the possibility of combining other desirable traits in the above hybrid progeny.

Unplanned/accidental Hybridization

Unplanned and accidental hybridization in hatchery stocks may cause a genetic deterioration into the aquaculture production and open water fisheries. During the production of Indian major carp seeds, different species often are induced to spawn in a common spawning tank thus providing the opportunity for unintentional hybridization. Silver carp (*Hypophthalmichthys molitrix*) and bighead carp (*Aristichthys nobilis*) sometimes are hybridized inadvertently because of their similar appearance and because of the shortage of "the correct" species at spawning time due to differences in maturation times between male and female. This hybridization often results in a fish that does not feed efficiently as its gill rakers are intermediate in shape between those of the silver carp that eats phytoplankton and those of the bighead carp that consumes zooplankton.

There is much anecdotal evidence of genetic deterioration of carp hatchery stocks in Bangladesh, through inbreeding, negative selection and hybridization. Stocks of exotic (i.e., non-indigenous) carps are particularly vulnerable to such degradation, given that the opportunities to go back to wild populations for broodstock replenishment are very limited. Furthermore, anecdotal evidence suggests that hybridization between silver carp and bighead carp is common, at least partly due to a shortage of mature bighead carp males towards the end of the breeding season.

Reported aquaculture production of the silver carp in Bangladesh in 2001 was 130,000 t, or 21.7% of freshwater aquaculture production, while there was no reported production figure for bighead carp. Bighead carp broodstock are present in many hatcheries, so presumably aquaculture production of bighead carp is present, but not high enough to be reported separately.

Hybridization between silver carp has also been reported to occur fairly frequently in commercial aquaculture hatcheries in Bangladesh. The consequences of hybridization for broodstock purity had recently been investigated. Allelic variation at three microsatellite DNA loci isolated from silver carp routinely distinguished between silver carp and bighead carp.

These markers were used in the analysis of samples collected from hatcheries in different regions of Bangladesh. Of 422 hatchery broodstock that were morphologically identified as silver carp, 8.3% had bighead allele(s) at one or more of the three microsatellite loci, while 23.3% of the 236 fish morphologically identified as bighead carp had silver carp allele(s) at one or more loci. The results suggested that while some of these fish might be F_1 hybrids, others had more complex genotypes, suggesting further generations of hybridization or introgression between the species in hatcheries, with potentially damaging consequences for the integrity of these stocks and their performance in aquaculture.

Interspecific hybridization in some carp species has recently been reported in Bangladesh. Either out of scientific interest or shortage of adequate hatchery populations (i.e., brood stock), introgressed hybrids are being produced intentionally or unintentionally by private hatchery operators and sold to hatchery and nursery owners. These hybrids are being ultimately stocked knowingly or unknowingly, either in grow-out ponds or in open water bodies like floodplains under the government's massive carp seed stocking program. There is widespread concern that mass stocking of such hybrids in the floodplains and other related open water might cause a serious genetic introgression problem that could adversely affect aquaculture and inland open water fish production. There is every possibility of segregation of genes with the result that some of the fish carrying the introgressed genes could not be easily distinguished from the pure species.

Hybrid introgression in major carp species is very likely to have negative consequences as a result of loss of distinct feeding strategies of the pure species, which are the basis of successful polyculture systems. If the introgressed hybrids reproduce in natural water bodies or are used as broodstock in hatcheries, they will not be true breeders; therefore, collection of carp seed from the pure species/strains will be difficult.

Hybridization with wild fish is especially prevalent in tilapia ponds connected to natural water bodies that contain indigenous or feral tilapia populations. Such uncontrolled and unintentional hybridization could undermine the performance of cultured stocks and make future use of the contaminated stocks as broodstock questionable. For example, wild three-spotted tilapia (*Oreochromis andersonii*) invaded ponds in Mozambique and produced hybrid tilapia that were less marketable than pure . Inadvertent hybridization at a chinook salmon hatchery was suggested as the probable explanation for the appearance of chinookxcoho salmon hybrids in a California stream. The level of unintentional or accidental hybridization has important considerations of aquatic biodiversity and will influence risk assessment on the use of hybrid fishes in aquaculture.

Recommendations

A number of hybridization studies in fishes have been reported but certainly not all of the hybrids are contributing to commercial aquaculture production. However, the contributions that hybrid fishes make to global aquaculture production are underestimated. Approximately, 80% of Thai catfish production from hybrids and there is a growing concern that these hybrids may be impacting native catfish.

The tilapia hybrids in Israel are the main tilapia produced, but the 6691 mt reported were not identified as hybrid. Production of 4257 mt of hybrid striped bass was reported from the USA, but production of no other fishes was reported, in spite of the fact that red tilapia and other tilapia hybrids are being produced and sold.

Accurate identification of hybrids is important not only for sustainable aquaculture development but also to allow for a better understanding of biodiversity and conservation issues. It would be unfortunate to experience wide-spread loss of pure species in aquaculture as happened with tilapia as a result of wide-spread loss of pure species and subsequent hybridization. It would be a significant cause for concern if hybrid Thai catfish or the hybrid Venezuelan characids pose more of a threat to local species than the pure species. The following points need to be addressed to overcome the above situations as well as to understand the role of hybrids in global aquaculture production:

- Genetic stock improvement through inter-generic or inter-specific hybridization of cultured fish species should be initiated under well-designed breeding plans at research institutes and lead central hatcheries under the guidance of fish breeding specialists/biologists
- Data on parental origins and stock identity should be recorded for each hybrid. When crosses are made, the female species should be listed first; random crosses in regards to sex of each parent should also be identified

8

GENE MAPPING, QUANTITATIVE TRAIT LOCUS MAPPING AND MARKER-ASSISTED SELECTION

INTRODUCTION

"Gene mapping is a method used to determine the place of a gene, the distance between the genes or identification of the location of a gene in a genome or on a chromosome."

Gene mapping is also known as "genome mapping" or "mapping of a genome" helps researchers navigate around the genome and create landmarks.

As with the landmarks for different addresses, here, the landmark for the different genes can be created using some genetic methods.

These landmarks help to find out regulatory sites, repeated DNA sequences or a gene, even it helps in discovering new genes related to some phenotype.

In the year 1911, Alfred H. Sturtevant created the first genetic map denoting the location of genes on a chromosome.

Working on the model organism *Drosophila melanogaster*, he found that the genes are arranged in a linear manner on a chromosome.

He also observed that the distance between any two genes is fixed if the genes are located on one particular chromosome and the length of the chromosome is the same in all organisms of that species.

Furthermore, he realized that the location of genes on a chromosome could be determined by calculating the frequency of crossing over.

Genetics is defined as the study of heredity and embodied in the definition of heredity is the transmission of genetic characters from parents to offspring. Among the several definitions that exist for a map, the most appropriate is the concept that a map is a two-dimensional reconstruction of a part of the earth or heavens. In its totality, a genetic map does portray the ordering and landmarks of the physical entities we know as chromosomes. But in and of itself any map, including a genetic map, can only direct the observer to a specific location or place. It cannot convey to the observer the nature of that location or the activities that occur at that location. To attain an appreciation for the location we recognize that is necessary to actually visit and observe the location. As a biological metaphor genetic maps are only the genesis for understanding the evolutionary complexities of any species.

To use an analogy based upon a world map, we may regard genetic mapping as a multi-layered process spanning the identification of individual landmarks that may be miles apart on several continents, down to the cataloging of individual grains of sand between these landmarks. By equating chromosomes to continents, landmarks to genetic markers, and grains of sand to individual nucleotide base pairs, a measure of the complexity inherent in the construction of genetic maps may be gained. Fortunately, the analogy to 'grains of sand' is somewhat of an exaggeration, as species have far fewer nucleotide base pairs than inferred.

Nonetheless, the concept does portray a feeling for the magnitude of organization that is expected at each level. We have yet to witness the completion of a gene map for any vertebrate species, even though this target is nearing completion for our own species, Homo sapiens. Studies of gene mapping in fishes are largely focused on the second level of organization (i.e., the addition of genetic markers to define linkage groups within the species).

However, in the zebrafish (*Danio rerio)* and pufferfishes (*Fugu rubripes*, *Tetraodon nigroviridis*), gene mapping studies are well advanced at the highest level of organization, and a large amount of raw sequence data exists for either species in public databases.1 In addition, plans are currently underway to sequence the entire genome of the pufferfish due to its relatively small size (~400 Mb) which would make it the first teleost for which complete genome information is available. It is expected that functional gene coding regions will be in a higher density in the pufferfish (given the compact nature of the genome) per number of contigs assembled and thus sequencing efforts are expected to yield more immediate rewards.

Gene mapping is a hierarchical process and utilizes a number of different biological techniques to gain information on the genome. In this paper, some of the techniques that are currently used in the study of genomics will be outlined, and an overview of various types of genetic maps will be presented with consideration given to the source material and markers that are needed to construct these maps. Based upon the knowledge of genetic marker map orders, it is possible to relate physical chromosome morphologies to the linear order of markers on the chromosomes via gene-centromere mapping distances. This methodology will be discussed in relation to its utility in defining the location of centromeres on genetic maps. In the latter sections of the paper, an overview of the applications of genetic mapping will be given. Included in those sections will be a discussion of gene transmission dynamics not only within species, but also between the sexes within a species. Highlighted, will be the use of a genetic map and its representative markers in investigating and understanding the process of adaptation, survival, and evolution (i.e., the raison d'etre of genetic maps).

TECHNIQUE AND APPROACH USED

Unlike transgenesis, which involves the transfer of a gene from one organism to another and is highly contentious, genome editing allows specific, targeted and typically minor changes to the genome of the species of interest. One of the most popular gene-editing tools is clustered regularly interspaced short palindromic repeats (CRISPR). It was initially used by scientists to target specific points on DNA strands to knock out certain genes and the tool can be applied to almost any organism. Not only have researchers refined their techniques and the range of organisms that have had their DNA sequences expanded but this also applies to the gene-editing options available to study.

In order for any modifications made by tools such as CRISPR to be heritable they need to take place when the organism is at a very early stage in its embryonic development, so that they can be incorporated into its gametes once it matures. This process is generally done by micro-injection into the ova in fish species like salmon. Through advances in genetics and genomic research, it is possible to identify specific genes where CRISPR editing is expected to give a significant boost to performance or health. For example, CRISPR was used to identify a gene affecting resistance to IPN in salmon in a recent study.

The use of CRISPR to bring in desirable traits from other strains or species is also possible. For example, a major international project called CrispResist aims to understand the genes causing differential resistance to sea lice in salmonid species, and then to use CRISPR to edit Atlantic salmon to show a response similar to other resistant salmonid species.

SCOPE AND SCALE OF APPLICATION

The mass use of CRISPR to edit individual production animals is unlikely to be feasible. However, editing broodstock populations or use of emerging methods of reproductive biotechnology including surrogate broodstock may offer avenues for commercial application. This could be relevant for a wide range of aquaculture species, as their high fecundity and external fertilization can facilitate genome editing at a scale that is not possible in farmed terrestrial animals.

Current peer-reviewed studies into gene-editing projects cover a growing range of aquatic species - including various salmonids, roho labeo (Labeo rohita), grass carp (*Ctenopharyngodon idellus*), common carp (*Cyprinus carpio*), channel catfish (*Ictalurus punctatus*), Pacific cupped oyster (*Magallana gigas*), Nile tilapia (*Oreochromis niloticus*) and red seabream (*Pagrus major*).

ACCESSIBILITY

The cost of using gene-editing tools has decreased dramatically in the last few years, but it is still a process that requires some specialist facilities and equipment. For example, it is necessary to have access to reproducing broodstock, micro-injection and molecular biology equipment, and a suitably biosecure hatchery facility. However, the reagents themselves are relatively cheap, perhaps only a few hundred dollars, and it is a technique that shows plenty of promise for a range of aquatic species and geographies.

Recently, one of the most significant projects relating to the use of CRISPR in aquaculture involved researchers from the Kyoto-based Regional Fish Institute, who partnered with Kyoto University and Kindai University to develop a red sea bream with 20 percent more meat. They succeeded by knocking out a protein that suppresses muscle growth. While genetically modified food that contains foreign genes must undergo safety screening in Japan, gene-edited products do not face the same rules, allowing the fish to be harvested and sold to Japanese consumers.

However, many countries are yet to pass legislation related to the sale of gene-edited animals for human consumption, and the position of the European Union to date is to consider gene-edited animals in the same manner as genetically modified organisms. Such restrictions are likely to drastically limit the possibility for the industry to benefit from the improvements in animal production, health and sustainability which CRISPR has to offer. Therefore, it is important that informed dialogue continues about the techniques and a reasoned risk–benefit pragmatic approach is taken for its regulation.

OUTCOME AND BENEFITS

Gene editing has the potential to improve a range of issues that currently affect aquaculture that include improving animal welfare, reducing environmental impact and increasing efficiency - and therefore the profitability - of production.

As some of the examples listed above illustrate, researchers have already shown the huge potential of gene editing for improving the performance of a wide range of traits in a growing selection of farmed aquatic species. It is a technique with huge promise, should regulations allow it to take place on a commercial scale.

Whether, when and how widely this is permitted remains to be seen. But given the raft of potential benefits, and that gene-edited red sea bream have already been legitimately sold for consumption in Japan, those involved in gene-editing projects believe that it is only a matter of time before it is widely adopted in commercial aquaculture production.

MAP CONSTRUCTION

Genomes may be mapped at three different levels ranging from physical maps that localize large DNA segments onto the cytological karyotype of a species, to genetic maps that depict linear marker orders along a chromosome, and finally, by cataloging and sorting raw DNA sequence along a chromosome. Genetic maps are the most common and are typically composed of linear marker arrangements depicting various linkage groups. Such marker linkage maps are generally produced using inbred/semi-inbred or outcrossed lines in an intercross or backcross design, where F1 parents are produced by mating two strains that are divergent for some character of interest in a quantitative trait locus (QTL) study. Haploid and diploid gynogenetic or androgenetic family lines have also been used for mapping in fishes.

Some of the programs also perform QTL analyses on the same data set but require that the phase of the markers be specified in the mapping families. When access to fully inbred lines is not available then the method of producing doubled haploid progeny provides an efficient way to generate a genetic map using fully homozygous progeny lines. By producing mitotic diploids from two lines and then hybridizing the lines, the F1 may be used to re-establish haploid progeny lines in the following generation. Linkage assignments will be completely in phase as only homozygous progeny genotypes are produced and the allelic origin can be traced to either donor strain (if the alleles differ between the strains).

Most mapping studies in fishes are currently focused at the second level of genetic map organization (i.e., the construction of marker linkage maps). A marker linkage map may be defined as a landmark map that is composed of anonymous DNA markers (e.g., most microsatellite markers, AFLP (amplified fragment length polymorphism), RAPD (randomly amplified polymorphic DNA), SNP (single nucleotide polymorphism) markers), or known gene sequence regions marking either introns or exons (e.g., SNP marker derived from an EST (expressed sequence tag) or designed from partially homologous DNA sequence information).

In general any marker that identifies a known expressed gene region (e.g., insertion/deletion (INDEL) polymorphism, microsatellite repeat, SNP) may be defined as an expressed sequence marker polymorphism (ESMP). If the marker identifies a known gene using the polymerase chain reaction (PCR), amplifying DNA through a known intron-exon boundary with subsequent detection of an ESMP, the marker may also be classified as an EPIC (exon primed intron capture) marker. Markers may also be developed as STS (sequence tagged sites) from cloned or PCR amplified DNA.

By reference back to our original analogy, we must recognize that a marker linkage map is largely incomplete. It can give us a reference to landmarks along the physical surface of a chromosome, but it is largely characterized by unlinked sequence data spanning 1 cM (a centiMorgan) or more. A centiMorgan is defined as 1% genetic recombination between a pair of markers, and the actual number of base pairs embodied in this distance is dependent upon the species being examined, and even the linkage group being examined. Certain regions of chromosomes may have higher recombination rates than others, and this will make the actual translation of estimated recombination distances into physical map distances variable throughout the genome. A sequenced linkage map of overlapping cloned fragments (contigs) that is characterized as containing large tracts of assembled sequence is generally referred to as a sequenced genome rather than a map. Therefore, the term map will be used to refer to a collection of genetic markers. The attainment of even a draft status of a sequenced genome requires a large investment of time and money to perform the necessary multiple sequencing runs on a large number of cloned fragments from the genome.

Genetic maps are simply built on the examination of recombination differences among pairs of genetic markers transmitted by a given parent. Implicit in this is the use of doubly heterozygous markers that uniquely mark four chromosome segments in the genome. Each genetic marker position will of course identify one homologous pair of chromosome segments that possess a different allelic marker on each chromosome arm.

If the other pair of marker alleles being compared are also derived from the same set of homologs (i.e., from an identical linkage group) then the recombination levels between any pair of linked markers will be reduced inversely to the physical distance they are separated on a chromosome arm. This will be detected as a significant deviation in the expected 1:1:1:1 segregation pattern of genotypic combinations, with parental coupling alleles being in excess to non-parental recombination alleles. In the absence of information on the phase of the markers transmitted by a parent, the phase may be inferred by comparing the genotypic combinations that occur in the highest frequency (i.e., if the genotypes being compared are Aa at locus 1 and Bb at locus 2, then either AB and ab genotypes must occur in the highest frequency, or Ab and aB gametes must occur in the highest frequency). Theoretically, only those markers that are physically close enough to one another to escape recombination via chiasmata cross-overs will map syntenically. Chiasmata cross-overs involve exchange between only one pair of non-sister chromatids that are present at meiosis I.

Markers that are physically close to one another have a reduced probability of having parental phases disrupted during meiosis, and thus, are expected to transmit a high frequency of parental (phase intact) genotypes to their progeny. In contrast, markers that are located far from one another on a linkage group may experience a cross-over event in 100% of the meioses generated. However, if chiasma interference is complete during meiosis (i.e., the formation of one chiasmata junction inhibits the subsequent formation of other junctions) then only two of the four chromatid strands present during meiosis I will experience a cross-over event.

Thus, 50% of the potential gametes formed at meiosis II should contain only non-recombinant parental marker orders. Assuming that cross-over gametes are transmitted in equal frequency to non-cross gametes, then total map length is expected to be fairly accurately represented by additive pairwise recombination distance estimates. For sets of markers that are located less than 50 cM from one another (i.e., less than 50% recombination for any pairwise combination) map distances are expected to be additive for any three point recombination estimate (e.g., A - B + B - C = A - C, if B is the intercalary marker). Double cross-overs involving the same pair of non-sister chromatids may also occur during meiosis I, which will disrupt a linear map order by shortening the recombination distance estimates among proximally and distally located markers along a linkage group. Non-random or preferential transmission of parental phase chromatids compared to recombinant chromatids will also shorten the apparent map distances between distally located markers and may mimic the results obtained from double crossover events.

Mapping functions such as Haldane's and Kosambi's mapping function account for the shortened map distances that may occur between any pair of linked markers due to double cross-over events. These mapping functions readjust observed recombination distances upward, thus generating a larger estimate of total genome size than would be obtained by directly summing recombination intervals among linked markers. The Haldane mapping function gives larger interval readjustments than the Kosambi function. Dense marker maps saturated with markers spanning only 1–2 cM intervals are largely equivalent in terms of their genome size estimates regardless of the mapping functions used.

Genetic map construction using information from the segregation of genetically variable markers is essentially a four stage process. The first stage is composed of scoring progeny genotypes in a source mapping family, and then entering the genotypes into a computer program for analysis of the segregation patterns obtained. The second step is to ascertain the linkage arrangements among the markers tested. Implicit in this analysis is the fact that only one contributing parent is scored for segregation and linkage. The use of androgenetic or gynogenetic haploids or diploids makes this a requisite, and when inbred lines are used, F1 hybrids between the inbred lines may be backcrossed to one of the homozygous inbred parents resulting in scored segregation of heterozygous loci from the single hybrid parent. Hybrids may also be intercrossed, although the phase assignment of heterozygous progeny genotypes in such crosses is not possible (i.e., the parental origin of the alleles cannot be ascertained). In outcrossed families, however, both parents may be heterozygous for different alleles at marker loci making phase assignments possible.

Three of the software programs (CARTHAGENE, JOINMAP, and LINKMFEX) currently accommodate linkage analyses using outcrossed pedigrees. When groupings of genetic markers have been statistically identified the markers belonging to any one group may be tentatively regarded as belonging to a linkage group within the species. It is not unusual to initially have more linkage groups identified in the species than may be accommodated by the actual known haploid number of chromosomes in the species. This is due to the fact that clusters of 2 or 3 markers may occur that remain unlinked due to low marker density. These clusters may occur at opposite ends of larger linkage groups, and they will eventually be joined as intervening markers in the linkage group are genotyped.

Physical mapping involves in situ hybridization of cloned segments of genomic DNA to their physical location on a chromosome. Although at the lowest level of genomic discrimination, physical mapping is extremely useful in terms of giving researchers insights into structural arrangements within the genome (i.e., localizations of repetitive DNA (e.g., distribution of specific SINE (short interspersed nuclear elements) families among chromosomes, identification of multi-gene family locations in the genome (e.g., rDNA clusters), localization of linkage group specific repetitive clusters (e.g., chromosomal paint probes). *In situ* hybridization methods can also be of great use in verifying the relative location of specific candidate genes that have been localized to large cloned DNA fragments (e.g., BAC clones).

Through the integration of genetic map information with the localization of selected markers from the genetic map, the confirmation of genetic map orders to specific chromosomal landmarks can be confirmed (i.e., the localization of terminal markers in a genetic map linkage group to telomeres and markers with small gene-centromere recombination distances to their relative location on the chromosome). Markers with apparent small gene-centromere map distances may in fact prove to be variable in their placement along chromosome arms dependent upon the frequency and localization dynamics of chiasmata formation among different linkage groups.

Markers

It is sobering to contemplate that little more than two decades ago, genetic maps in a wide variety of organisms were largely built on information obtained from phenotypic mutant lines and a few biochemical marker loci. Currently, genetic maps are constructed with a wide variety of molecular markers that may broadly be classified as sequencespecific and sequence-independent markers. The latter category is represented by anonymous DNA markers such as AFLP and RAPD markers. These markers have proven to be of great utility in establishing initial framework maps (e.g., in zebrafish) and in rainbow trout, since they are generated with relative ease by the researcher and appear to be highly polymorphic in most of the species that have been examined. Since no knowledge of a targeted DNA region is required prior to their utilization it is possible to implement these markers in virtually any species of fish from which genomic DNA may be extracted in a pure form that is amenable to restriction enzyme digestion, ligation, and PCR amplification.

While sequence-independent DNA markers may be beneficial in establishing an initial framework map, it is unlikely that these markers will be of great utility in population-based studies examining gene function and action. This class of marker is characterized by a lack of tractable inheritance across family lines since these markers are predominantly expressed in a dominant/recessive fashion. Thus, the genotypes of individuals expressing a given fragment may not be accurately classified as being either homozygous or heterozygous. This is possible if previous pedigree information confirms that the individual is descended from an individual expressing the fragment and one that is lacking the fragment, but this is tempered by the need to assess the segregation patterns in the progeny of such a mating in the previous generation. If both classes of phenotype are present in the progeny, the individual may accurately be classified as being heterozygous, whereas if all siblings in the family expressed the fragment, then the inferred genotype is that of a homozygote. The labour required to establish inheritance modes in each generation will severely limit the utilization of such markers.

Additionally, AFLP homologies are often difficult to infer between families given the large number of fragments that may be generated with this technique, and reproducibility is a concern with RAPD fragments in certain species of fish. AFLP fragments may aid in the establishment of sequence-specific markers if they can be cleanly purified and sequenced. Thus by virtue of their ubiquitous distribution, AFLP markers may be more readily localized to QTL regions in the genome, facilitating the sequence characterization of these regions. Whether the reported localization of AFLP markers to centromeric regions in salmonid fishes is a general finding among all fish species remains to be established.

Undoubtedly, comparative evolutionary studies will be enhanced greatly as details of syntenic marker arrangements among species becomes available (e.g., between human and zebrafish). Anonymous markers will provide the greatest amount of information by virtue of the fact that such markers will be most widespread throughout the genome. However, these markers are likely to only be utilized among relatively closely related species where the homology of the marker generation can be confirmed between the pair of species being compared (e.g., microsatellite markers in salmonid fishes). Comparisons on the homologous relationships among functional genes will be possible across more divergent taxa. Information on the organization of multi-gene families and repetitive elements such as LINEs (long interspersed nuclear elements), and SINEs, and any retroposon-like elements in the genome may also be garnered from homologous sequence information among closely related species.

Both the genes themselves and their nature of expression may be studied once the genes have been characterized and sequenced. Thus, if a researcher is interested in studying a given physiological function, and has a priori knowledge that a given gene is likely to be of importance in this physiological process, then a genome search for the candidate gene may be initiated.

The influence of this candidate gene on the physiological function may be studied in greater detail once the gene has been located. An alternative method, termed a genome scan, utilizes a marker linkage map to localize chromosomal regions (and putative QTL) that may influence the trait. The advantage of the latter scenario is that it may detect multiple regions (in addition to identified candidates) influencing a given trait. New chromosomal regions (devoid of existing candidate loci) may be discovered with this approach.

QUANTITATIVE TRAIT LOCUS MAPPING

In order to begin a QTL analysis, scientists require two things. First, they need two or more strains of organisms that differ genetically with regard to the trait of interest. For example, they might select lines fixed for different alleles influencing egg size (one large and one small). Second, researchers also require genetic markers that distinguish between these parental lines. Molecular markers are preferred for genotyping, because these markers are unlikely to affect the trait of interest. Several types of markers are used, including single nucleotide polymorphisms (SNPs), simple sequence repeats (SSRs, or microsatellites), restriction fragment length polymorphisms (RFLPs), and transposable element positions. Then, to carry out the QTL analysis, the parental strains are crossed, resulting in heterozygous (F_1) individuals, and these individuals are then crossed using one of a number of different schemes. Finally, the phenotypes and genotypes of the derived (F_2) population are scored. Markers that are genetically linked to a QTL influencing the trait of interest will segregate more frequently with trait values (large or small egg size in our example), whereas unlinked markers will not show significant association with phenotype.

For traits controlled by tens or hundreds of genes, the parental lines need not actually be different for the phenotype in question; rather, they must simply contain different alleles, which are then reassorted by recombination in the derived population to produce a range of phenotypic values. Consider, for example, a trait that is controlled by four genes, wherein the upper-case alleles increase the value of the trait and the lower-case alleles decrease the value of the trait.

Here, if the effects of the alleles of the four genes are similar, individuals with the *AABBccdd* and *aabbCCDD* genotypes might have roughly the same phenotype. The members of the F_1 generation (*AaBbCcDd*) would be invariant and would have an intermediate phenotype. However, the F_2 generation, or the progeny from a backcross of an F_1 individual with either parent, would be variable. The F_2 offspring would have anywhere from zero to eight upper-case alleles; the backcross progeny would have anywhere from four to eight upper-case alleles.

A principal goal of QTL analysis has been to answer the question of whether phenotypic differences are primarily due to a few loci with fairly large effects, or to many loci, each with minute effects. It appears that a substantial proportion of the phenotypic variation in many quantitative traits can be explained with few loci of large effect, with the remainder due to numerous loci of small effect. For example, in domesticated rice (*Oryza sativa*), studies of flowering time have identified six QTL; the sum of the effects of the top five QTL explains 84% of the variation in this trait. Once QTL have been identified, molecular techniques can be employed to narrow the QTL down to candidate genes. One important emerging trend in these analyses is the prominent role of regulatory genes, or genes that code for transcription factors and other signaling proteins. For instance, in rice, three flowering time QTL have been identified at the molecular level, and all of these loci encode regulatory proteins known from studies of *Arabidopsis thaliana*.

A meta-analysis of extensive data in pigs and dairy found that QTL effects were skewed towards fewer QTL with large effects. Orr (2001) addresses the question of defining and distinguishing between "large" and "small" effects. As with all statistical analyses, sample size is a critical factor. Small sample sizes may fail to detect QTL of small effect and result in an overestimation of effect size of those QTL that are identified. This is known at the "Beavis effect". Researchers suggested a method for comparing detected QTL to a distribution of expected values in order to estimate how many loci might have been missed. Recent studies have taken these biases into account.

Another consistent trend in looking at QTL across traits and taxa is that phenotypes are frequently affected by a variety of interactions (e.g., genotype-by-sex, genotype-by-environment, and epistatic interactions between QTL), although not all QTL studies are designed to detect such interactions. Indeed, several complex traits in the fruit fly Drosophila melanogaster have been extensively analyzed, and this research has indicated that the effects of such interactions are common.

For example, detailed examination of life span in *D. melanogaster* has revealed that many genes influence longevity. In addition, significant dominance, epistatic, and genotype-by-environment effects have also been reported for life span. Similarly, QTL studies examining plant architecture differences between maize and teosinte have repeatedly shown significant epistatic interactions. These same types of interactions have additionally been demonstrated in soybeans.

It is also possible to perform QTL analysis on unmanipulated natural populations using hybrids, sibships (half-sibling or full-sibling families), and/or pedigree information. Diverse ecological and evolutionary questions have been addressed using these tools. For example, Researchers identified multiple QTL associated with differences in male calling song between two closely related species of the Hawaiian cricket, a trait involved in rapid speciation.

Similarly, researchers addressed the question of possible gene flow between domesticated crops and their wild relatives in contrasting environments using crop-sunflower hybrids. Environmental and conservation questions have also been explored. For instance, researchers examined various loci that influence invasive success by exotic species, while researchers reviewed questions surrounding QTL for tolerance to heavy metal exposure in plants that could contribute to phytoremediation of polluted soils.

Caveats and Qualifications of QTL Analysis

Like most methods, QTL analysis is not without limitations. For instance, QTL studies require very large sample sizes, and they can only map those differences that are captured between the initial parental strains. Because these strains are unlikely to contain segregating alleles of large effect at every locus contributing to variation in natural populations, some loci will remain undetected. Furthermore, the specific alleles that do segregate, particularly in inbred lines, may not be relevant to natural populations. Other alleles at these same loci are likely to be of interest, however. Thus, the goal for many studies is to identify loci rather than particular alleles. (One notable area of exception involves applied studies in medicine and agriculture, which are often interested in specific segregating alleles).

The number of times that individual genes have been identified following a QTL mapping study remains small. Indeed, researchers lists examples of quantitative traits in which single genes have major effects and their molecular basis has been studied, and he notes that this number is modest relative to the effort invested in QTL studies. One reason for this discrepancy is that many QTL map to regions of the genome of perhaps 20 centimorgans (cM) in length, and these regions often contain multiple loci that influence the same trait.

Moreover, identifying the actual loci that affect a quantitative trait involves demonstrating causality using techniques like positional cloning followed by targeted gene replacement. Frequently, the quest for individual genes within a QTL is assisted by the identification of *a priori* candidate genes using classical reverse genetics or bioinformatics. A functional relationship between the candidate gene and the QTL must then be demonstrated, such as by using functional complementation (the addition of wild-type complementary DNA from the gene in question into the nucleus to rescue a loss-of function mutation or to produce an alternative phenotype). Other techniques, such as deficiency mapping (deletion mapping), are available for specific organisms, including *Drosophila*.

The Future of QTL Mapping

New permutations of QTL mapping build upon the utility of the original premise: locus discovery by co-segregation of traits with markers. Now, however, the definition of a trait can be broadened beyond whole-organism phenotypes to phenotypes such as the amount of RNA transcript from a particular gene (expression or eQTL) or the amount of protein produced from a particular gene (protein QTL or PQL). QTL mapping works in these contexts because these phenotypes are polygenic, just like more traditional organismal phenotypes, such as yield in corn.

For example, transcript abundance is controlled not just by cis-acting sequences like the promoter, but also by potentially unlinked, trans-acting transcription factors. Similarly, protein abundance is controlled by "local" variation at the coding gene itself, and by "distant" variation mapping to other regions of the genome. Local variation is likely to be composed of *cis* variants controlling transcript levels (though the correlation between transcript level and protein abundance is often quite low, so this may represent a minority of cases). Other local mechanisms might include polymorphisms for the stability or regulation of the protein. In contrast, distant variation could include upstream regulation control regions. Beyond these examples, further extension of QTL analysis includes mapping the contribution of imprinting to size-related traits, and other adaptations of QTL mapping will no doubt follow.

Historically, the availability of adequately dense markers (genotypes) has been the limiting step for QTL analysis. However, high-throughput technologies and genomics have begun to overcome this barrier. Thus, the remaining limitations in QTL analysis are now predominantly at the level of phenotyping, although the use of genomic and proteomic data as phenotypes circumvents this challenge to some extent.

Genome-wide association studies (GWAS) are becoming increasingly popular in genetic research, and they are an excellent complement to QTL mapping. Whereas QTL contain many linked genes, which are then challenging to separate, GWAS produce many unlinked individual genes or even nucleotides, but these studies are riddled with large expected numbers of false positives. Though GWAS remain limited to organisms with genomic resources, combining the two techniques can make the most of both approaches and help provide the ultimate deliverable: individual genes or even nucleotides that contribute to the phenotype of interest.

Indeed, combining different QTL techniques and technologies has great promise. For example, researchers used data on gene expression in fat and kidney tissue from two previously generated, recombinant rat strains to study hypertension. Alternatively, samples adapted to different environments may be compared, or other populations of interest might be selected for expression analysis. This approach permits measurement of hundreds or even thousands of traits simultaneously. Differences in expression may be co-localized with phenotypic QTL that have been previously determined to create manageable lists of positional candidate genes.

Other interesting questions concerning gene regulation can be addressed by combining eQTL and QTL, such as the relative contributions of *cis*-regulatory elements versus *trans*-regulatory elements. Regarding hypertension, researchers identified 73 candidate genes deemed suitable for testing in human populations, and many of the most highly linked eQTL were regulated in *cis*. These integrated approaches will become more common, and they promise a deeper understanding of the genetic basis of complex traits, including disease. Integrating phenotypic QTL with protein QTL can also give investigators a more direct link between genotype and phenotype via co-localization of candidate protein abundance with a phenotypic QTL. Still more kinds of data can be integrated with QTL mapping for a "total information" genomics approach (e.g., eQTL, proteomics, and SNPs).

QTL studies have a long and rich history and have played important roles in gene cloning and characterization; however, there is still a great deal of work to be done. Existing data on model organisms need to be expanded to the point at which meta-analysis is feasible in order to document robust trends regarding genetic architecture. Data generated by lab-based QTL studies can also be used to direct and inform other efforts, such as population genomics, wherein a large number of molecular markers are scored in the attempt to identify targets of selection and thus genes underlying ecologically important traits. Furthermore, QTL studies can inform functional genomics, in which the goal is to characterize allelic variation and how it influences the fitness and function of whole organisms. Thus, although the map between genotype and phenotype remains difficult to read, QTL analysis and a variety of associated innovations will likely continue to provide key landmarks.

MARKER-ASSISTED SELECTION IN FISH

The desirable phenotypic variations in the performance traits of fishes are used to increase the aquacultural yield, improve incomes of farmers and enhances food security through selective breeding by choosing better-performed individuals. However, phenotype-based selection needed considerable time to optimise the traits, so researchers are now moving from phenotype based selection to genotype-based selection. The lacking of a molecular marker is the main limiting factor for the realization of genotype based selection potentials in fishes. However, with the advent of DNA-based genetic markers in the late 1970s and now the ease of the marker discovery through the next generation sequencing allowed the researchers to identify large numbers of markers spreads throughout the genome of any species of interest.

The markers are used to detect linkage with the traits of interest, thus allowing MAS finally to become a reality. In aquaculture, genetic mapping has been completed for tilapia, tiger shrimp, kuruma prawn, rainbow trout, catfishes and atlantic salmon. Now genetic map needs to be developed for the other essential species for successful development of aquaculture industry by increasing overall production through MAS. The traits, difficult and expensive to measure or showing low heritability and recessive traits, can easily be optimised using MAS.

However, lack of the resources such as dense linkage maps, needed to locate QTL accurately in fishes is one of the major problem in aquaculture. Recently, expressed sequence tag (EST) databases based sequencing approach has been used for developing molecular markers such as simple sequence repeats (microsatellites) and single nucleotide polymorphisms (SNPs). In aquaculture, SNPs are mostly used for genomic studies and identification of diagnostic markers for various diseases. Marker-assisted selection method (MAS) or genome-wide marker-assisted selection method (G-MAS) are not widely used in aquaculture, but nowadays its use is increasing due to its ease of use and quicker than traditional phenotype-based selection. Now it becomes a fertile field of research for the aquaculture researchers to discover novel genetic marker that can be used to link with the QTLs in selective breeding programmes.

Most of the genetic improvement in fish and shellfish species to date has been made through the use of traditional selective breeding. Well-designed breeding programmers have shown substantial response to selection for body weight, e.g. Atlantic salmon, 10 to 14 percent, reported rates of genetic gain 8 percent for indirect selection for body weight at sea in rainbowtrout but it is 13 percent for direct selection.

Recently MAS become very popular method of indirect selection for production of the genetically improved offspring's in aquaculture breeding programme. As most of the performance traits such as growth or disease resistance are controlled by multiple genes and therefore inherited as quantitative traits, analysis of their associated quantitative trait loci (QTL) is an essential part of aquaculture genomics. QTLs are largely unknown genes that affect performance traits (such as growth rate and disease resistance) and these are important to breeders. Relative chromosomal positions of QTL in a genome can be identified in a two-step process that begins by constructing a genetic linkage map. Genetic linkage maps are made by assigning (mapping out) polymorphic DNA markers (such as microsatellites, SNP or AFLPs) to chromosome configurations based on their segregation relationships. SNPs and other markers that are linked to QTLs form the basis of marker assistance selection (MAS) in a selective breeding program for improving aquaculture production.

Importance of MAS in selective breeding programme

The rate of the genetic gain through MAS can increase when there is a continuous identification of new QTL for the targeted traits. The extra genetic gain through the MAS may decreases very quickly in successive generation of selection for the same QTL and also the rate of identification of new QTL is hard to predict. The genetic gain due to MAS for a targeted QTL is higher when the characteristicis measured after the selection.The frequency of the favourable QTL allele can be quickly increased in a population using marker assisted selection method during the first generations compared to traditional selection based on BLUP (Best linear unbiased prediction). The information generated from the detected QTL in the selection requires developing some selection criteria to connect this molecular information with phenotypic information.

MAS can be efficiently used to identify the major QTLs for a traitand for a phenotypic which is expensive to measure. According to researchers, in more than 20 aquaculture fish species, QTL for traits like cold and salinity, sex determination, growth and disease resistance have been mapped. Such studies can provide necessary information about patterns and the rate of evolutionary changes besides may provide tools for marker assisted selection. Mapping typically results in differing crosses and segregation pattern analysis in the offspring of hybrids between the divergent crosses. The applied goal of mapping understood so far is to identify the position of specific phenotypic traits, which could be helpful in breeding programs. Mapping genes and QTLs is of fundamental and applied interest in aquaculture species. Researchers mentioned polymorphic markers, genotyping platforms, reference families and software programs are prerequisite for constructing a linkage map of any species. The purpose of QTL mapping is to assist in selective breeding by understanding the effect and quantity of genes determining a trait.

QTL analysis for important traits like thermal tolerance, spawning time, embryonic development, growth traits, stress, salinity tolerance, infectious hematopoietic necrosis and bacterial cold-water disease have been conducted for more than 20 aquaculture species including finfish, mussels and crustaceans. QTL mapping is the practical application of marker-assisted selection in aquaculture. Among the 20 species studied so far, tilapia and rainbow trout are among most important freshwater food fish species where salmons are the major cultured marine fish species.

Advantages of marker-assisted selection

MAS is profitable for the aquaculturist over conventional selection methods when the traits show low heritability. It helps to discover the existing genetic diversity in breeding populations and can be used to improve diserable characters in aquaculture species. Improvement of performance traits through traditional selection integrated with molecular tools is fast and more accurate and allows us to understand the genetic mechanism affecting performance traits. The dramatic development of molecular genetics laid the groundwork for genomics that has introduced a new generation of molecular markers for use in the genetic improvement of farm animals. These markers provide more accurate genetic information and better understanding of the animal genetic resources.

Molecular markers have been used in QTL mapping for better genetic management in aquaculture. It is beneficial for sex-limited traits such as sperm production, egg production or traits that are a poor predictor of breeding value like fertility. The traits which aredifficult or expensive to measure (disease resistance), expressed late in life is very difficult to optimize using conventional breeding programme. Selection programme needs several progenies to optimise traits and in some case the progeny testing scheme requires a long generation interval and also very costly. There the MAS is very useful tool to reduce generation interval through early selection, even before maturity of the fish and to select those traits which are restricted to only one sex.

Limitation of marker-assisted selection

The main restriction in MAS is the increasing sample collection cost for genotyping and lack of complete genotype information of the fishes. There are very limited study has been conducted on MAS in fisheries in India. Genotyping the whole population is also difficultin commercial fish populations. To decrease genotyping costs, researcher are concern on the identification of the most informative individuals based on phenotypic information, segregation analysis or combining the phenotypic and genotypic information in breeding. There is a limited number of genefully studied for a performance trits and also lack of confidence of users to associated this gene which the molecular marker leads to lower the accuracy of QTL detection for fishes.

In case fish there is a lack of the complete genotype information in the population used in MAS schemes and lack of markers for the species make it difficult to link a QLT with a molecular marker. The selection of the marker information is not entirely reliable because of the possibilities of over estimation of QTL effects in QTL position in the fish population. The inconsistency is also a major problem in QTL estimation in fish because the QTL effect is not expressed similarly over the years orwhen it is used in a different population or stock. The main targeted traits in aquaculture for MAS are body growth, disease resistace, meat quality, feed efficiency and maturation.

9

GENE EXPRESSION, ISOLATION AND CLONING

INTRODUCTION

Fish is a healthy diet which is more demand in global market. Due to over exploitation, capture fish production has been decreased. To curtail this problem, aquaculture is the solution to fulfill the healthy food. To focus on exploited wild stocks should be admitted to biodiversity and genomic study to find out genetic variation. Population genetic study is a useful tool to identification of fish stocks. Genomic study deals with the evolution, structure and function of the gene. Gene is defined as a sequence of nucleotide/ segment of chromosome/ unit of heredity. Genes encode proteins and proteins dictate cell function. Therefore, thousands of genes expressed in a particular cell function. The genetic information stored in DNA which is expressed by the process of transcription and translation.

Transcription (DNA to mRNA) process occurs in the nucleus of the cell which genetic information encoded in DNA which is transferred to RNA. RNA carries this information out of the nucleus and into the cytoplasm where it becomes directly involved with protein synthesis via translation (Birnstiel et al., 1985). Translation follows the movement of mRNA to the cytoplasm where it interacts with structures called ribosomes to synthesize a protein. Translation of the mRNA at the ribosomes in the cytoplasm then involves several components including other types of RNA including tRNA and rRNA.

Functional genomics scrutinizes the biochemical and physiological role of gene products and their interactions and expression. Comparative genomics examines entire genomes, structure, evolution and taxonomy. Environmental genomics comprises molecular variation in natural or artificial populations of different taxa and their response to environmental conditions. Present review paper explores the concept and application of gene expression techniques and genomics study with special reference to fisheries and aquaculture.

Gene expression analysis plays an important role in identifying differentially expressed genes under different environmental conditions and gene expression regulation, shedding light on gene functions. EST analysis has been demonstrated effective for detection of differential expression and regulation of certain genes. Without normalization or subtraction in library construction, the number of the sequenced ESTs for a given gene reflected the abundance of the gene expression at the corresponding scenario (e.g. environmental conditions, developmental stages, treatments).

Direct EST sequencing is inefficient in discovery of rarely expressed genes. To solve this problem, the method to construct normalized cDNA libraries was developed. The basic principle is using hybridization to reduce redundant genes and increase the representation of rarely expressed genes. However, with the adoption of RNA-seq technology, as described below, it is unlikely that additional EST analysis will be conducted because of the low efficiency as compared with RNA-seq.

Initial annotation of ESTs (Expressed sequence tags) can be conducted by simple sequence similarity comparisons. Further annotation analysis can be carried out after obtaining the consensus sequences (putative unigenes), such as determination of gene identity based on homology search, open reading frame identification, gene ontology annotation and gene-enrichment analysis.

In order to assign gene identity to contigs and singletons, homology search is widely used. Such an approach is especially helpful for newly studied species. BLAST is the most widely used program to obtain high throughput EST analysis and annotation results. The BLAST package provides different flavours of algorithms for sequence similarity searching. BLASTX is used to search against protein database by translated consensus EST sequences, while BLASTN is used to search against nucleotide sequence databases.

The National Center for Biotechnology Information (NCBI), ENSEMBL and Swiss-Prot are three important databases for BLAST search. For instance, the Swiss-Prot database has a fully manually curated and annotated unigene database, Uniprot, which can be used for identifying putative function for unigene by BALSTX.

The NCBI provides a dbEST database that can be used to search novel transcripts by BLASTN. The dbEST is a main EST resource database, including ESTs for over 200 aquaculture species. The ENSEMBL database can provide chromosome location information of genes, which is a useful tool for comparative genome analysis. However, the BLAST sequence similarity comparison provides only sequence homology information, and one cannot purely rely on BLAST for gene identification. Detailed phylogenetic analysis and/or orthology analysis is needed to determine the identities of genes.

STUDY OF GENE EXPRESSION BY USING VARIOUS TECHNIQUES

Nowadays, the role molecular techniques in fisheries research have increased. The genetic research including stock identification and differentiation study, source of pathogen and their confirmation study and gene expression with reference to growth and gonadal maturation.

Northern blotting is used to detect specific RNA molecules present within a RNA mixture, employed in the analysis of RNA sample from a cell type or tissue so as to determine the RNA expression of certain genes. Western blotting is used to detect the specific protein molecules present within a protein mixture. This mixture might include all the proteins that are associated with a certain cell type or tissue.

The technique also to determine a protein's size, and how much of it is expressed. Fluorescent in situ hybridization (FISH) is a cytogenetic technique that can be used to recognize and locate specific gene sequences. FISH can be used to visualize copy number aberrations such as the deletion, translocation or amplification of chromosomes. Reverse transcription polymerase chain reaction (RT-PCR) is the most sensitive technique available for detecting and quantifying mRNA. Extremely small sample sizes can be used in the quantification of mRNA. Serial Analysis of Gene Expression (SAGE) is a technique used to generate a library of short sequence tags which can be used to detect a transcript. The transcript expression level can be determined by evaluating how many times each tag is detected.

Growth is a trait of moderate to high heritability that displays a complex development and physiology. It involves both the increase of muscle cell number (hyperplasia) and cell size (hypertrophy), and the balance between both processes depends on the developmental stage, but it is also influenced by the environment. Growth is regulated by the hypothalamic-pituitary axis hormones which also control feeding behaviour, yet local interactions between different muscle cell types and with other nearby tissues also condition muscle growth. Additionally, growth rate is regulated by other physiological and environmental factors, such as food intake, temperature, diet composition, age or sex.

Furthermore, identifying key growth-related genes is more complex in teleost than in other vertebrates, since this group of fish has undergone an additional round of genomic duplication6 involving neo- or sub-functionalization processes. Yet, although a complex task, understanding the genetic basis of growth variation might aid to achieve more efficient breeding programs. Despite its good response to traditional selection, the sole application of phenotypic and relatedness information to improve growth rate might determine the loss of relevant genetic variation affecting the medium- and long-term performance of breeding programs.

In this sense, knowing the molecular mechanisms, genes and pathways involved in growth can have multiple benefits. Beyond increasing our basic knowledge with important implications in different research areas related to growth, identification of allelic variants of moderate to large effect could be useful in marker-assisted selection (MAS), provided a cost-benefit balance. Finally, genome-editing techniques may allow recovering or even introducing variants not present in the population of interest, increasing genetic gain. As such, different studies have been focused on understanding growth differences in different fish species in the last decade.

Turbot is the flatfish with highest aquaculture production, mainly held by Spain and China. Females reach sexual maturity at ~3 years (~50 cM length) and can spawn between 5 and 10 million eggs per season. Reproduction can be achieved along the whole year under a variable controlled photoperiod. Thus, production can benefit from the high fecundity of a single selected couple over a relatively long period. Several studies have been carried out in this species to identify quantitative trait loci (QTL) associated with growth, but the information provided, though useful, is still of limited application.

Knowing the genes or even the allelic variants behind growth differences is important, not only for a more efficient selection, but also to guide future studies in turbot and other fish species. However, functional studies on differential growth have not been performed in turbot yet.

To gain knowledge on the key genes involved in growth, we carried out a transcriptomic study in fast- and slow-growing turbot to identify differential expression profiles and genetic markers between extreme growth phenotypes. Samples of muscle and brain tissues were taken from family-matched fast- and slow-growth fish selected from eight unrelated full-sib families, and their transcriptomes compared by RNA-Seq. These organs play a key role in the regulation of somatic growth, and muscle also represents the main edible part of fish.

In order to evaluate a high number of individuals and to avoid family bias, a pooling strategy was followed. Differential gene expression and marker association between high- and low-growth turbot were obtained. The detected candidate genes and markers were mapped using the recently assembled genome of the species and their positions were compared with previously reported QTLs. The most interesting genes and markers were validated using individual Real-Time PCR (qPCR) and SNP genotyping to disentangle general and family-specific factors involved in turbot growth.

GENE EXPRESSION IN AQUACULTURE MANAGEMENT

Gene expression techniques significantly contributing in sex determination system, fish farmers interested in monosex culture to obtain higher production at the same time avoid the breeding (e.g. triploidy). Several microsatellite markers are linked to the sexdetermining factor for linkage analysi.

Aquaculture genomic studies mainly focus on improved growth and development rates, food efficiencies, increased resistance to disease and pathogens and molecular phenotypic selection for improved brood stock characteristics. Differential gene expression has been used to solve production problems in commercially valuable fish species, such as Atlantic halibut (*Hippoglossus hippoglossus*) during the early stages of aquaculture development by using microarray analyses.

Culture-developed molecular tools can switch vital information and technologies ragards to conservation and management of hatchery and wild fish populations. Gene regulation and expression are thought to play significant roles in adaptive evolution in fishes. Genetic mechanisms underlying adaptive differences in important physiological traits.

Gene expression profiles and physiological genomics applied can be applied to aquaculture and fisheries by developing new technologies and significantly change the way culture management uses genetic information. An innovative cost effective feeds can be developed by using nutrigenomics. For example, plant based protein for carnivorous fish or low fish meal protein feed can be developed.

An implementation of genomic approaches should be encouraged in the fields of fisheries and aquaculture by supporting the development of genomic resources such as linkage maps, expressed sequence tag databases and expression profiling. Genomic and gene expression studies significantly help to nativity of stock identification, to promote the fish genetics & fisheries biotechnology sector.

RNA-seq technologies

RNA-seq is a technology to sequence transcriptomes using next-generation sequencing technologies. It has been widely used for analysis of gene expression profiling and identification of differentially expressed genes. RNA-seq can be done with a number of sequencing platforms, including the Illumina sequencing platform, ABI Solid Sequencing, and less efficiently, the Life Science's 454 sequencing. Of these, the Illumina HiSeq sequencers are the most popular because of their very high throughput and accuracy of sequencing reads.

RNA-seq analysis starts with RNA samples. Before RNA-seq, considerations need to be made to allow statistical analysis of the results with proper biological and technical replica. For instance, a control sample can be compared with the infected samples at various times after infection. In order to allow statistical analysis of the results, a minimal of three biological replications is required for each condition (treatment).

The biological issues of RNA-seq are actually extremely simple. The researchers are interested in which genes are expressed, how much they are expressed, and how the samples in different conditions compare. Before answering these questions, the immediate task is to assemble the short reads into reference transcriptome. In general, one of the two types of assembly methods can be used for the assembly of RNA-seq sequences, depending on the existing genome resources. If a reference genome sequence is available, reference-guided assembly methods can be used. In contrast, de novo RNA-seq assembly methods must be used in the absence of a reference genome sequence.

RNA-seq has been extensively used for the identification of gene-associated markers. In catfish, hundreds of thousands of gene-associated SNPs have been identified by deep sequencing of RNA from many individuals of both channel catfish and blue catfish, which will be used in the development of high-density catfish SNP chips for genome-wide association studies. In the study to understand the adaptive divergence between dwarf and normal lake whitefish species, 454 sequencing was used with the aim to generate a set of SNP markers; 89 SNPs showed pronounced allele frequency differences between sympatric normal and dwarf whitefish.

RNA-seq data sets are being deposited to the NCBI's Sequence Read Archive (SRA) at www. ncbi.nlm.nih.gov/sra. The RNA-seq data sets can be searched, retrieved and downloaded from the SRA database. For instance, searching of SRA using "catfish" as the keyword results in 109 records, as of March 2016. Researchers can download any of the records for additional analysis. Such data sets are most often used for meta-analysis.

Analysis of differentially expressed genes using RNA-seq

RNA-seq has many applications. However, the two popular analyses using RNA-seq are the identification of differentially expressed genes during development, under a specific physiological condition, or after certain treatments. The basic principle for the identification of differentially expressed genes is to calculate the ratio of counts of short reads aligned to each gene before and after treatment. For instance, if 50 RNA-seq reads align to gene X after infection while only 5 RNA-seq reads align to the same gene before infection, then the infection induced the expression of this gene is tenfold. This is to demonstrate the concept. In practice, however, detailed normalization and statistical analysis must be conducted.

With RNA-seq, the higher the expression level a gene has, the more reads are sequenced from this gene. However, the read count is correlated with the size of the gene exons. Therefore, the sizes of the gene must be normalized to compare their expression. Reads per kilobase per million (RPKM) has been widely used as a reasonable normalizer. RPKM is the number of reads per kb long of a transcript per 1 000 000 reads of RNA-seq sequences. RPKM = 109 × C/NL, where C is the total number of reads mapped onto the gene; N is the total number of mapped reads; and L is the sum of the genes in base pairs. For instance, the RPKM of a 2 kb transcript with 3 000 alignments in a sample of 10 million of mapped reads is calculated as:

$$RPKM = 10^9 \times 3\ 0002 \times 10\ 000\ 000 = 150$$

Once all the read counts are normalized into RPKM, expression levels among genes and before and after treatment can be calculated.

Analysis of co-expressed genes using RNA-seq data sets

In addition to the identification of differentially expressed genes, RNA-seq can be used to determine correlated or coordinated expression. Such analysis can be quite insightful. For instance, a single treatment such as high temperature treatment may induce a common set of genes, and these genes may all have similar functions in dealing with the heat shocks.

The correlation of expression patterns among various genes, referred to as "co-expression", can be revealed by network analysis. An increasing number of studies have demonstrated associated behaviour of genes with related biological functions. Given that most biological processes cannot be carried out by a single gene, analysis of co-expressed genes from RNA-seq data sets may be quite informative.

Analysis of allele-specific expression

A diploid organism has two sets of chromosomes and thereby two alleles at a given locus. Allele-specific expression (ASE) refers to the phenomenon that the two alleles are not equally expressed, up to the exclusive expression of only one of the two alleles. A number of recent studies have demonstrated that allele-specific gene expression is common. Interest in the existence of ASE in non-imprinted autosomal genes has increased with awareness of the important role that variation in non-coding DNA sequences can play in determining phenotypic diversity.

A number of approaches have been used for the detection of ASE. In early studies, ASE was detected by a single-base extension of a primer adjacent to the variable single nucleotide polymorphism. Several recent studies applied a variety of technologies to scale up the tested genes, of which the array-based approach was the most widely used. A number of array-based ASE studies have been published in the past decade. Owing to rapidly increasing throughput and decreasing costs, next-generation sequencing is rapidly replacing array-based technology for functional genomic assays.

In addition, the ability to resolve single-base differences, digital quantification, and comprehensive genome-wide coverage provides information on the abundance and the allelic biases in transcripts or regulatory DNA, which otherwise could not be achieved using hybridization-based arrays. RNA-seq technology using high-throughput sequencing platforms allows for relatively unbiased measurements of expression levels across the entire length of a transcript. This technology has several advantages, including the ability to detect transcription of unannotated exons and measure both overall and exon-specific expression levels and assay allele-specific expression. Notably, RNA-seq is the only technology that provides concurrent allelic and total expression data.

Measuring ASE is vital to better understanding global mechanisms of genetic variations. ASE analysis has been widely employed in mammals, insects and plant systems. In aquaculture species, despite many reports regarding expression of specific genes, very little is known about ASE. Recently, researchers reported that ASE is highly enriched with ribosomal protein genes. With more and more application of RNA-seq in aquaculture species, it is inevitable that it will be used for the analysis of ASE.

EXPRESSED SEQUENCE TAGS

Genetic maps are essential tools to locate genomic regions associated with productive characters, which can eventually be applied in marker-assisted selection programs or used to identify genes related to specific traits through fine mapping and/or positional cloning strategies. Additionally, they provide the support to study genome organization and evolution through comparative mapping, and provide useful landmarks for genome assembly. A first generation turbot consensus map (242 anonymous microsatellites; 26 linkage groups (LG)) was reported. It has been used to identify QTL for sex determination, growth rate and resistance to pathogens. Recently, a new microsatellite genetic map has been reported by researchers using 158 anonymous markers.

Genomic resources have greatly increased in aquaculture species especially after the arrival of NGS, and several genome projects are underway in several fish species. However, most comparative genomic studies still rely on model species. Genome sequences with high coverage are available in zebrafish (*Danio rerio*), fugu (*Fugu rubripes*), Tetraodon (*Tetraodon nigroviridis*), medaka (*Oryzias latipes*) and stickleback (*Gasterosteus aculeatus*). Since gene-associated markers are much more conserved than anonymous ones, they constitute the preference target to go further on comparative mapping and evolutionary genomics. Comparative mapping also represents the best strategy to capture candidate genes at genomic regions associated with productive characters in aquaculture species.

DEVELOPMENT OF EST RESOURCES OF AQUACULTURE SPECIES

Expressed sequence tags are single-pass sequences of random complementary DNA (cDNA) clones from cDNA libraries. They are traditionally generated using Sanger sequencing and therefore the resultant sequences are approximately 500 to 800 base pairs in length. Several years ago, because sequencing was relatively cheap, large numbers of ESTs can now be generated at a reasonably low cost from either the 5' or 3' end of a cDNA clone to get an insight into transcriptionally active regions. ESTs were used as a primary resource for human gene discovery. Thereafter, there has been an exponential growth in the generation and accumulation of EST data in public databases for various organisms, with approximately 74 million ESTs now available in these databases.

EST analysis is an effective genomic approach for rapid identification of expressed genes, and has been widely used in genome-wide gene expression studies in various tissues, developmental stages and under different environmental conditions. In addition, the availability of cDNA sequences has accelerated further molecular characterization of genes of interest and provided sequence information for microarray construction and genome annotation.

Table 1. Some examples of aquatic species with major expressed sequence tag (EST) resources (>10 000) using zebrafish (1,488,275) as a reference

Species	No. of ESTs
Danio rerio (zebrafish)	1 488 275
Ciona intestinalis	1 205 674
Xenopus laevis (African clawed frog)	677 911
Oryzias latipes (Japanese medaka)	666 891
Salmo salar (Atlantic salmon)	498 245
Ictalurus punctatus (channel catfish)	354 516
Oncorhynchus mykiss (rainbow trout)	287 564
Crassostrea gigas (Pacific oyster)	206 388
Litopenaeus vannamei (white shrimp)	161 248
Ictalurus furcatus	139 475
Oreochromis niloticus (Nile tilapia)	120 991
Petromyzon marinus (sea lamprey)	120 731
Sparus aurata	79 216

EST resources provide sequence information for microarray development. For instance, in a recent study, researchers developed a large-scale oligonucleotide microarray platform containing 20,000 features (20K), which was used to study immune response of the Atlantic cod spleen with stimulation of formalin-killed atypical Aeromonas salmonicida. Similarly, oligo microarray for gilthead sea bream (*Sparus aurata*) was developed based on ESTs, and the microarray was used to identify 1 050 differentially expressed genes between two developmental stages.

Although EST analysis has been important for transcriptome characterization, it is now becoming expensive, relative to several of the recently developed approaches, as detailed below.

However, EST resources still have a great value to serve as reference for RNA-seq analysis. It has been found that ESTs are useful for high-quality reference-guided assembly of next-generation sequencer-generated short reads.

OSMOREGULATION

Osmoregulation is a process that regulates the osmotic pressure of fluids and electrolytic balance in organisms. In animals, this process is brought about by osmoreceptors, which can detect changes in osmotic pressure. Humans and most other warm-blooded organisms have osmoreceptors in the hypothalamus. Besides the brain, osmoregulators are also found in the kidneys.

Freshwater fish and marine fish osmoregulate in different ways. The environments which they have varying levels of salinity, hence the process of osmoregulation is different.

Osmoregulation in Freshwater Fish

Freshwater fishes are hypertonic to their surrounding environment, which means that the concentration of salt is higher in their blood than their surrounding water. They absorb a controlled amount of water through the mouth and the gill membranes. Due to this intake of water, they produce large quantities of urine through which a lot of salt is lost. The salt is replaced with the help of mitochondria-rich cells in the gills. These cells absorb salt into the blood from the surrounding water.

Osmoregulation in Marine Fish

Compared to freshwater fish, marine fish face the opposite problem. They have a higher concentration of water in their blood than their surrounding environment. Consequently, it results in the tendency to lose water and absorb the salt. To get around this problem, marine fish drink large quantities of water and restrict urination. Another additional energy expenditure also arises as these organisms actively need to expel salt from the body (through the gills).

Osmoregulation in Bacteria

Bacteria use a transport mechanism to absorb electrolytes when osmolarity around it increases. The osmotic stress activates certain genes in bacteria that synthesize osmoprotectants.

Fish Maintain Osmotic Balance by Osmoconforming or Osmoregulating

Osmoconformers maintain an internal solute concentration—or osmolarity—equal to that of their surroundings, and so they thrive in environments without frequent fluctuations. All osmoconformers are marine animals, although many marine animals are not osmoconformers.

Most fish are osmoregulators. Osmoregulators maintain internal osmolarity independent of the environment, making them adaptable to changing environments and equipped for migration.

Osmoregulation Requires Energy

Osmosis tends to equalize ion concentrations. Since fish require ion levels different from environmental concentrations, they need energy to maintain a solute gradient that optimizes their osmotic balance.

The energy required for osmotic balance depends on multiple factors, including the difference between internal and external ion concentrations. When osmolarity differences are minimal, less energy is required.

Alternative Osmotic Strategies

The bodily fluids of marine sharks and most other cartilaginous fish contain TMAO; this enables them to store urea and internally surpass the external osmolarity, allowing them to absorb water through osmosis.

Most animals are stenohaline—unable to tolerate large external osmolarity fluctuations. Euryhaline species, like salmon, can change osmoregulatory status. When salmon migrate from freshwater to the ocean, they undergo physiological changes, such as producing more cortisol to grow salt-secreting cells.

EPIGENETIC AND ITS MECHANISM

Before discovering DNA as inheritance component, scientists were concluded that all the gens in all time are not active in all organisms, although carry the same data. Then Epigenetic was introduced as a mechanism that controls the gene expression in an inheritable pathway. In a molecular level the modification of cytosine and histone alkaline modification and changes in nucleosome location are known as the typical mechanisms of epigenetic. These mechanisms are essential for regulating cellular process such as: difference, transposon movements, genomicimprinting, × deactivation and Embryogenesis. In a multi cellular organism the ability of preservation and retention of Epigenetic during genesis and transition to next generation is essential to make a lot of phenotypes that originates from the same genotype. In human, although monozygotic twins are alike in DNA sequence they are different in DNA methylation and the profiles related to histone modification. This matters even in one single cell. How stem cell can become any cell will be answered by epigenetic. The importance of epigenetic came out when their inappropriate outbreak leads to a number of diseases such as cancer.

Epigenetic modification and related complexes

Epigenetic modifications divided into three branches:

1. DNA methylation
2. Histone DNA
3. Nucleosome positioning

DNA methylation

DNA methylation generally occurs in CpGdinucleotide. These nucleotides are gathered in CpG islands- which about 60% of human gene promotersare in contact with them and usually they are not methylated and have the permission to be copied and make a suitable chromatin formation for gene expression but their malapropos hypo methylation leads to deactivation of copping, but some of them will be methylated during the tissue differentiation. Generally CpG islands methylation is correlated with gene silence. DNA methylation doesn't occur just in CpG Island. The idiom of CpG Island Shores refers to the areas with low concentration of CpG which are in vicinity of CpG Islands, and methylation in these areas are completely in contact with copping suppress. It seems that most of tissue methylation occurs in CpG Island Shores instead of CpG Island. DNA methylation has a key role in genomic imprinting; it means hyper methylation in one or two maternal alleles is related to mono allele gene expression.

DNA methylation has several mechanisms to suppress gene expression

1. Methylated DNA increases use of MBD protein. The family member of MBD protein calls the histone modifier complexes and chromatin remodeling's to methylated sites.
2. DNA methylation suppresses transcription directlyby blocking the use of joining protein (such as E2F, NF-KB, AP2) to DNA in Cis binding element areas.

It is so rare that DNA methylation be along with transcription activation but in plants and vertebrates, gene bodies' methylation (the exon of some genes which is under transcription and translation) is positively related to gene expression. It is suggested that this matter be along with elongation step blockage of spurious initiation of transcription. Gene bodies tend to DE methylation in disease that leads to transcription initiation in several wrong sites.

Histone modification

Histones are key elements in epigenetic. H2B, H2A (two dimmer form) and H3, H4 (one tetra dimmer form) form nucleosome.

H1 histone is copulative and will join the DNA which enters to nucleosome. All histones will be revised after translation and most of this correction is done on histone tail including: stilation, methylation, phosphorylation, obi coientination and ADP rebozilation.

Refer to the transcription form, genomes are divided into two form: a) Eu-chromatin active form b) de active hetero chromatin form. In Eu-chromatin form high levels of H3K4, H3K36, H3K79 estilasion and methylation is detected but in hetero chromatin form low estilasion and high methylation levels will be detected. Totally transcriptionally active gene be spotted in promoters by high level of H4K20 me11, H2BK5ac, H3K27ac, H3K4me3 and in gene bodies by H4K20 me1, H3K79me1. Several Non coding RNAs become syntheses from some hetero chromatin areas. For instance a SiRNA expressed from hetero chromatin pombe saccharomyces centromere area that will jointranscription silence inductor RNA complex and affect it.

Epigenetic in aquatic animals

During embryonic development, epigenetic modifications of DNA occur through various processes (e.g. DNA methylation and histone acetylation) and are assumed to facilitate differentiation into specific cell types. An epigenetic alteration can be defined as a mitotically and/or meiotically heritable change in the function of a gene without alterations in the gene sequence. Genes that are to be silenced from one of the parental alleles (i.e. expressed by only one allele) become methylated during the embryonic development in a process called imprinting. There is only limited evidence of genomic imprinting in oviparous species, such as fish.

However, DNA methylation reprogramming was observed during the early embryonic development of zebrafish (*Danio rerio*) in a recent study using an anti-5-methylcytosine antibody in immunohistochemistry and southwestern immunoblotting. This reprogramming of DNA methylation found in zebrafish is similar to the reprogramming during mammalian development.

Vitellogenesis is the production of yolky eggs in oviparous species, and involves transport of gene products from the liver to the ovary where proteins are deposited in the maturing oocytes. Researchers shown that exposure to EE2 results in an alteration of DNA methylation levels in zebrafish which warrants studies of epigenetic changes in fish toxicological studies.

They have for the first time shown sex and tissue differences in the level of DNA methylation in adult zebrafish in a franking region of the vitellogenin I gene. Further, following exposure to 100 ng EE2/L during 14 days the DNA methylation levels they decreased in liver of both females and males. The results of them demonstrate the usefulness of the zebrafish as a model organism for studying epigenetic processes and possible epigenetic alterations following exposure to chemicals.

Pyrosequencing technology represents a tool to determine methylation levels of multiple CpG sitesin specific genes of interest and this study shows its applicability to investigate pollutant-induced alterations of methylation levels in fish.

Genetic and epigenetic interactions between redundant genes in polyploid fish have probably influenced their evolutionary fate, leading to their current impressive biological diversity. Spontaneous polyploids have been observed in several phylogenetically distant orders, including both wild and farmed fish species.

In the vertebrates, polyploid species are not exclusive to fish, since they have been reported in different groups, from amphibians to occasionally even in mammals. Polyploids can originate either from alterations of meiotic or mitotic processes in specimens within a species (autopolyploidy) or by reproductive contact among species (allopolyploidy).

Regulatory changes in gene expression following tetraploidisation may result in epigenetic instability, because they are more likely to be deleterious than advantageous.

The genetic sources of variation can be associated with the presence of the extra maternal set of chromosomes and can involve simple gene dosage (additivity) between chromosome sets or positive or negative dosage compensation effects (heterosis), epigenetic mechanisms, and transcriptional co-suppression (negative gene dosage compensation).

In zebrafish modulation of parental abiotic environment and nutrition confers increased resistance to the environmental stressor and alterations in cardiovascular parameters (stroke volume, heart rate, cardiac output and red blood cell concentration) to the subsequent generation.

TRANSPOSABLE ELEMENTS

Transposable elements (TEs) represent the largest genomic component of most eukaryotic organisms. Their contribution to genome size reaches up to two third in vertebrate invertebrate, and plant genomes. They play a central role in the structural organization and plasticity of genomes allowing the establishment of evolutionary new cellular functions. The distinctive parasitic and selfish behavior of TEs provides a rational mechanism for their recruitment as genome building blocks. Genome sequencing projects and genome-wide functional assays showed that in several instances TEs facilitate the generation of new *cis*-regulatory elements by inserting themselves in close proximity to genes. They modulate and coordinate eukaryotic gene expression by supplying transcription factors binding sites, transcription start sites, and enhancer and silencers elements.

Convergent domestication of transposons (i.e. adaptation of a transposon to be used for a host function) were recently uncovered. SINE family species-specific expansions remodel the repertoire of CTCF-binding sites in highly diverse mammalian lineages. Likewise the recruitment of different and unrelated transposable elements control the extrapituitary expression of Prolactin in various mammal lineages. The third example might stem from a predisposition of *pogo*-like transposases to encode centromeric proteins. Members of this family of transposons were tamed in mammalian and fission yeast genomes to encode centromere-associated protein B (CENP-B) and CENP-B-like proteins, respectively. These examples of convergent evolution illustrate well previous findings of the ENCODE consortium that many functional elements are seemingly unconstrained constituting a "warehouse" for natural selection.

DNA-based transposable elements

Four major classes of transposable elements are known: viral family retrotransposons, long interspersed nuclear elements (LINEs), short interspersed nuclear elements (SINEs) and DNA-based transposable elements. The last class, also called terminal-inverted-repeat elements, is different from the others in that elements move directly from DNA to DNA, that is, without forming RNA intermediates. The enzyme that is central or solely required for their transposition reaction is called a transposase and is considered to act mainly on their terminal and/or subterminal regions. Transposable elements are present in multicopies in host genomes and, in many cases, albeit not all, autonomous copies exist including a gene for their own transposase. Copies that do not carry the transposase gene or carry an imperfect gene are called nonautonomous copies, these being thought to be produced from autonomous copies by errors during transposition or subsequent mutational sequence changes.

As long as the transposase cannot distinguish autonomous from nonautonomous copies, which so far appears to be true, the evolutionary scenario they would be expected to follow is first increase in frequency of nonautonomous elements, subsequent loss of transposition activity, and ultimate extinction from the host genome.

Several DNA-based transposable elements have been identified in vertebrates, analogous to many other organisms. However, autonomous copies are exceedingly rare, which suggests the situation where most vertebrate DNAbased elements are in the final or nearly the final stage. For studying transposition mechanisms, as well as evolution, the few elements in teleost fishes demonstrated or suggested to be still active offer unique materials. The authors have been working on identification and examination of active elements of the medaka fish *O. latipes*. The present review describes results so far obtained on various aspects of medaka fish transposable elements, such as structure, transposition mechanisms, evolution and application to biotechnology.

Medaka fish

The medaka fish is a freshwater teleost native to Eastern Asia, including China, Japan and Korea. It is, together with zebrafish, a superior model animal for vertebrate genetics and developmental biology because of its small genome size (800 Mb), small body size (4 cm), short generation time (3 months), high reproductive rate (10 to 20 eggs per female every day) and transparent egg chorion. In addition, a strain having completely transparent bodies has recently been generated by accumulating deficient alleles at several pigmentation loci.

Another remarkable advantage is the in-depth knowledge of closely related fishes. Fourteen species are so far identified in the genus *Oryzias* and their detailed phylogenetic relationship is known. This provides a strong basis for evolutionary studies comparative to that in the melanogaster species subgroup of Drosophila. A great amount of genetic variation is known to exist not only among but also within species and many inbred strains reflecting geographic differences have been prepared.

The Tol1 element

There are several different albino mutants in Tomita's collection. He was convinced that some of them are due to defective tyrosinase genes. Tyrosinase is the key enzyme in melanin biosynthesis. Because structural analysis of spontaneous mutant genes has long been the main procedure for identifying new transposable elements, we first cloned the wild-type tyrosinase gene of the medaka fish and then examined its structure in candidate albino strains. The Tol1 element (transposable element of *Oryzias latipes*, 1; GenBank accession number D42062) was found as a result of this analysis. It had been inserted in the first exon of the tyrosinase gene, making the gene nonfunctional.

The particular Tol1 copy found in the tyrosinase gene was denoted Tol1-tyr. It is 1.9 kb in length, carries terminal inverted repeats of 14 bp, and is flanked by an 8-bp target site duplication. No meaningful open reading frame (ORF) is evident in Tol1-tyr. Mainly for this reason, it is not clear which other transposable elements Tol1 is related to. We have estimated the copy number of Tol1 to be about 80 per diploid genome.

Tol1-tyr — tgg**cctttagc**CAGTAGCGGTTCTAGGCACGGG

ATTCAGGGTAGAACCGCCACTG**cctttagc**ttt —

Tol2-tyr — ggg**tcaagaac**CAGAGGTGTAAAGTACTTGAGT

TTGAGTACTTTTTACACCTCTG**tcaagaac**tcc —

Fig. 1. Nucleotide sequences of Tol1 and Tol2 terminal regions. Tol1-tyr and Tol2-tyr are specific copies of Tol1 and Tol2, respectively, inserted in the tyrosinase gene. Nucleotide sequences of their terminal regions are shown in uppercase letters. Tol1 has terminal inverted repeats of 14 bp each and Tol2 carris terminal inverted repeats of 17 bp and 19 bp, indicated by underlining. The sequences of the tyrosinase gene are shown in lowercase letters. Both elements are flanked by 8-bp target site duplications indicated by boldface letters.

While the human tyrosinase gene is more than 70 kb in length, that in the medaka fish is as small as 5 kb, primarily because of the smaller sizes of introns. The entire defective gene of the albino strain, containing the 1.9-kb Tol1 copy, is within the carrying capacity of a lambda phage vector. We obtained such a clone and replaced a small part of the first exon, in which Tol1 is included, with the corresponding region of the wild-type gene.

Introduction of this modified clone into fertilized eggs of the albino strain resulted in melanin pigmentation in the skin of hatched fish. There was no nucleotide sequence difference in the manipulated region between the wild-type and the defective genes, except for the Tol1 insertion and its flanking target site duplication. These results indicate that the particular Tol1 insertion event in the tyrosinase gene is not so old as to lead to accumulation of other mutations on the already defective tyrosinase gene. In other words, Tol1 is, or was until recently, active with respect to transposition.

More direct evidence for its transposition activity has recently been obtained. We exposed embryos of the albino strain to UV, and conducted PCR with a pair of primers encompassing the Tol1 insertion point. The length of a major PCR product was longer by 1.9 kb than the product with a wild-type fish. However, in some cases, minor products of a length identical, on agarose gels, to the wild-type form were observed.

Cloning and sequencing of these products revealed excision footprints, remnants of Tol1 terminal regions or loss of nucleotides originally belonging to the tyrosinase gene. Such footprints are observed with many DNA-based transposable elements.

From the above results, an autonomous copy that is presumed to carry a transposase gene would be expected to be present somewhere in the genome. However, no such copy has been detected among more than 20 copies we have so far cloned and sequenced. Tol1 occurs in seven of the nine *Oryzias* species we have examined by genomic Southern blot analysis (Fig. 2).

The probe used was the 1.9-kb entire Tol1-tyr copy cloned from the medaka fish *O. latipes*. Hybridization signals became weaker with the phylogenetic distance from *O. latipes*, with the exception of *O. mekongensis*. This result suggests that Tol1 was present in the common ancestor of the Oryzias species. We have also conducted more analytical Southern blot experiments, in which different parts of the 1.9-kb Tol1 copy were used separately as probes. We observed several species-specific structures.

For example, a central 0.4-kb region of Tol1-tyr hybridized to virtually all Tol1 copies in *O. latipes* but did not produce any hybridization signals with *O. javanicus*, indicating that all the Tol1 copies of the latter species lack this portion (unpublished). These results suggest that, although Tol1 appears to be an ancient resident in the genus *Oryzias*, its transposition occurred after species divergence.

The Tol2 element

Tol2 was identified by the same approach, as adopted for Tol1, in a different albino strain, that is, a different tyrosinase allele. It had been inserted in the last, fifth exon of the tyrosinase gene. It is 4.7 kb in length, carries terminal inverted repeats of 17 bp and 19 bp, and is flanked by an 8-bp target site duplication (Fig. 1). This element belongs to the hAT transposable element family that includes hobo of Drosophila, Activator of maize and Tam3 of snapdragon. Tol2 was directly demonstrated to be active soon after its discovery. We carried out PCR with primers encompassing the Tol2 insertion point, and PCR products with the same length as the product from the wild-type fish were readily obtained without any treatment such as exposure to UV. In these PCR products, the element was found to have been removed and footprints were observed.

Tol2 has four ORFs that roughly correspond to four exons making up the mRNA for the Tol2 transposase, which catalyzes both excision and insertion of Tol2. Microinjection of fish eggs with *in vitro* synthesized mRNA for the transposase triggers transposition of a co-injected Tol2 clone. This opened a way to the development of a gene transfer vector. We embedded the green fluorescent protein (GFP) gene in Tol2 and injected this clone into fertilized eggs together with the transposase mRNA. The GFP gene was successfully introduced into the fish genome and transmitted to subsequent generations. Sequencing analysis demonstrated the entire Tol2 element, including the GFP gene, to have been integrated into the chromosomes through its transposition.

The Tol2 clone, modified to carry the GFP gene and other necessary components, was 9.0 kb in length. Therefore, Tol2 still retains transposition activity even when it is as large as 9.0 kb. Larger versions are now being tested to determine the limitation in size.

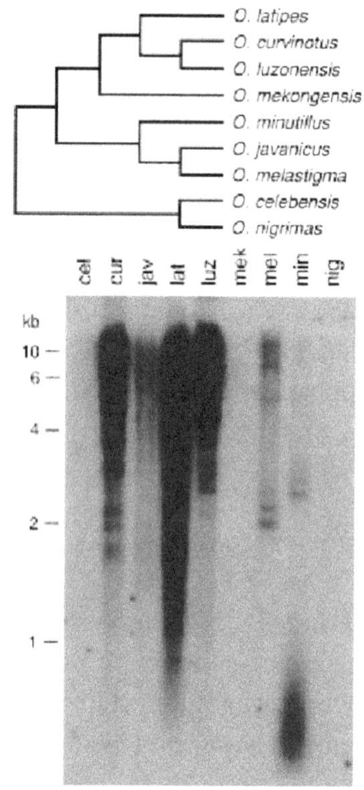

Fig. 2. Distribution of Tol1 among species.

Genomic Southern blot hybridization to examine the distribution of Tol1 among species was conducted as described, using the entire Tol1-tyr element as probe. The species names are abbreviated with their first three letters. Hybridization signals are not evident in the the lane for O. mekongensis on this autodiagram, but longer exposure revealed about ten bands. Along the left margin are indicated the sizes and mobilities of the size marker DNA fragments.

Regarding variation among copies in the genome and the distribution among species, Tol2 presents a striking contrast to Tol1. We examined the presence/absence of Tol2 in ten Oryzias species and found it only in two, the medaka fish O. latipes and the Hainan medaka fish O. curvinotus, the latter being an inhabitant of Southern China. These two species are relatively closely related to each other but nucleotide sequence variations exist on their chromosomes. Despite this variation, Tol2 copies exhibit virtually no nucleotide sequence differences between the species. From results of quantitative analysis, we have proposed that Tol2 was recently introduced into one of these two species from the other or into both of them from an unknown source.

RIBOSOMES

Ribosomes are the complex molecular machines that synthesize proteins as instructed from the genetic information on messenger RNAs (mRNAs). Most of the observed phenotypes in cells and organisms arise from the function of polypeptides that the ribosomes produce.

Hence, ribosomes are at the critical junction of the genotype-phenotype relation in all species. Fully assembled ribosomes have large and small subunits.

The small, and large, subunits in eukaryotes are referred to as the 40S, and 60S subunits, respectively, based on their sedimentation properties. 80S refers to fully assembled ribosomes. Each subunit is a ribonucleoprotein particle, composed of one (in the 40S), or three (in 60S), ribosomal RNA (rRNA) molecules, and many (79 in yeast, 80 in animals) proteins in 80S ribosomes.

The ribosomal proteins are structural, non-catalytic components of ribosomes. Bacterial ribosomes have similar architecture, but they are smaller and have fewer proteins.

The majority of ribosomal proteins are essential for ribosome function and life. In budding yeast, 15 out of a total of 79 cytoplasmic ribosomal proteins are not essential. In several cases, more than one gene may encode a ribosomal protein. For example, in the budding yeast *Saccharomyces cerevisiae*, 59 ribosomal proteins are encoded in each case by a pair of highly similar paralogous genes.

As will be described (Fig. 3), cells carrying mutations in ribosomal proteins display a broad spectrum of phenotypes, depending on the locus and alleles involved. It is the objective of this review to systematically go over these phenotypes and examine how they might come about.

The focus in this review is on six eukaryotic organisms, three invertebrate ones (budding yeast, worm, fly) and three vertebrate (fish, mouse, human). The next sections will describe the following:

species	mutant genes	phenotypes	phenotypes (most common)
	137	111	decreased: competitive fitness, resistance to chemicals, vegetative growth
	77	151	larval arrest, embryonic lethality, maternal sterility
	75	43	lethal-all die during larval stage, partially lethal, Minute
	41	223	head decreased size, extension decreased thickness, eye decreased size
	23	289	decreased body weight, kinked tail, prenatal lethality
	24	25	Diamond-Blackfan anemia (multiple types)

(Data collected from SGD, WormBase, FlyBase, ZFIN, MGI, OMIM)

Fig. 3. Summary of the most common phenotypes that arise from loss-of-function mutations in ribosomal proteins, in each of the organisms examined in this review. The number of mutant genes for which phenotypes have been described is shown in the second column, while the total number of phenotypes detected by mutations in any ribosomal protein gene in that organism is shown in the third column.

i. the generation of a complete matrix of ribosomal protein mutant loss-of-function phenotypes in each of the six organisms;

ii. a computational approach to define and group those phenotypes and the genes that underpin them;
iii. a similar analysis for gain-of-function phenotypes of ribosomal proteins in yeast;
iv. a discussion of the evidence linking in some cases ribosomal protein mutants to increased proliferation, including cancer; and
v. an examination of the observed phenotypes in the context of changes in gene expression, especially at the translational level, which may bridge the genotype–phenotype relationships.

Lastly, it is worth pointing out that all the datasets generated here are provided in the attached files, with the hope of stimulating further analyses of the remarkable properties and consequences of ribosomal protein perturbations.

PROTEIN EXPRESSION PROTEOMICS.

The quantitative study of protein expression between samples that differ by some variable is known as expression proteomics. In this approach, protein expression of the entire proteome or of subproteomes between samples can be compared. Information from this approach can identify novel proteins in signal transduction or identify disease-specific proteins.

Structural proteomics

Proteomics studies whose goal is to map out the structure of protein complexes or the proteins present in a specific cellular organelle are known as "cell map" or structural proteomics. Structural proteomics attempts to identify all the proteins within a protein complex or organelle, determine where they are located, and characterize all protein-protein interactions. An example of structural proteomics was the recent analysis of the nuclear pore complex. Isolation of specific subcellular organelles or protein complexes by purification can greatly simplify the proteomic analysis. This information will help piece together the overall architecture of cells and explain how expression of certain proteins gives a cell its unique characteristics.

Functional proteomics

"Functional proteomics" is a broad term for many specific, directed proteomics approaches. In some cases, specific subproteomes are isolated by affinity chromatography for further analysis. This could include the isolation of protein complexes or the use of protein ligands to isolate specific types of proteins. This approach allows a selected group of proteins to be studied and characterized and can provide important information about protein signaling, disease mechanisms or protein-drug interactions.

Technology of Proteomics

An integral part of the growth of proteomics has been in the advances made in protein technologies. Twenty-six years ago, when 2-DE was introduced, very few tools existed for proteomics. Since that time, new technologies have emerged and old ones have been improved in areas from protein separation to protein identification. However, it is also clear that it is still not feasible to conduct many types of proteomics because of limitations in technology. These problems will have to be solved and new technologies must be developed for proteomics to reach its full potential. A typical proteomics experiment (such as protein expression profiling) can be broken down into the following categories:

i. the separation and isolation of proteins from a cell line, tissue, or organism;
ii. the acquisition of protein structural information for the purposes of protein identification and characterization; and
iii. database utilization.

SEPARATION AND ISOLATION OF PROTEINS

By the very definition of proteomics, it is inevitable that complex protein mixtures will be encountered. Therefore, methods must exist to resolve these protein mixtures into their individual components so that the proteins can be visualized, identified, and characterized. The predominant technology for protein separation and isolation is polyacrylamide gel electrophoresis. Unlike the breakthroughs in molecular biology that eventually enabled the sequencing of the human genome, some aspects of protein science have shown little progress over the years. Protein separation technology is one of them. Since its inception some 32 years ago, protein electrophoresis still remains the most effective way to resolve a complex mixture of proteins. In many applications, it is at this stage where the bottleneck occurs. This is because 1- or 2-DE is a slow, tedious procedure that is not easily automated. However, until something replaces this methodology, it will remain an essential component of proteomics.

One- and two-dimensional gel electrophoresis

For many proteomics applications, 1-DE is the method of choice to resolve protein mixtures. In 1-DE, proteins are separated on the basis of molecular mass. Because proteins are solubilized in sodium dodecyl sulfate (SDS), protein solubility is rarely a problem. Moreover, 1-DE is simple to perform, is reproducible, and can be used to resolve proteins with molecular masses of 10 to 300 kDa. The most common application of 1-DE is the characterization of proteins after some form of protein purification. This is because of the limited resolving power of a 1-D gel. If a more complex protein mixture such as a crude cell lysate is encountered, then 2-DE can be used. In 2-DE, proteins are separated by two distinct properties. They are resolved according to their net charge in the first dimension and according to their molecular mass in the second dimension. The combination of these two techniques produces resolution far exceeding that obtained in 1-DE.

One of the greatest strengths of 2-DE is the ability to resolve proteins that have undergone some form of posttranslational modification. This resolution is possible in 2-DE because many types of protein modifications confer a difference in charge as well as a change in mass on the protein. One such example is protein phosphorylation. Frequently, the phosphorylated form of a protein can be resolved from the nonphosphorylated form by 2-DE. In this case, a single phosphoprotein will appear as multiple spots on a 2-D gel. In addition, 2-DE can detect different forms of proteins that arise from alternative mRNA splicing or proteolytic processing.

The primary application of 2-DE continues to be protein expression profiling. In this approach, the protein expression of any two samples can be qualitatively and quantitatively compared. The appearance or disappearance of spots can provide information about differential protein expression, while the intensity of those spots provides quantitative information about protein expression levels. Protein expression profiling can be used for samples from whole organisms, cell lines, tissues, or bodily fluids. Examples of this technique include the comparison of normal and diseased tissues or of cells treated with various drugs or stimuli.

Another application of 2-DE is in cell map proteomics. 2-DE is used to map proteins from microorganisms, cellular organelles, and protein complexes. It can also be used to resolve and characterize proteins in subproteomes that have been created by some form of purification of a proteome. Because a single 2-DE gel can resolve thousands of proteins, it remains a powerful tool for the cataloging of proteins. Many 2-DE databases have been constructed and are available on the World Wide Web.

A number of improvements have been made in 2-DE over the years. One of the biggest improvements was the introduction of immobilized pH gradients, which greatly improved the reproducibility of 2-DE. The use of fluorescent dyes has improved the sensitivity of protein detection, and specialized pH gradients are able to resolve more proteins. The speed of running 2-DE has been improved, and 2-D gels can now be run in the minigel format.

In addition, there have been efforts to automate 2-DE. Hochstrasser's group has automated the process of 2-DE from gel running to image analysis and spot picking. The use of computers has aided the analysis of complex 2-D gel images. This is a critical aspect of 2-DE because a high degree of accuracy is required in spot detection and annotation if artifacts are to be avoided. Recently, a molecular scanner was developed to record 2-DE images.

Software programs such as Melanie compare computer images of 2-D gels and facilitate both the identification and quantitation of protein spots between samples. A recent exciting advance in 2-DE was developed by Minden and coworkers. This technology is called difference gel electrophoresis (DIGE) and utilizes fluorescent tagging of two protein samples with two different dyes. The tagged proteins are run on the same 2-D gel, and postrun fluorescence imaging of the gel is used to create two images, which are superimposed to identify pattern differences.

The dyes are amine reactive and are designed to ensure that proteins common to both samples have the same relative mobility regardless of the dye used to tag them. This technique circumvents the need to compare several 2-D gels. In their original paper, DIGE was used to detect differences between exogenous proteins in two *D. melanogaster* embryo extracts at nanogram levels. Moreover, an inducible protein from *Escherichia coli* was detected after 15 min of induction.

However, a number of problems with 2-DE still remain. Despite efforts to automate protein analysis by 2-DE, it is still a labor-intensive and time-consuming process. A typical 2-DE experiment can take two days, and only a single sample can be analyzed per gel. In addition, 2-DE is limited by both the number and type of proteins that can be resolved. For example, the protein mixture obtained from a eukaryotic cell lysate is too complex to be completely resolved on a single 2-D gel. Many large or hydrophobic proteins will not enter the gel during the first dimension, and proteins of extreme acidity or basicity (proteins with pIs below pH 3 and above pH 10) are not well represented.

Some of these problems can be overcome with different solubilization conditions and pH gradients. Another limitation of 2-DE is the inability to detect low-copy proteins when a total-cell lysate is analyzed. In a crude cell extract, the most abundant proteins can dominate the gel, making the detection of low-copy proteins difficult. It was determined in the analysis of yeast proteins by 2-DE that no proteins defined as low-copy proteins were visible by 2-DE.

Yet it is estimated that over half of the 6,000 genes in yeast may encode low-copy proteins. In mammalian cells, the dynamic range of protein expression is estimated to be between 7 and 9 orders of magnitude. This problem cannot be overcome by simply loading more protein on the gel, because the resolution will decrease and the comigration of proteins will increase. Because of these limitations, the largest application of 2-DE in the future will probably involve the analysis of protein complexes or subproteomes as opposed to whole proteomes.

Alternatives to electrophoresis.

The limitations of 2-DE have inspired a number of approaches to bypass protein gel electrophoresis. One approach is to convert an entire protein mixture to peptides (usually by digestion with trypsin) and then purify the peptides before subjecting them to analysis by MS. Various methods for peptide purification have been devised, including liquid chromatography, capillary electrophoresis, and a combination of techniques such as multidimensional protein identification or cation-exchange chromatography and reverse-phase (RP) chromatography.

The advantage of these methods is that because a 2-D gel is avoided, a greater number of proteins in the mixture can be represented. The disadvantage is that it can require an immense amount of time and computing power to deconvolute the data obtained. In addition, considerable time and effort may be expended in the analysis of uninteresting proteins. One of the most exciting techniques to emerge as an alternative to protein electrophoresis is that of isotope-coded affinity tags (ICAT).

Profiling Proteomes

Gene expression is a highly regulated, multistep process, and it is impossible to predict the exact protein concentration or activity from the measurement of mRNA levels. Proteomics has therefore become a key tool in systems biology because it provides quantitative and structural information about proteins, which are the major functional determinants of cells. Phenotypic alterations associated with genetic perturbations often result from changes in protein accumulation or stability, or changes in protein posttranslational modifications, which can disrupt protein-protein interactions and network connectivity. Quantitative protein information complements data from transcriptional profiling and metabolomics. It represents a key link between different levels of gene expression regulation and provides insights into their causal relationships. Unlike transcriptional profiling, however, comprehensive proteome analysis remains challenging, and information about proteome complexity and dynamics is far from complete.

Moreover, the rate of metabolite synthesis is often controlled by regulatory posttranslational modifications of enzymes and not only by their abundance. Information about quantitative relationships between RNA and protein accumulation, posttranslational protein modifications, and metabolite levels is therefore required to fully understand regulatory circuits that control systems behavior and function.

Protein quantification can be absolute or relative. While relative protein quantification mostly depends on stable isotopes, absolute quantification of comprehensive protein sets is much more difficult. Recent improvements in statistical data evaluation and increasing accuracy of mass spectrometry instruments allow quantifying large numbers of proteins in shotgun-type experiments on the basis of spectral counting.

This method is reliable and comparable to most other quantification methods, including two-dimensional PAGE-based protein staining; however, the protein dataset must be very large. More accurate information about the exact in vivo concentration of individual proteins requires specialized targeted approaches.

Current methods for absolute protein quantification include isotope dilution strategies using isotopically labeled peptides as internal standards. Signature peptides for internal standardization are characteristic for a protein of interest, and are often referred to as proteotypic peptides (PTPs).

In AQUA, PTPs are added to analytical protein samples in known concentrations. The protein samples are subsequently scanned for PTPs of interest. Using the extracted ion chromatograms the native peptide can then be quantified relative to the added PTP.

A modification of this strategy accounts for quantification errors derived from incomplete tryptic digest of the analytical sample. In QconCAT (for quantification concatamer), a synthetic protein with concatenated, isotopically labeled PTPs is expressed as recombinant protein in a biological system, added to the sample prior to Trypsin treatment and carried through the digestion procedure, such that losses from incomplete tryptic digestion will also affect the quantity of the PTPs.

Both the AQUA and the QconCAT strategies are incompatible with upstream fractionation techniques, which is a potential problem in biomarker quantification.

A way around this constraint is offered by the protein standard absolute quantification strategy, which uses isotopically labeled protein standards that are added to the sample prior to fractionation. Several prediction tools exist that help to define the most suitable PTPs for the detection and quantification of specific proteins.

However, only experimental data provide the necessary reliability for PTP selection because in practice PTP prediction often deviates from experimental observations. Therefore, efforts are under way to catalogue PTPs for model organism proteomes. Proteome maps for Arabidopsis generated PTPs for 4105 proteins, many of which may be optimal for the detection of proteins in different organs.

Similar quantitative approaches are also used for metabolites, because in addition to RNA and protein levels, understanding the function and behavior of metabolic networks requires global information about metabolite concentrations and fluxes as well.

10

GENE-TRANSFER TECHNOLOGY

INTRODUCTION

Gene transfer technique is used very widely both in basic research and applied biology. The delivery of DNA into animal cells is a fundamental and established procedure. It has become an indispensable tool for gene cloning, the study of gene function and regulation and the production of small amounts of recombinant proteins for analysis and verification. Gene transfer experiment helps to express the introduced genetic construct (or transgene) in the recipient cells or to disrupt or inactivate particular endogenous genes (resulting in a loss of function.

There are many applications of gene transfer like large-scale commercial production of recombinant antibodies and vaccines and gene medicine or gene therapy. They range from the use of mammalian and insect cell cultures to transfer of DNA into human patients for the correction or prevention of disease.

In the organisms or genetically modified whole animals created by Gene transfer every cell or a specific target population of cells carries a particular alteration. Such animals are used to study gene function and expression, model human diseases, produce recombinant proteins in their milk and other fluids, and to improve the quality of livestock herds and other domestic species. The technology has contributed to understand the functions of the many genes discovered in the genome projects (functional genomics).

Examples of such experiments include systematic DNA-mediated mutagenesis and gene trap programs in the mouse and in the fruit fly, Drosophila melanogaster, genome-wide RNA interference experiments in the nematode, Caenorhabditis elegans, and novel protein interaction screens based on the yeast two hybrid system but performed using mammalian cells.

Genetic engineering of food is the science which involves deliberate modification of the genetic material of plants or animals. Introduction of DNA into plants is of great agricultural potential and medical importance. The gene transfer results can be transient and stable transfection.

Gene therapy can be defined as the deliberate transfer of DNA for therapeutic purposes. Gene transfer is one of the key factors in gene therapy, and it is one of the key purposes of the clone. Gene transfer can be targeted to somatic (body) or germ (egg and sperm) cells. In somatic gene transfer the recipient's genome is changed, but the change will not be passed on to the next generation. In germline gene transfer, the parents' egg and sperm cells are changed with the goal of passing on the changes to their offspring.

HISTORY

The concept of gene transfer between cells was first demonstrated in bacteria, which are capable of at least four natural forms of genetic exchange. The first mechanism was discovered in 1928 by Frederick Griffith named as transformation in bacterium Streptococcus pneumoniae. However, he was unable to determine the nature of the transforming principle. In 1944 Oswald Avery established that the substance transferred between cells was DNA. A second form of gene transfer was named conjugation discovered by Joshua Lederberg and Edward Tatum in 1946 in Escherichia coli. This process involved the transfer of DNA through a direct link between the bacterial cells. In E. coli this conduit between cells took the form of a proteinaceous tube known as a pilus. The ability of the cells to construct the pilus and pass DNA through it was encoded on a large plasmid known as the F (for fertility) factor.

In most cases, the act of conjugation involved transfer of the plasmid alone, which became established in the recipient cell thereby converting it from an F" to an F+ phenotype. In some cases, however, the F plasmid could integrate into the bacterial chromosome, and conjugation could result in the transfer of chromosomal genes. This process, which was used to construct the first genetic map of E. coli, was termed sexduction. Then in 1951, transduction, a new type of gene transfer mediated by bacteriophage was discovered by Joshua Lederberg and Norton Zinder in Salmonella. They found that newly formed phage could occasionally package some of the host cell's DNA and then transfer it to a second host cell in a subsequent infection.

Two forms of transduction were identified – generalized and specialized transduction. In generalized transduction the phage head was mistakenly stuffed completely with host cell DNA. Whereas in specialized transduction the phage genome integrated into the bacterial chromosome and became linked to host DNA. The fourth mechanism of gene transfer in bacteria is mediated by complete cell fusion, and occurs in several genera of bacteria including Bacillus and Streptomyces. Under natural conditions, the viral capsid is needed to introduce the oncogene and the rest of the genome into an animal host cell. The unusual process of gene transfer without the viral capsid was named transfection to distinguish it from normal infection.

In bacterial genetics, the term transformation continued to be used to describe the uptake of naked plasmid or genomic DNA (essentially any DNA which had the potential to transform the phenotype of the recipient cell) while transfection was used specifically to describe the uptake of naked phage DNA (or RNA), i.e. nucleic acid which had the potential to initiate a phage replication cycle. The term transfection became generally accepted to mean the introduction of any sort of DNA— phage, plasmid, genomic or otherwise—into an animal cell, in the absence of a biological vector.

DIFFERENT GENE TRANSFECTION TECHNIQUES

There are two types of transfection transient and stable. In transient transfection, the transfected DNA is not integrated into host chromosome. DNA is transferred into a recipient cell in order to obtain a temporary but high level of expression of the target gene. Stable transfection is also called permanent transfection. By the stable transfection, the transferred DNA is integrated (inserted) into chromosomal DNA and the genetics of recipient cells is permanent changed.

Regardless of the delivery method, gene transfer into animal cells must accomplish three distinct goals. First, the exogenous genetic must be transported across the cell membrane. In physical transfection methods, transport across the membrane is achieved by direct transfer, where the membrane is breached during delivery through which DNA and RNA can diffuse.

In other delivery methods, the nucleic acid must form some sort of complex which binds to the cell surface before internalization. For example, in chemical transfection methods the complex is formed between nucleic acid and a synthetic compound, while in transduction methods the complex comprises nucleic acid packaged inside a viral capsid.

Once across the cell membrane, the genetic material must be released in the cell and transported to its site of expression or activity. Again, the nucleic acid is passive at this stage. In most transfection methods, DNA or RNA complexes are deposited in the cytoplasm, following escape from the endosomal vesicle. DNA must be transported to the nucleus, while RNA can function directly in the cytoplasm. In methods such as particle bombardment and microinjection, it is possible to deliver DNA directly into the nucleus, so intrinsic transport pathways are not required.

Many viruses also deliver their nucleic acid cargo to the nucleus as part of the infection cycle, often after interaction with cell surface receptors and either internalization within endosomes or direct fusion with the plasma membrane. However, few exceptions are there like poxviruses (e.g. Vaccinia virus) and alphaviruses (e.g. Sindbis virus) which replicate in the cytoplasm. In the final stage of gene transfer, the exogenous genetic material must be activated. It must be released from its complex and rendered competent for expression and/or interaction with the host genome. Exogenous RNA exist only transiently in the host, whereas exogenous DNA can exist transiently or permanently.

The gene transfer methods normally include three categories:

1. transfection by biochemical methods;
2. transfection by physical methods;
3. virus-mediately transduction.

The first gene transfer protocols used naked DNA which was mixed with particular chemicals to form synthetic complexes. These synthetic complexes either interact with the cell membrane and promote uptake by endocytosis, or fuse with the membrane and deliver the DNA directly into the cytoplasm. Such chemical transfection methods have been widely used. However they are generally inefficient for gene transfer in vivo. In contrast, physical transfection methods are efficient for both in vitro and in vivo gene transfer.

Physical transfection methods involve breaching the cell membrane and introducing the nucleic acid directly into the cell or nucleus. Although there are advantages and disadvantages to both sets of procedures, some of the most efficient transfection methods in use today involve a combination of chemical and physical processes. Chemical and physical transfection methods were first used for the transfer of naked, wild-type viral DNA into animal cells. Now both techniques are more widely used for the introduction of plasmid vectors and recombinant viral genomes carrying specific transgenes of interest.

METHODS IN THE PRODUCTION OF TRANSGENIC FISH

Microinjection method

Microinjection method has been used successfully in the production of transgenic fish and is a commonly used technique due to its simplicity and reliability. The use of the microinjection method results in higher survival rates for manipulated fish embryos than the electroporation method. The most established method for gene transfer in fish is microinjection. Microinjection that allows delivery of the transgene directly into the nucleus, Transgene is directly microinjected into the male pronuclei of fertilized eggs. Transgenic technology through DNA microinjection into zebra fish embryos has made great gain in the last decade. It is shown that the DNA injected into the cytoplasm of fertilized zebra fish eggs could integrate into the fish genome and be inherited in the germ line. The frequency of germline transmission of a microinjected DNA could be as a high as 20% in zebra fish.

Disadvantage

This method is not only time-consuming procedure for animals such as fish that produce a large number of eggs and labour intensive but is also limited by the physiology of fish eggs. The nuclei of fish eggs are small and difficult to visualize, the chorion, hardens soon after fertilization and in many fish species the pronuclei of fertilized eggs is not visible and transgenes are usually injected into the egg cytoplasm. This technology, however, still has as the major constrains the low efficiency generation of transgenics. To improve the efficiency of selection of transgenics, genetic markers are co-injected with the transgene to monitor for transformed zygotes. The Green Fluorescent Protein (GFP) from Jellyfish (Aequorea victoria) has been used for this purpose in zebrafish.

Electroporation method: Electroporation has been shown to be the most effective means of gene transfer in fish since a large number of fertilized eggs can be treated in a short time by this method.

Electroporation utilizes a series of short electric pulses to permeate the cell membrane that make possible the formation of temporary pores on the surface of the target cells through which the transgene is introduced into the cytoplasm where it is then delivered to the nucleus by the cellular machinery. Electroporation has been preferred in many laboratories for gene transfer in fish systems because of its efficiency, speed and simplicity.

Disadvantage

The presence of a tough chorion layer around the fish eggs reduces efficiency, removal of the chorion is a tedious procedure and introduces additional stress on newly fertilized eggs. In traditional electroporation methods, either levels of transfer have been low with few studies showing germ-line transmission and expression or the device used is no longer commercially available.

Sperm mediated gene transfer method

Spermatozoa are capable of binding DNA and carrying it into an egg. Fish spermatozoa can be stored in seminal plasma with little loss of viability for long periods. Therefore, this technique appears very promising for gene transfer in fish.

Retroviral infection method

The successful use of Retroviruses for gene delivery in fish has been reported.

Disadvantage

The preparation of retroviral particles including the transgene of interest is a very laborious process, increases costs and requires technology.

Advantages of transgenic fish

The growth rate of Transgenic fish can be increased by 400% to 600% while simultaneously reducing feed input by up to 25% per unit of output, thereby improving food conversion ratios.

Transgenic fish have been developed for applications such as the production of human therapeutics, experimental models for biological research, environmental monitoring, ornamental fish and aqua cultural production. Growth enhanced transgenic fish have improved feed-conversion efficiency, resulting in economic and potential environmental benefits such as reduced feed waste and effluent from fish farms. Transgenic strains provide useful model systems for studying the consequences of growth enhancement from genetic, physiological and ecological standpoints. The trait of growth rate has received the greatest attention because of its importance in aquatic animal production systems.

Two to threefold increase relative to non-transgenic fish have been reported for tilapia and Atlantic salmon and up to twofold increases in common carp. Commercial production of transgenic fish able to transmit desirable characteristics, such as enhanced growth or disease resistance to their progeny. Growth acceleration in fish has been one of the targets of gene transfer experiments.

Zebra fish are a good model for the rapid study of GH-transgenes, before selecting the construct to use in economically important species. AquAdvantage salmon, a modified Atlantic salmon reach commercial size in one-third of the time required for nontransgenic salmon. The faster growth of each generation can lead to increased production per unit time along with savings on total food per pound of meat produced. Use of transgenic fish as bioreactors for the large-scale production of rare human therapeutic proteins or novel foods for specific dietary requirements. Transgenic lines of tilapia engineered to produce human clotting factor VII, which is used in liver transplants and in treating injuries.

Zebra fish genetically modified to manifest a defective aortic valve development has made it possible to identify the role of an enzyme, UDP-glucose dehydrogenase, in the process of the embryonic development of this valve. Transgenesis is proposed as a mean for removing allergenic substances in seafood. Fish are recognized as test organisms with distinct and superior benefits in providing insights to disease processes. The development of transgenic animal models represents a revolutionary advance in the study of a variety of disease processes. Fish as cost-effective and important animal models in genetics, developmental biology and toxicology.

Development of transgenic fish as a model in reducing or replacing selected mammals used in toxicity testing. Medaka fish has several characteristics that are especially well suited for environmental toxicology, including well-characterized histopathology. They have been used extensively in chemical hazard testing, carcinogenesis bioassays and germ cell mutagenesis studies.

Tilapia expressing extra copies of trout growth hormone gene grow much faster than their non transgenic counterparts which do not possess extra copies of the growth hormone gene. Transgenic fish have been used to study developmental regulation of genes, improve cold tolerance, increase growth rate, and improve feed utilization.

Fast growing transgenic coho salmon, medaka, rainbow trout and tilapia containing novel growth hormone genes reached sexual maturity sooner than unmodified counterparts. The efficiency of growth and feed conversion can also be increased in finfish by creating transgenic fish that incorporate a gene construct encoding growth hormone. Genetically modified fish have also been developed as experimental models for biomedical research, especially in studies involving embryogenesis and organogenesis as well as in the study of human diseases, Xenotransplantation and recombinant protein production for producing important therapeutic agents.

Disadvantages of transgenic fish

Inducing transgenesis is a relatively transgene. Mosaicism, the condition in which not all cells in embryos have the targeted transgene, can occur. The primary ecological concerns regarding utilization of transgenic fish are the loss of genetic diversity and loss of biodiversity, and reduction in species richness. Transgenic fish are those related to their involuntary escape into environment.

Concerns range from interbreeding with native fish populations to ecosystem and effects resulting from heightened competition for food and prey species. The sterilization of transgenic fish will help in reducing the interbreeding risks associated with the escape of transgenic fish. Introduction of transgenic fish into natural communities is a major ecological concern. Reduced sperm production in the GH transgenic Nile tilapia, Oreochromis niloticus.

Extremely fast-growing transgenic salmon and loach have low fitness and die inefficient process. Only about one out of every 100 eggs microinjected will stably incorporate the recombinant DNA sequence into its genome and subsequently transmit the transgene to its progeny. Transgenesis is unnatural in the sense that the insertion of DNA material in transgenic animals would not take place without human intervention. Transgenic fish could produce new or modified proteins that could be toxic to humans, they generally show reduced swimming ability and lower reproductive performance than non-transgenic fish.

The transgenic fish are also more active and aggressive when feeding, and more willing to risk exposure to predation. Changes in cognitive abilities or brain function and structure have occurred in transgenic fish. The transgenic salmon pose serious ecological threats to wild populations and are not engineered for natural environments. Transgenesis can result in an unpredictable number of copies and site of integration of the early.

GENE TRANSFER TECHNOLOGY IN FISH

Principle of gene transfer

Transgenics are organisms into which heterologous DNA (transgene) has been artificially introduced and the transgene stably integrated into their genomes. Although transgenics have been produced since the early 1980s, the production of transgenic fish was accomplished only a decade ago, in the early 1990s. The main principle of gene transfer is to facilitate the transfer of heterologous DNA into the nucleus of a target cell where integration into the host genome takes place. This can be accomplished using a number of techniques, such as microinjection that allows delivery of the transgene directly into the nucleus, and electroporation that facilitates the formation of temporary pores on the surface of the target cells through which the transgene is introduced into the cytoplasm where it is then delivered to the nucleus by the cellular machinery.

Fish species used for gene transfer studies

To date, a range of fish species, such as Japanese medaka (*Oryzias latipes*), zebra fish (*Branchydanio rerio*), trout (*Onchorynchus mykiss*), Atlantic salmon (*Salmo salar*), common carp (*Cyprinus carpio*), channel catfish (*Ictalurus punctatus*), tilapia (*Oreochromis*), loach (*Misgurnus anguillicaudatus*) and goldfish (*Carassius aurata*), have been used for gene transfer studies. The species of fish to be used for gene transfer studies is determined according to the purpose of the study.

Fish species that are small in size, such as Japanese medaka (*Oryzias latipes*), are chosen to test the efficiency of a method or to establish a model system for basic or applied studies. However, larger species such as tilapia are chosen if the transgenic fish will be used as a bioreactor or for aquaculture purposes.

Methods used for gene transfer

A variety of gene transfer techniques, such as calcium precipitation, microinjection, lipofection, retrovirus infection, electroporation, embryonic stem cells, spermmediated gene transfer and particle gun bombardment, have been used to produce transgenics. Different gene transfer techniques present various advantages and disadvantages depending on the species of animal to be used for gene transfer studies. For instance, methods such as microinjection and electroporation are currently used for gene transfer in egg-laying fish, whereas others, such as retroviral vector-mediated gene transfer, are used in live-bearing fish species. The methods currently used or under investigation for gene transfer in fish are summarized below:

Electroporation

Electroporation utilizes a series of short electric pulses to permeate the cell membrane and allow the transgenes to enter the cells. The presence of a tough chorion layer around the fish eggs reduces efficiency, but removal of the chorion layer, physically or chemically, increases the efficiency of this technique. However, removal of the chorion is not applied routinely because it is a tedious procedure and introduces additional stress on newly fertilized eggs.

Electroporation has been shown to be the most effective means of gene transfer in fish since a large number of fertilized eggs can be treated in a short time by this method. There are a number of commercially available electroporaters, such as Beakon 2000 (Baekon Co.), that can be effectively used for gene transfer studies in fish.

Sperm-mediated gene transfer

Sperm-mediated gene transfer has the potential to be one of the most favorable gene transfer techniques because of the nature of the procedure. It has been reported that in mice, spermatozoa are capable of binding DNA and carrying it into an egg. However, the spermmediated gene transfer procedure in mammalian systems is quite controversial at present, since the viability of sperm outside the reproductive tract is very short (a few minutes at most). On the other hand, fish spermatozoa can be stored in seminal plasma with little loss of viability for long periods (days). Therefore, this technique appears very promising for gene transfer in fish.

Lipofection

Lipofection enables encapsulation of DNA within a phospholipid bilayer and subsequent delivery into a cell. Liposomes have been used successfully as gene delivery systems in mammalian cell cultures (3). Although no attempts have been reported for gene transfer in fish using lipofection, this technique may also be applied for gene transfer in fish.

Retroviral infection

Retroviruses are ideal gene delivery vehicles due to their efficient integration and single copy insertion of genetic material into the host genome. Their succcessful use for gene delivery in higher vertebrates and in fish has been reported. However, the preparation of retroviral particles including the transgene of interest is a very laborious process, increases costs and requires technology. Therefore, use of this method is limited. Nevertheless this technique makes gene transfer possible for species such as live-bearers and crayfish, from which newly fertilized eggs cannot be readily obtained.

Among the techniques mentioned above, electroporation has been preferred in many laboratories for gene transfer in fish systems because of its efficiency, speed and simplicity.

Egg Collection and Embryo Culture

Currently, the preferred way of producing a transgenic fish is to introduce transgene into a newly fertilized egg. For this purpose, fertilized eggs can be collected immediately after fertilization and maintained in ice-cold Yamamato's saline to slow down embryonic development for a short period of time before gene transfe.

GENE TRANSFER

Gene transfer can be carried out using one of the techniques for generating transgenic fish mentioned above. For gene transfer in an egg-laying fish species such as medaka a Baekon 2000 electroporation device can be used under the following conditions: pulse frequency, 2^6 pulse per cycle; burst time, 0.4 s; cycle number, 5; pulse width, 160 μS and distance of electrode from surface of buffer, 1 mm; voltage, 2.5-5.0 kv. Yamamoto's saline can be used as the electroporation medium.

Gene transfer via retroviral vectors or lipofection can be accomplished by incubation of newly fertilized eggs or injection of gonads with retroviral vector or lipofection reagent packaged with the transgene of interest. In sperm-mediated gene transfer, the sperms are first incubated with transgene and then the eggs are fertilized with these treated sperms.

After the gene transfer, treated embryos can be incubated under standard conditions established previously for each fish species until hatching.

Screening of Putative Transgenic Fish

It has been reported that only a part of gene tranfer trials are successful. Therefore, following the gene transfer, each individual needs to be analyzed for the presence of the transgene. For this purpose, following hatching, the treated individuals are reared to a stage at which a fin clip can be obtained without harming the individual to check for the presence of transgene. Genomic DNA from each surviving treated fish is prepared from a small piece of fin tissue fin clip.

A variety of methods, such as dot blot and Southern blot hyridization, can be utilized to identify putative transgenic individuals using genomic DNA samples. However, Polymerase Chain Reaction (PCR) based assay and Southern blot analysis of the PCR products are currently commonly used in order to economically handle a large number of samples. A transgene-specific oligomer set is used to amplify the transgene from the genomic DNA sample using a thermocycler. PCR products are then run on agarose gels and blotted to a nylon membrane. A radiolabeled transgene probe is used to identify the PCR products containing the transgene sequence blotted to the nylon membrane.

Positive PCR products show the transgenic fish, and these are referred to as founder (P1) individuals. P1 individuals are then crossed with nontransgenic individuals to investigate the transmission of the transgene and to establish the transgenic fish lines. Progeny produced from the cross of a P1 transgenic and nontransgenic individuals are referred to as the F_1 generation. F_1 transgenic individuals are identified by the same method used to identify P1 transgenic individuals. A number of previous studies demonstrated that some of the P1 generation transgenic fish are germline mosaics, meaning that these individuals contain the transgene in only a part of their germ cells; these therefore, produce transgenic offspring in limited numbers. F_1 transgenics are usually crossed with nontransgenic counterparts or with one another to obtain an F_2 generation and to determine whether there is a Mendelian inheritance of the transgene from F_1 progeny. F_2 generation individuals are identified by the same method used to identify former generations.

GENE TRANSFER TECHNOLOGY FOR DEVELOPMENT OF TRANSGENIC FISHES

The most commonly used methods in fish biotechnology are chromosome manipulation and hormone treatments, which can be produced triploid, tetraploid, haploid, gynogenetic and androgenetic fish.

Other popular methods of gene transfer in fish are microinjection, electroporation of sperms, electroporation of eggs and incubation of sperms. Following are the main steps in gene transfer for development of transgenic fish.

Preparation of DNA Construct:

Desired transgene should be a recombinant gene or DNA construct, which is constructed in plasmid that contains an appropriate promoter-enhancer element and a structural DNA sequence.

The foreign genes are typically introduced with strong genetic signals, promoters and/or enhancers, which enable the foreign genes to be expressed at very high levels continuously (or constitutively), effectively placing those genes outside the normal metabolic regulation of the cell, and of the transgenic organism resulting from the transformed cell.

There are three main types of transgenes:

Gain-of-Function:

These transgenes are able to increase particular function in transgenic individual after their expression. For example growth hormone genes from mammal and fish linked to appropriate promoter-enhancer element and a structural DNA sequence to produce GH transgene.

This GH transgene when express in transgenic individuals increases production of growth hormone leading to enhanced growth of transgenic animal.

Reporter Function

These transgenes are able to identify and measure the strength of promoter-enhancer element.

Loss of Function

This transgene is not yet used for modification of transgenic fish. Such transgenes are used for interfering with the expression of host genes. The promoter-enhancer elements of transgenes are linked to a growth hormone gene of fish.

Hence transgenic fish contain extra DNA sequences that are originally derived from same species. Gene construct is then introduced into fertilized egg or embryo, so that transgene be linked to genome of each cell of egg or embryo.

Gene Transfer by Microinjection

Microinjection is most successfully and widely used technique for gene transfer in fish. One method of microinjection technique involves the use of fine injection needle for introducing DNA into cut site in the cell. While doing so it destroys those cells that are in direct contact with the injected DNA.

To ensure the integration of the DNA it should be injected to intact cells close to the cut site. The injection apparatus consists of a dissecting stereomicroscope and two micromanipulators, one with a glass micro-needle for delivering transgene and other with a micropipette for holding fish embryo in place.

The success of microinjection technique depends on the nature of egg chorion. The soft chorion facilitates the microinjection while the thick chorion limits the ability to visualize the target for injection of DNA. In many fishes (Atlantic salmon and rainbow trout) the egg chorion gets tough and hard just after the fertilization or to contact with the water and provides a difficulty in injecting the DNA.

But using the following methods can solve this problem:

1. By using the micropyle (an opening on the egg surface for sperm entry during the fertilization) for inserting the injecting needle.
2. By using microsurgery for making an opening on the chorion.
3. By digesting the chorion with enzymes.
4. By using 1mM glutathione for initiating fertilization and reducing hardness of chorion.
5. By direct injection to the unfertilized eggs.

Another technique of gene transfer is intra-nuclear microinjection, which involves direct physical approach using a fine needle to deliver DNA into cell or even nuclei.

To facilitate rate of microinjection protoplast with partially reformed cell wall may be attached to a solid support with artificially bound substrate -without damaging the cells. Solid support may be of either glass cover slips or slides.

Steps of Microinjection Technique

1. Desired eggs and sperms are stored separately at the optimum conditions.
2. Add water and sperms and initiate the fertilization.
3. Ten minutes after the fertilization, eggs are dechorionated by trypsinization.
4. Fertilized eggs are microinjected with desired DNA just within a few hours of fertilization. DNA is released into the centre of the germinal disc to the first cleavage in dechorionated eggs. The time available for microinjection is first 25 minutes and that too between fertilization and first cleavage.
5. After microinjection the embryos are incubated in water until hatching takes place.

Survival rates of microinjected fish embryos is seem to be about 30-80% depending the fish species.

Advantages of Microinjection Technique:

This technique has the following merits:

1. Optimum quantity of DNA can be delivered per cell, increasing chances for integrative transformation.
2. The delivery of DNA is precise, even into nuclei of target cell again improving chances for integrative transformation.
3. The small structure can be injected.
4. It is a direct physical approach, hence it is a host range independent.

Disadvantages of Microinjection Technique:

1. A single cell can be injected at a time, hence it is time consuming process.
2. It requires sophisticated instruments and specialized skills.
3. Limited embryonic time restricts injection to more eggs and a low transformation rate.

Gene Transfer by Electroporation

It is a simple, fast, efficient and convenient method for transferring gene. This method involves an electrical pulse to deliver DNA into cells. The cells are exposed to a short electrical shock, which make the cell membrane temporarily permeable to DNA.

The desired DNA fragment is placed in direct contact of protoplast membrane, which enters into the cell upon electric shock. Hole may be created as a result and stabilized by a favourable dipole interaction with electric field.

Electroporation involves a chain of electrical pulses for permeation of cell membrane, thereby allowing the entry of DNA into fertilized eggs. The rate of DNA integration in electroporated embryo is more than 25% is the surviving rate, which is slightly higher in comparison of microinjected ones.

Advantages of Electroporation Technique:

1. It allows simultaneous entry of DNA constructs.
2. It is more suitable method for those species, which has very small eggs for microinjection.
3. This method does not require specialized skill.

Antifreeze Protein Gene Transfer

Many teleost inhabiting icy marine water in the Polar Regions produce antifreeze glycoproteins (AFGPs) or antifreeze proteins (AFPs) in their sera to protect them from freezing. This protein lowers the freezing temperature of solution without altering its melting temperature.

Thermal hysteresis, the difference between the freezing and melting temperature, is a unique property of these proteins. AFPs and AFGP have been demonstrated to bind to ice crystals and inhibit ice crystal growth.

Despite their similar antifreeze properties, these proteins are quite different in their protein structures. There are one type of AFGP and three types of AFP. Recently fourth type of AFP has also been reported in longhorn sculpin.

The Atlantic salmon Salmo salar, lacks any of these AGFPs or AFPs gene(s) and are unable to survive in sub-zero sea water temperature. An inability to tolerate temperature below – 0.6 °C to – 0.80 °C is one of the major problems of sea cage farming in Northern Atlantic coast. Hew and his co-workers developed antifreeze-resistant Atlantic salmon containing the AFP or AFGP genes using gene transfer technology.

They used genomic clone (2A-7) encoding the major liver-type AFP (wflAFP-6, previously known as (HPLC-6) from the winter flounder (Pleuronectus amaricanus) was used as a candidate for gene transfer.

Flounder AFPs belonged to the type I AFPs that are small polypeptides and high in alanine and helical content. Flounder AFPs is multi-gene family of 80-100 copies encoding two different isoforms, namely the liver type and skin type AFPs.

The liver type AFPs such as wflAFP-6 or wflAFP-8 (HPLC-8), are synthesized exclusively in the liver as prepro AFPs. In contrast the skin-type AFPs, including wfsAFP-2 and wfsAFP-3, are expressed widely in many peripheral tissues as intracellular mature AFPs.

Growth Hormone Gene Transfer:

Recently scientists have developed an "all fish" growth hormone model. They have cloned and sequenced the grass carp and common carp carbonic anhydrase (CA) gene and growth hormone gene. The grass CA gene (beta-actin) promoter has been linked to a grass carp growth hormone cDNA to form a high efficiency expression vector called pCAZ.

Using the CAT gene as receptor gene, a pCA grass carp growth hormone was microinjected into fertilized, non-activated common carp via the micropyle, generating "all fish" transgenic carp. The presence of transgene was detected by reverse transcriptase PCR and Northern blotting. These transgenic fish showed about 137% high growth rate of the control.

Disease-resistance Gene Transfer

In China scientists piloted a gene contributing resistance to the grass carp haemorrhagic virus (GCHV). Eleven different gene fragments encoding protein was cloned and isolated from translation in vitro using GCHV genomic single gene fragments.

Based on the information of capsid protein SP6 and SP7 gene cDNA, 3 oligonucleotides were synthesized and fused with SV40 MT promoter and transferred into grass carp cytokine-induced killer (CIK) cells via a constructed expression vector and transfected with GCHV. The result indicated that the mortalities were reduced by one order after challenge with the virus.

PERFORMANCE OF TRANSGENIC FISH

A transgenic fish is one that contains genes from another species. A transgenic fish is an improved variety of fish provided with one or more desirable foreign gene for the purpose of enhancing fish quality, growth, resistance and productivity.

Typically, genes of one or more donor-species are isolated, and spliced into artificially constructed infectious agents, which act as vectors to carry the genes into the cells of recipient species. Once inside a cell, the vector carrying the genes will insert into the cell's genome.

A transgenic organism is regenerated from each transformed cell (or egg, in the case of animals), which has taken up the foreign genes. And from that organism, a transgenic variety can be bred. In this way, genes can be transferred between distant species, which would never interbreed in nature.

The application of genetic engineering to animals, like potatoes with built-in insecticide, could provide numerous benefits, including the possibility of a safer, cheaper food supply and the creation of new sources for inadequate pharmaceutical resources.

With the advancement in the field of genetic engineering, the application of its commercial use has also increased. Aquatic animals are being engineered to increase aquaculture production.

The use of genetic engineering and rDNA technology has done miracles in medical and industrial research. The transgenic fish are being promoted as the first marketable transgenic animals for human consumption.

One of the most important aspects between fish and other terrestrial animals for cultivation and genetic improvement is that, usually, fishes have higher levels of genetic variation and hence more scopes for selection than most mammals or birds.

Using the gene transfer technology, scientists now have created a genetically engineered variety of Atlantic salmon that grow to market size in about 18 months, otherwise the fish takes about 24-30 months for becoming market size fish. It is also hoped that we can now modify a large number of fishes with fast growing characteristics and bring Blue Revolution.

The following are the important points needed for genetic engineering (gene transfer) to produce transgenic fish:

1. A gene sequence is to isolate for the particular characteristics; for example, growth hormone gene.
2. These genes (gene sequence) are then inserted into a circular DNA known as plasmid Vector (enzymes endonucleases and ligases are used).
3. Plasmids are harvested in the bacteria to produced billions of copies.
4. Plasmids are introduced into linear DNA. The linear DNA is sometimes called a gene cassette because it contains several sets of genetic material in addition to new inserted gene; for example, growth hormone gene. The technology is available to integrate genes in germ line of developing individual (fish) and finally transmitted into further generations.
5. Making the cassette a permanent part of fish's genetic makeup.

DEVELOPMENT OF TRANSGENIC FISHES

Development of transgenic fish has focused on a few species including salmon, trout, carp, tilapia and a few others. Salmon and trout are cash crops while the others primarily provide sources of protein. Currently, about 40 or 50 labs around the world are working on the development of transgenic fish.

About a dozen of them are in the U.S., another dozen in China, and the rest in Canada, Australia, New Zealand, Israel, Brazil, Cuba, Japan, Singapore, Malaysia, and several other countries. Some of these labs are associated with companies that expect to commercialize their fish in a few more years.

Many of the fish under development are being modified to grow faster than their wild or traditionally bred aquaculture siblings.

Faster growth is usually accomplished by transferring a fish growth hormone gene from one species of fish into another. The faster growing fish not only reach market size in a shorter time, they also feed more efficiently. Trout growth hormone (GH) was used to produce transgenic carp with improved dressing properties. Such transgenic carp are recommended for production in earthen ponds.

Transgenic Salmon

The Atlantic salmon is engineered with a pacific salmon, growth hormone driven by the arctic antifreeze promoter gene. The rapid growth of that transgenic salmon is achieved, not so much by the transgenic growth hormone as by the antifreeze gene promoter that functions in the cool water desirable for salmon flavour.

Researchers research scientists with Fisheries and Oceans, Canada, in West Vancouver, British Columbia has modified the growth hormone gene in Coho salmon by developing a gene construct in which all the genetic elements are derived from sockeye salmon.

The transgenic Coho grew on average 11 times faster than unmodified fish and the largest fish grew 37 times faster. The growth hormone levels in the transgenic fish are high year-round, rather than falling off in the winter as occurs in ordinary salmon. The modified salmon are large enough to be marketed after one year, in contrast to standard farmed salmon that do not reach market size for at least three years.

Transgenic Tilapia

Tilapia fish, native to Africa, are cultured world-wide as "poor man's food", second only to carp as warm water food fish, and exceeding the production of Atlantic salmon (whose market value is twice that of tilapia). Tilapia has been extensively genetically modified and promoted as a transgenic fish exclusive for isolated or contained production. Transgenic tilapia, which is modified with pig growth-hormone, has three times larger than their non-transgenic siblings. Tilapia genetically modified with human insulin grew faster than non-transgenic siblings, and could also serve as a source of islet cells for transplantation to human subjects.

Transgenic Medaka Fish

Purdue animal scientist used tiny Japanese fish, *Oryzias latipes* called medaka to examine what would happen if male med akas genetically modified with growth hormone from Atlantic salmon. Inserting a gene construct consisting of the human growth hormone driven by the salmon growth promoter into medaka produced the transgenic medaka.

The viability of groups of modified and conventional fish was measured at three days of age, and 30 percent fewer transgenic fish survived to that age. The researchers calculated that large males had a four-fold mating advantage, based on observations of wild-type medaka. In another experiment Silk moth genes were introduced into Medaka fish to create resistance to bacterial pathogens.

Transgenic Zebra Fish

The tiny zebra fish (Bmchydanio rerio) that lives in aquariums, was genetically modified to produce a fluorescent red pigment, and is being promoted for sale as a household aquarium pet, the "goldfish".

The goldfish caused a stir in the United States because regulation of such transgenic pets is murky and none of the major regulatory agencies: Food and Drug Administration (FDA), United States Department of Agriculture (USDA) or Environment Protection Agency (EPA), has been willing to take the lead in regulating the goldfish (even though USDA does deal with pet animals).

Researchers developed novel varieties of the Zebra fish. Three "living colour" fluorescent proteins, green fluorescent protein (GFP), Yellow fluorescent protein (YFP), and red fluorescent protein (RFP or dsRed), were expressed under a strong muscle-specific mylz2 promoter in stable lines of transgenic zebra fish.

These transgenic zebra fish with vivid fluorescent colours (green, yellow, red or orange) fluorescent proteins can be seen with naked eyes under both daylight and ultraviolet light in dark. The green fluorescent protein (GFP) is originally isolated from the jellyfish (*Aequorea tictoria*).

Auto-Transgenesis

Indian scientists are concentrating on developing transgenic fish through auto-transgenesis which involves just increasing the copies of growth hormone genes present in a fish as opposed to allotransgenesis which amounts to transfer of genes from different species.

The increase in growth homone genes leads to an increase in flesh content. Indian scientists feel that auto-transgenesis is safer and less controversial. According to researchers, the generation time of most fish species is shorter and breeding frequency is relatively higher.

A single female can produce several hundred or thousand eggs and thus provide a larger number of genetically identical eggs. Besides, the most important advantage is that the fertilization is external and can be readily controlled by experimental manipulation.

According to Pandian, "the limited availability of transgenes of piscine origin had been the major hurdle in production of transgenic fish. However, with advancements in molecular biology, more than. 8500 genes and cDNA sequences of piscine origin have been isolated, characterized and cloned in the world."

Controlled Culture of Transgenic Fish and Feed

Pond commercial culture is effective for carp and tilapia, but more difficult with salmon and trout. Currently, pond culture is suitable for carp and tilapia because the fish are vegetarians, carnivorous salmon and trout depend on a diet of fish and fishmeal but the worldwide stock of feed fish has diminished and suitable vegetable meat substitutes must be found.

Atlantic salmon (as typical cold water carnivores) cannot thrive on a diet of rapeseed oils but the fish can achieve maturity if finished with fish oils at least 20 weeks near the end of their maturity cycle.

GM oil rapeseed with enhanced production of long chain fatty acids are proposed to serve as feed for pond cultured fish. And glyphosate- tolerant GM canola meal has been pronounced substantially equivalent to non-GM canola as feed for rainbow trout.

Applications of Transgenic Fish

Transgenic Fish may be better used for the following purposes:

1. For increasing fish production to meet the growing due to demand of food due to increase in world population.
2. For production of pharmaceutical and other industrial products from piscine origin.
3. For development of transgenic native glow fish varieties for aquarium.
4. As fish biosensors for monitoring aquatic pollution.
5. For isolation of genes, promoters and synthesis of effective gene constructs.
6. For researches in embryonic stem cells and in-vitro embryo production.
7. For production of anti-freeze protein.

Environmental Concerns about Transgenic Fish

The primary environmental concerns about releases of transgenic fish, for example, include competition with wild populations, movement of the transgene into the wild gene pool, and ecological disruptions due to changes in prey and other niche requirements in the transgenic variety versus the wild populations.

Transgenic Fish could Threaten Wild Populations

Researchers have found that releasing a transgenic fish to the wild could damage native populations even to the point of extinction. Transgenic fish could present a significant threat to native wildlife.

Transgenic fish are typically larger than the native stock, and that can confer an advantage in attracting mates", Muir says. "If, as in our experiments, the genetic change also reduces the offspring's ability to survive, a transgenic animal could bring a wild population to extinction in 40 generations.

Although at Canadian research facilities, elaborate precautions are being taken to prevent the release of transgenic fish into the environment. The fish are often raised in ponds covered with nets to keep birds out; enclosed by electric fences to keep muskrats, raccoons, and humans out; and the outlets are fitted with screened drains to prevent the loss of small fishes or eggs.

GENE FLOW

One of the larger environmental concerns raised by transgenic fish is the possibility that a transgenic species raised in open water pens will escape and spread novel traits into the ecosystem by breeding with wild relatives, a biological process known as "gene flow."

Gene flow between transgenic or conventionally bred fish and wild populations is an environmental concern, because it may present a threat to natural biodiversity.

Some researchers believe that the genetic differences introduced to a transgenic fish may impact its net fitness, a scientific term meaning an organism's ability to survive and pass its genes to future generations.

The concept, which factors in characteristics such as the juvenile and adult viability of a fish, the number of eggs produced by a female, and the age at which a fish reaches sexual maturity, provides a useful barometer for discussing some gene flow scenarios. According to one scientific model, if a transgenic fish escapes and mates with a wild fish, gene flow could follow one of three scenarios:

Purge Scenario

When the net fitness of a transgenic fish is lower than that of its wild relatives, natural selection will quickly purge from the wild population any novel gene(s) introduced by the transgenic fish. In theory, evidence of the novel trait will disappear from subsequent generations.

Spread Scenario

When the net fitness of a transgenic fish is equal to or higher than the net fitness of a wild mate, gene flow is likely to occur and the genes of the transgenic fish will spread through the wild population. This means evidence of the transgenic genome would persist in subsequent generations.

Trojan Gene Scenario

When the net fitness of a transgenic fish is altered such that the fish has enhanced mating success but reduced adult viability (i.e., chances of surviving long enough to mate), introduction of that fish into the wild population could result in a rapid decline of the wild population.

Essentially mating success would ensure the spread of the novel gene throughout the population, but the inability to survive would reduce the population size of subsequent generations and potentially lead to extinction.

A declining fish population would also have secondary impacts on other aquatic species that feed on, or otherwise depend on it. Populations unable to' successfully switch over to another food source, or those whose survival or reproduction depends directly on the declining population, would also suffer.

TRANSGENIC FISH INVASIVE SPECIES

Even if they do not breed with wild relatives, transgenic fishes that escape into natural ecosystems could be an environmental nuisance by becoming an invasive species.

This danger mainly arises for those transgenic fishes endowed with new genes that improve such fitness traits as breeding capabilities and the ability to withstand harsh conditions. The establishment of a thriving transgenic fish population in an ecosystem where it has never existed could crowd out native fish populations.

RISK MITIGATION

It is important to note that developers of transgenic fish are attempting to reduce or eliminate both gene flow and invasive species risks by sterilizing transgenic fish. Sterilization is relatively easy and inexpensive but success rates are highly variable.

In addition, sterilization does not necessarily neutralize environmental risks. Academic scientists note that an escaped, sterile fish might still engage in courtship and spawning behaviour, disrupting breeding in wild populations. Waves of escaped sterile fish could also create ecological disruptions as each group is replaced by another equally strong group of transgenic sterile fish.

FOOD SAFETY ISSUES

One important food safety issue involves the extent to which fish absorbs and stores environmental toxins, such as mercury, high levels of which could pose a danger to humans who eat the contaminated fish.

Some scientists worry that discrete biological changes induced by the genetic engineering process might enable transgenic fish to absorb a toxin that conventional fish cannot absorb or to better tolerate higher levels of a toxin already known to cause concern.

Some scientists have expressed concern that the genetic engineering process could increase the allergic potential of fish, particularly through the introduction of novel proteins that never before existed in the food chain.

However, it is equally possible that genetic engineering will form their diet. Genetically engineered plant crop had faced protest in various countries regarding safety for food and environment. There is a need for the regulating transgenic animals for debate.

11

COMBINING GENETIC ENHANCEMENT PROGRAMMES

Because fish lay large numbers of eggs, in which both fertilisation and development is external, they are rather suitable organisms for transgenic induction. To date fish of many different species have been used in such work - rainbow trout, atlantic salmon, channel cat fish, tilapia, medaka, zebrafish, goldfish, carp and loach. Normally the induction of transgenics has involved injection of multiple copies of the chosen novel gene into the egg cytoplasm soon after fertilisation, followed by assays for subsequent gene integration, expression and transmission to progeny.

Genes which have been utilised for transgenic induction in fish include somatotrophin (growth hormone) of both mammalian and piscine origin, metallothionein, 'antifreeze' protein, and crytallin coding sequences. A number of laboratories have demonstrated by Southern blotting of restricted DNA fragments that genomic incorporation has been achieved, although in others there appears to be a continuing problem of long term persistance of extrachromosomal DNA material. Expression of the injected sequences tends to be at a low level and in some cases not to be detectable. This is probably due to the inefficiency of mammalian gene promoters in fish cells and will hopefully be overcome by the use of genes of piscine origin. Germ line integration and transmission to progeny of the novel sequence has been established in a few laboratories.

GENETIC DIVERSITY AND THE NEED FOR GENETIC RESOURCE MANAGEMENT

Management of FiGR is necessary for more than just increased production. Besides being essential for genetic improvement programmes in aquaculture, genetic resources are the necessary raw ingredients that allow species to adapt to short-term and long-term changes in their environment; they provide species, populations and individuals with the flexibility of dealing with and adapting to changes to their environment, changes both from humans and from natural causes. That is, genetic diversity is necessary for the continued evolution of species. Genetic diversity interacts with environmental variation to produce the variety of shapes, sizes, life-history characters, behaviour, and colours that make aquatic species so valuable and interesting. Some of these differences show up as different colours of fish or as different scale patterns, whereas other differences show up as different migration patterns or reproductive behaviour. Without genetic diversity, there would be no species diversity, no adaptation, no breeds, and no evolution; there eventually would be extinction as climate and habitats change as a result of natural or human actions.

The common carp has by far the longest history of domestication and genetic improvement for aquaculture. Farmed Atlantic salmon, channel catfish and Nile tilapia have been genetically improved more recently. However, with the success of these breeding programmes (i.e. changing the genetic structure of a wild fish) and the inevitable use of these improved breeds in many farming systems, comes the problem of interaction between genetically improved aquaculture stocks and their wild relatives. These wild relatives often support viable fisheries and will provide new genetic material that can be useful to aquaculture. The aquaculture sector is in an advantageous position to minimize extinction of the wild relatives of farmed species, as was allowed to happen to many in the livestock and crop sectors.

Management of aquatic genetic resources must have defined objectives in order to plan programmes and to judge success and impact. These objectives will depend on the purpose of the aquaculture facility: whether it is maximizing production, maximizing efficiency, reducing inputs, releasing fish for culturebased fisheries, or helping restock threatened or endangered species. Each of these objectives will require different management programmes for aquatic genetic resources.

Broodstock management: inbreeding, genetic drift and domestication

Aquaculture is not only a critical sector of food production, it is also a necessary component of recreational and commercial fisheries and a required management tool for conservation programmes. As is the case for all types of animal husbandry, aquaculture means that humans must intervene in and manage a species' life cycle. The moment this occurs, we usually produce irreversible changes in the population's gene pool. These changes can be desired, which occurs when selective breeding programmes are conducted to improve growth or when domestication creates fish that are better adapted to the hatchery environment.

Unfortunately, we also produce undesired, damaging changes in the genome through inbreeding and genetic drift, which lower viability and growth and increase developmental instability. While domestication is beneficial in food fish culture, it is harmful for fish that are stocked in the wild, because fish that are well adapted for a hatchery may be maladapted in the wild. In this chapter, broodstock management involves the control of inbreeding and genetic drift and the process of domestication.

Inbreeding

Inbreeding is the mating of relatives. Inbreeding is one of the three traditional breeding programmes, and it has been used to develop new breeds and can be used in combination with cross-breeding to produce uniform, outstanding individuals. Even though planned and directed inbreeding can be beneficial, unintentional and unplanned inbreeding will cause problems.

Genetically, inbreeding increases homozygosity in the offspring (i.e. genetic similarity) which means it also decreases heterozygosity (i.e. genetic diversity). Homozygosity is also produced when non-relatives mate and, genetically, the two forms of homozygosity are identical. Even though the two forms of homozygosity are identical, a distinction is made because of the way homozygosity is produced and its consequences.

Relatives are more alike genetically than non-relatives. Consequently, when relatives mate, they produce offspring that are more homozygous than is the case when non-relatives mate; the closer the relationship between mates, the more homozygous the offspring.

This is of concern, because all animals contain a small number of harmful or deleterious recessive alleles.8 In most cases an individual is not affected and survives because it has only one copy of the harmful allele, inherited from one of its parents (it is heterozygous); two copies of the allele are needed (one from both parents) to produce the harmful or lethal effect (it is homozygous). Because relatives are more alike than non-relatives, they tend to share the same deleterious recessive alleles. Two non–relatives might only share one or two in common, while relatives usually share more in common; the closer the relationship, the more that are shared.

When relatives mate, the pairing and subsequent expression of these deleterious recessive alleles in their offspring produces inbreeding depression–lower growth rate, viability, and fecundity and an increase in the number of abnormalities. Studies in fish have shown that inbred fish have these classical clinical signs of inbreeding depression, as well as a decreased return rate when stocked in the wild.

The negative effects of inbreeding usually do not occur immediately. Inbreeding depression is often delayed (i.e. they might not occur until several generations after inbreeding has begun). How quickly inbreeding depression occurs depends on the amount of inbreeding that has been produced and the trait.

The ideas that were described above can give farmers the erroneous idea that inbreeding is a major reason behind many of their production problems, so they come to the erroneous conclusion that inbreeding has occurred and their stock is no longer of good quality when they observe a deformed individual or yield has declined. Deformities and a decrease in yield are often due to nongenetic factors such as developmental errors, toxins, nutritional deficiencies, or weather and may not be due to inbreeding.

Because inbreeding is the mating of relatives, if individuals can be given unique tags, it is rather easy to prevent inbreeding or to minimize it by preventing parent-offspring, brother-sister, and half-sib matings. If the closest mating allowed is between second cousins, inbreeding will never become a problem.

When conducting selective breeding programmes, inbreeding is inevitable, because when you only allow the best to reproduce, you often mate relatives. Minimizing inbreeding during a selective breeding programme is important, because you do not want to use the genetic gain produced via selection simply to counteract inbreeding depression. To prevent this, a number of breeding programmes have been designed to minimize inbreeding during a selective breeding programme.

While it is important to prevent the systematic mating of close relatives in selective breeding programmes, incidental (random) matings of close relatives (e.g., brother-sister matings) in large-scale breeding programmes is not as big a problem as it is in small populations, because it is likely that the offspring produced by these matings will be mated to non-relatives in the following generation, which will produce fish with no inbreeding.

While it is easy to give livestock individual marks and thus prevent relatives from mating, it is rather difficult for fish. Therefore, aquaculturists must manage the population as a whole to minimize the accumulation of inbreeding. To do this, aquaculturists must manage the effective breeding number (N_e)

One way N_e is determined is by the number of males and number of females that reproduce and leave viable offspring:

$$N_e = \frac{4\,(\text{number of females})\,(\text{number of males})}{(\text{number of females}) + (\text{number of males})}$$

Thus, Ne is determined by the number of males that leave viable offspring, the number of females that leave viable offspring, and by the sex ratio of the brood fish. This means that N_e is maximized by increasing both the number of males and females that are spawned and by bringing the sex ratio as close to 1:1 as possible. Skewed sex ratios, which are often used in aquaculture, make N_e far smaller than the number of brood fish that are spawned. N_e is most strongly influenced by the least represented sex (e.g. when few males are used, N_e approximates the number of males rather than the total).

The reason aquaculturists and fisheries biologists need to manage N_e is that it is inversely related to inbreeding:

$$F = 1/2N_e$$

where F is the amount of inbreeding produced (0-100%) in a single generation; F is the percent increase in homozygosity. This formula shows that as Ne decreases, F increases; N_es.

In a closed population, once inbreeding has occurred, it lowers the N_e of the next generation:

$N_eF = N_e /1 + F$

where N_eF is the effective breeding number in a closed population with F >0%. For practical purposes, the total F that is produced over a series of generations can be calculated by summing the F that is produced in each generation, without considering previous inbreeding.

How big should N_e be to minimize inbreeding? Unfortunately, there is no universal value of F that aquaculturists or fisheries biologists want to avoid, which means there is no universal N_e. N_es ranging from 30-500 have been recommended, 50 being the most common. To calculate a desired N_e, one must determine what level of genetic risk is acceptable; in this case, it is the maximum amount of inbreeding that is desired after a given number of generations. In addition, the concern about inbreeding, whether N_e needs to be managed, and how large N_e must be depends on the purpose/goal of the hatchery or fish farm, if fish are spawned, and how broodstock will be managed.

SEX REVERSAL AND TRIPLOIDY

In simple terms, a triploid fish is merely a fish that is sterile. Triploid fish have three setsof chromosomes, unlike a fertile fish that have two sets of chromosomes (a diploid fish).Triploids are common in many industries; Seedless watermelons are triploids, as are bananas. Triploids are also naturally occurring.

A triploid fish is not a genetically modified organism. The genes of the fish have not been manipulated or changed in any way and do not result in the expression of any foreign or novel proteins or tissues. Triploid fish simply have 3 sets of chromosomes instead of 2 and for this reason are not able to reproduce, but are otherwise normal fish.

Recent legislation codified in the Fish and Game Code describes the rational to use sterile fish for recreational purposes. These include the conservation of native fish genotypes (genetics) and the protection of endangered or threatened species. For these and other reasons the California Department of Fish and Wildlife is producing triploid trout eggs that hatch into fish, are raised, and ultimately result in sterilized troutreleased for recreational purposes.

Triploid trout are created by forcing the egg to retain a chromosome that is normallyejected during egg development. There are many ways to do this.

As you can see in Fig. 1 below, in fertilized trout eggs, normally a chromosome (N) is kicked out of the egg as a polar body at some stage of development. Using pressure treatment at a specific time in the egg development, the polar body and chromosome is retained. With 3 chromosomes the fish is sterile and cannot reproduce.

Triploidy

Diploid (normal fish): 2 chromosome sets (2n)

Triploid: 3 chromosome sets (3n)

- larger cells

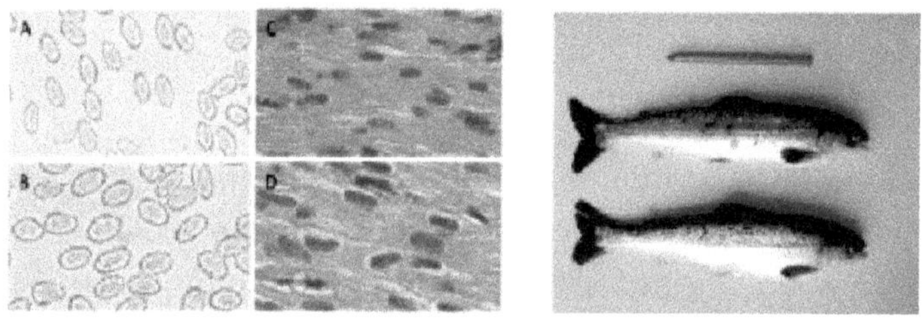

Fig. 1. Triploidy

Applications

The major consequence of triploidy is sterility.

Triploid fish are functionally sterile

Benefits of Triploids

- Sterility mitigates the problems associated with:
- Early sexual maturation
- Genetic interactions with wild fish
- Early sexual maturation
- Early sexual maturation is a major production problem for many commercially
- important aquaculture species, including salmonids, sea bass, Atlantic cod.
- Early sexual maturation results in:
    - **Reduced growth:** energy is diverted from somatic to gonadal growth
    - Reduction in flesh quality
    - Reduced immunocompetence

Mitigates the problems associated with escaped fish

- Farmed fish escape at every level of the production cyclempact of farmed fish on wild fish: is a major environmental concern
- Genetic introgression results in reduced fitness of wild fish and potentially can accelerate extinction of local populations
- Farmed fish escape at every level of the production cycle
- Impact of farmed fish on wild fish: is a major environmental concern

- Genetic introgression results in reduced fitness of wild fish and potentially can accelerate extinction of local populations
- Sterility would not solve all the problems associated with escapees but would mitigate the more serious environmental concerns (genetic introgression) Fishery Management (stock enhancement)

How to produce triploids?

The key to the original process of triploidy induction involves what is referred to as the second polar body, an excess (and expendable) set of chromosomes present in the egg prior to fertilisation. Once a sperm cell enters the egg, internal mechanisms are triggered to expel the second polar body prior to the fusion of chromosomes from the sperm and the egg nucleus. In many aquatic species, triploids were first produced by fertilising eggs under controlled conditions and applying a shock at just the right time to interfere with the expulsion of the second polar body, resulting in an organism with three chromosome sets.

Unfortunately, it became apparent that, in many species, the shocks utilised on newly fertilised eggs could result in reduced viability of the resulting fry. However, an alternative strategy for producing triploids soon emerged, also based on applying temperature or pressure shocks to newly fertilised eggs, but a bit later after fertilisation. These shocks were timed to interrupt the first cell division of a developing embryo, resulting in a tetraploid animal with four sets of chromosomes. In some cases, these organisms can be raised to maturity and mated successfully with normal (diploid) individuals to generate all-triploid offspring, referred to as interploid triploids.

This approach avoids the practical limitations of shocking individual batches of eggs and allows for the production of triploid offspring in far greater numbers. While tetraploids themselves often tend to underperform compared to diploids, their interploid offspring generally exhibit few or no physiological problems. Many species lend themselves to the interploid strategy, but it has proven problematic (or virtually impossible) in some cases, because viable tetraploid breeding stock is required.

Let's consider the application of interploid triploidy in some of the major aquaculture species groups. Carps remain the most important group of cultured fishes around the world, but the interploid approach to triploid production has not been widely adopted by carp culturists. Tetraploid bighead carp and grass carp were produced in Florida in the late 1980s, but the fry exhibited low, or no, survival. Researchers in India reported on the successful induction of tetraploids in the Indian major carps rohu and catla around the same time, apparently producing viable fry. Early efforts to produce tetraploid common carp had some success, but once again the resulting tetraploids died in their early life stages.

Some interesting and unexpected ploidy levels can result when certain carp species are hybridised. One line developed from a cross between red crucian carp and common carp has been shown to be all tetraploid, producing interploid triploids when crossed with either of the original parental species. Researchers in China have utilised this approach, in combination with other strategies, to achieve mass production of all-female interploid triploids.

While catfish have become the second most cultured group of fishes, the interploid triploid strategy has not been utilised to any great extent for these species either. Tetraploid *Clarias batrachus* were produced using cold shock at Kasetsart University in Thailand in 1988, and tetraploid *C. gariepinus* were produced in Hungary a decade later. Some early trials with channel catfish in the mid-1980s demonstrated useful protocols to produce tetraploids, but the industry never really followed up on this finding. Further work two decades later used different shock treatments, but success was still limited.

In channel catfish, harvestable size is attained prior to the onset of maturity, so sterility is not an important issue in terms of feed conversion or product quality. Additionally, concern over escapees is pretty much non-existent anywhere the species is currently cultured. This is also the case for many other species of catfish.

Much of the early research involving interploid triploids in aquaculture was focused on salmonids. Although many studies have been published on the use of interploid strategies in trout and salmon, shock-induced triploids are still the commercial approach for Atlantic salmon farmers. As a result, animal welfare concerns relating to the use of triploid salmon have been raised in several countries. Triploid salmon have been shown to suffer from elevated levels of skeletal deformities and cataracts, although diet and management can offset these problems to some extent. Some studies, however, have shown them to display disease resistance equal to that of their diploid counterparts.

Cultured molluscs and crustaceans offer huge potential for the adoption of commercial-scale interploid triploid culture, but the situation varies significantly between these major invertebrate categories. Although the mechanisms for producing tetraploid molluscs are much more complicated than those developed for fishes, the production of interploid seed is expanding rapidly around the planet, especially in oyster culture. In contrast, polyploidy is extremely difficult to attain in shrimp, and after some initial research the concept has largely been abandoned.

That leaves one last major group of farmed fishes to consider: tilapia. Tilapia are arguably the species most in need of techniques for producing sterile fingerlings. Since they were first cultured centuries ago, uncontrolled reproduction and early maturation have sullied the productive potential of various tilapia species and their hybrids. It is no surprise, then, that for several decades researchers have pursued the holy grail of tilapia breeding: tetraploid broodstock that could produce all-triploid offspring on a commercial scale. But results have been disappointing.

Sex reversal is a biological process whereby the pathway directed towards the already determined-sex fate is flipped towards the opposite sex, creating a discordance between the primary sex fate and the sex phenotype expressed. The process of sex reversal occurs during embryonic development or before gonad differentiation.

Fig. 2. Tilapia

In GSD species sex reversal means that the sexual phenotype is discordant with the genetic/chromosomal sex. In TSD species sex reversal means that the temperature/conditions that usually trigger the differentiation towards one sexual phenotype are producing the opposite sexual phenotype.

Sex reversal can occur naturally, by mutations, or can be induced artificially. Sex reversal can be genetically or hormonally induced in laboratory. It can also occur artificially by exposure to endocrine disruptors such as pollutants, including herbicides, which can act as estrogen promoters or inhibitors, for instance by altering aromatase expression.

In gonochoristic fishes, the sex can be determined genetically, environmentally or by a combination of both. In fishes, primary sexual fate can be susceptible to alteration by hormones exposure and multiple environmental factors, such as population density, water pH, or temperature. Those conditions can affect the gonad development and differentiation, which can lead to sex reversal. In medaka fish, where sex reversal has been documented show a shared gene related to normal male development, the *dmy* gene. In wild populations, genetically female fate can be phenotypically reversed to males if they carry the *dmy* gene or a mutated *dmy* gene and genetic males can be reversed to females if they lack the *dmy* gene.

GENETIC ENGINEERING AND CROSSBREEDING

Genetic engineering (also called genetic modification) is a process that uses laboratory-based technologies to alter the DNA makeup of an organism. This may involve changing a single base pair (A-T or C-G), deleting a region of DNA or adding a new segment of DNA. For example, genetic engineering may involve adding a gene from one species to an organism from a different species to produce a desired trait. Used in research and industry, genetic engineering has been applied to the production of cancer therapies, brewing yeasts, genetically modified plants and livestock, and more.

Several terms are used to describe genetically engineered animals: genetically modified, genetically altered, genetically manipulated, transgenic, and biotechnology-derived, amongst others. In the early stages of genetic engineering, the primary technology used was transgenesis, literally meaning the transfer of genetic material from one organism to another. However, with advances in the field, new technology emerged that did not necessarily require transgenesis: recent applications allow for the creation of genetically engineered animals via the deletion of genes, or the manipulation of genes already present.

To reflect this progress and to include those animals that are not strictly transgenic, the umbrella term "genetically engineered" has been adopted into the guidelines developed by the Canadian Council on Animal Care (CCAC). For clarity, in the new *CCAC* guidelines on: genetically-engineered animals used in science the CCAC offers the following definition of a genetically engineered animal: "an animal that has had a change in its nuclear or mitochondrial DNA (addition, deletion, or substitution of some part of the animal's genetic material or insertion of foreign DNA) achieved through a deliberate human technological intervention."

Those animals that have undergone induced mutations (for example, by chemicals or radiation — as distinct from spontaneous mutations that naturally occur in populations) and cloned animals are also considered to be genetically engineered due to the direct intervention and planning involved in creation of these animals.

Genetic engineering applied to aquaculture and fish in particular aims at increasing growth rates and feed utilization, increasing environmental tolerance and resistance to diseases, controlling reproduction, and improving food quality characteristics. This development is entirely in line with the development that has taken place using traditional breeding methods. However, new aspects have been added by genetic engineering. First, transfer of genes is possible between species which do not hybridize naturally. Second, genetic traits from one species can be introduced into another species 'in one shot', rather than by the elaborate breeding schemes of traditional breeding.

Techniques for gene transfer

The methodology for gene transfer to fish has been reviewed by various researchers. What follows is an outline of the techniques reviewed in those papers; the reader is referred to the original publications for a more technical treatment of the subject.

Microinjection

Direct microinjection of cloned DNA into fish eggs is by far the commonest method for introducing genes into the germ—line of fishes. Using this method, transgenic individuals have been generated in several fish species, including rainbow trout, Atlantic salmon, common carp (*Cyprinus carpio*), loach (*Misgurnus anguillicaudatus*), channel catfish (*Ictalurus punctatus*), tilapia (*Oreochromis niloticus*), goldfish (*Carassius auratus*), zebra fish (*Brachydanio rerio*), medaka (*Oryzias latipes*), walleye (*Stizostedion vitreum*) and northern pike (*Esox lucius*).

In mammals, the foreign DNA is microinjected into the male pronuclei of the fertilized eggs. In fishes, the male pronuclei are not visible (except in medaka) and the foreign DNA has to be microinjected into the egg cytoplasm. Usually, 106 copies of the gene construct are injected into each egg. The injection is carried out shortly after fertilization, usually between the one—cell and four—cell stages.

The eggs of many fish species have a tough outer membrane (chorion) which makes microinjection with a glass needle difficult. In order to tackle this problem, an opening in the chorion has been made by microsurgery before the glass needle is inserted, or the glass needle has been inserted through the micropyle, which is the opening through which sperm enters the egg during fertilization. Other methods include removing the chorion using a forceps or digestion by trypsin, or performing the microinjection before the fertilization and subsequent water hardening of the chorion take place.

Cross-breeding

Cross-breeding is another well-known means of genetic improvement which has application in aquaculture. Heterosis or hybrid vigour enables an offspring to surpass its parents for one or more traits. On the other hand, inbreeding depression caused by mating of closely related parents has a deleterious effect on the progeny.

The inbreeding measure is the coefficient of inbreeding, incorporating the degree of the animal's homozygosity. It shows what part of the genes in a group of individuals are in the homozygous state. Close inbreeding, especially sib mating (brothers and sisters and parents with offspring), causes homozygosity to increase rapidly, to as much as 0.9–0.95 (expressed as fractions of one) or even more. In most animals, inbreeding results in inbreeding depression, characterized by a drop in viability, growth rate and often fertility. Such depression has been observed by many workers in common carp, brook trout and other fish species. Outbreeding, on the other hand, is accompanied by heterosis in growth rate and viability, especially when fish from different high inbred groups are crossed. The main types of crossings undertaken are:

a. Commercial crossing, directed towards breeding of the first-generation hybrids for commercial purposes. Only the first generation, that has the heterosis of productive qualities or incorporates the advantageous characteristics of both the parental forms, is used. They are maintained for further reproduction.
b. Synthetic or distant crossing, in which distant parents, including those of inter-generic origin, are crossed to develop a new breed, in the course of

long selection. It may attempt to combine the qualities of parents of several breeds, species or even genera. Such crossings to produce new breeds should ensure the preservation and perfection of the productive qualities of the breed, the preservation of genetic variability and the prevention of inbred depression.

Distant outbreeding is indispensable in the selection of aquaculture species. The aims of such crossings are as follows:

1. an overall increase in genetic variability, resulting in an increase in selection response;
2. achievement of a combination of charac teristics of two or three breeds or two (rarely three) species;
3. improvement of the productive quality of the local breed by making use of the few valuable traits of another breed;
4. increase in the viability of the breed by intro ducing genes responsible for resistance to environmental factors and diseases.

Researchers describes different cross- breeding patterns to achieve these aims. Reproductive crossing is suggested when valuable properties from both parents are to be combined in the hybrids. It can be done with complete fertility of the hybrid and requires only meticulous selection in the subsequent generations.

Introductory crossing will be advantageous when one or only a few characteristics from a breed have to be incorporated in the hybrid. Each generation of the hybrid has to be crossed with the local breed and so there is the risk of losing the useful characteristics of the improved breed in back crossing, particularly in the case of polygenic inheritance of properties selected.

This type of crossing is of considerable use in selection for increased resistance to certain dis- eases, which is often dependent on the presence of one or a few genes. These genes can be pre served by means of proper selection in each generation.

Absorptive crossing differs from introductory crossing only in that its purpose is a nearly com plete substitution of the local breed of geno- type by the genotype of the improved breed. Only some features of the local breed, such as viability, are preserved.

Alternate crossing is the most complicated system which is most useful when a combina tion of many characteristics from two breeds with polygenic inheritance is required. It allows the preservation of high genetic variability through a number of generations. Selection effi ciency is kept at a high level owing to this variability and does not reach a plateau. The mainproblem of obtaining new hybrid breeds by means of crossing (interspecific or intergeneric) is their complete or partial sterility, which takes a lot of time-consuming work to overcome.

A number of breeding systems have been proposed to utilize completely the advantages associated with heterogeneous crossings. Paral lel breeding of two or more groups within a breed is possible when working with slowly maturing species, without intermingling, allow ing a moderate inbreeding among each and carrying out selection in each generation. In breeding in groups for family selection, a large number of crossings from different groups are carried out for each generation. The parents producing the best offspring are used for subsequent commercial crossings. This systemsuffers from the drawback that the genetic variability gradually decreases during family selec tion. Researchers recommend that a reserve group of a sufficient number of individuals should be kept for each group when selecting two groups marked by certain genes.

In case of a drop in genetic variability, addi tional gene pools can be introduced into the exhausted group.

Another possible system is alternate in- breeding and outbreeding. After two or three generations of close inbreeding, the evaluation of hybrids from different inbred lines is per formed. The best combinations are used for commercial rearing and among the offspring new inbred lines are established. Linear selec tion involving inbreeding for superior ancestors and top cross, where crossing is done between the best inbred individuals (say males) and individuals from the outbred population (say females) to preserve the genetic variability, are other methods that have applications in aquaculture.

One of the most complicated techniques of breeding is reciprocal recurrent selection (RRS), where the combining capacity of the parents from each of the two groups is evaluated by means of a cross with parents from the other group. The individuals thus selected are reproduced without recrossing and their off- spring again tested for combining potential. The basic feature of all the systems described is the utilization of heterosis in crossing individuals from different groups, lines and breeds.

Along with this, moderate to very close inbreeding is employed. The most appropriate system would naturally depend on the species and the traits that are of importance in commercial culture. Researchers , however, proposed a cross-breeding scheme for fish farming as summarized below:

a. Test all possible crosses between different strains or species for the economic traits in question and select crosses that are likely to give useful results. It may be better to use strains with very different origins and which, in combination, have favourable traits.
b. Develop inbred lines and test the crosses under natural conditions to find the most valuable cross for farming.
c. Start an RRS programme to ensure continuous genetic improvement, utilizing both general and specific combining abilities simultaneously.

The pros and cons of crossbreeding and hybridisation in aquaculture

Heterosis is based on what we call dominance genetic effects, and these effects result from interactions between the pairs of alleles at all of the genes that influence a characteristic of interest. Indeed, the utilisation of dominance genetic effects is fundamentally based on combinations at every level – of species, of lines within species, and occasionally even of individuals within a population.

While utilising heterosis can be an attractive strategy for improving traits that exhibit low heritabilities, any advantage that might be derived is only temporary. Since these interactions depend on specific combinations of alleles, they cease to exist during the formation of eggs and sperm. So heterosis is, unfortunately, not something that can be inherited or passed along from generation to generation. As explained in more detail below, sometimes the effort is more trouble than it's worth, but at times it can result in great gains in productivity.

HYBRIDISATION VS CROSSBREEDING (A FUZZY DISTINCTION)

Hybridisation, in the strictest sense, involves the crossing of related species. Some examples include crossing striped bass with white bass, channel catfish with blue catfish, or Nile tilapia with blue tilapia. In contrast, when distinct stocks, lines or varieties of the same species are crossed, the most appropriate term is crossbreeding, with the resulting offspring referred to as crossbreds. Nonetheless, some researchers use the term hybrid even when both parental lines are from the same species. In fairness, the distinctions between strains, races and species are often blurred to a point where crossbreeding or hybridisation are essentially indistinguishable.

Sometimes, producing these crosses is easily accomplished, but occasionally a lot of work is involved. Often, progeny must be evaluated on a cross-by-cross basis to determine how the alleles involved will interact to impact production performance.

And, while most tilapia species and synthetic strains have little or no inhibitions about mating outside of their parental lines, many hybrids can only be produced through induced spawning and/or artificial fertilisation. Depending on the cross one is pursuing, it may be necessary to capture wild broodstock or resort to hormonal injections or implants.

Fig. 3. A hybrid sunfish

Whenever hybridisation or crossbreeding is to be practiced season after season, the need to maintain access to each parental line is unavoidable. With this need comes the requirement for adequate facilities to house sufficient numbers of individuals. Care must also be taken to avoid inadvertent mixing of distinct lines within species. So why bother? The resulting benefits can include not only heterotic gains but also the combination of desirable traits from parental lines. Occasionally the production of monosex offspring is also possible by crossing closely related species that exhibit distinctly different sex-determining systems.

Experimental design

Once they are produced, evaluating hybrids or crossbreds of aquatic species can usually be a fairly straightforward process. The best approach to these evaluations involves a design that includes both pure lines and their reciprocal crosses (line 1 females by line 2 males, and line 2 females by line 1 males). Even more complex designs, involving multiple parental lines, can be accommodated if suitable incubation and rearing facilities are available. Mating designs involving more than two lines are typically referred to as diallel crosses.

In many hybrids, significant differences are apparent between reciprocal crosses. These can involve cytoplasmic inheritance from mitochondrial DNA, or more direct influences which may have genetic bases, such as egg size or quality. Inherent skill in caring for fertilised eggs (eg catfish, tilapia, macrobrachium) or fry also comes into play when evaluating maternal effects, and maternal effects themselves can exhibit heterosis.

Some recent research

In a recently published study, researchers in Cameroon evaluated reciprocal hybrids between their native catfish *Clarias jaensis* and the non-native *C. gariepinus*. Crossing female *C. jaensis* with male *C. gariepinus* resulted in the highest fertilisation and hatching rates, as well as the highest larval survival rate and lowest rate of deformities, while the reciprocal cross demonstrated poor performance. Growth of the *C. jaensis* × *C. gariepinus* hybrid (the female parent should always be listed first) was comparable to that of pure *C. gariepinus* from day 4 through day 32 but survival was slightly lower, probably as a result of greater size variation within the offspring.

Similarly, research in China documenting heterosis in grouper hybrids from a cross between *Epinephelus awoara* females and *E. tukula* males was recently reported. At 11 days post-hatch, hybrid survival was roughly 40 percent, while that of pure *E. awoara* was only 5 percent (pure tukula was not available for the analysis). At 4 months of age, hybrids weighed 24.1 g on average compared to 21.3 g for *E. awoara*. By 13 months, however, this modest advantage had increased to 409 g vs 129 g.

At times, heterosis is expressed as simple hardiness, and tolerance of marginal conditions. My colleague Dr Amina Zuberi was part of a team of researchers that recently examined the thermal tolerance of hybrids between the major carps *Labeo rohita* and *Catla catla* in Pakistan. *L. rohita* exhibited the highest maximum critical temperature and highest acquired thermal tolerance, followed by hybrids, while *C. catla* exhibited the highest minimum critical temperature and the lowest acquired thermal tolerance. Although hybrids exhibited limited heterosis (< 11 percent), the authors recommended replacing *C. catla* with hybrids in culture environments with relatively high or low temperatures.

Hybrid vigour is not just expressed in fish, it is also commonly seen in aquatic invertebrates. A group of researchers from Guangzhou examined the performance characteristics of two populations of the boring giant clam (*Tridacna crocea*) and their reciprocal hybrids. The two parental lines used in the study are separated by some 600 km of open sea, originating from Zhaoshu and Huangyan Islands.

Results from three trials, each including both parental lines and crosses, indicated that growth heterosis was apparent in both the spat and adult phases of culture. On day 15, heterosis for metamorphosis rate was approximately +26 percent and by day 360 heterosis for survival was 6.2 percent.

Combining traits

At times, the goal of crossing species or lines is not to capitalise on heterosis but rather to combine certain traits, such as the fast growth of pure striped bass with the hardiness of their smaller relatives, the white bass. Or, in the historic Taiwanese red tilapia, the red colouration of Mozambique tilapia with the faster growth and superior dressout of Nile tilapia.

Hybrids of the carps *Labeo fimbriatus* and *Catla catla* have been produced for decades to combine the relatively small heads of the former with the deep bodies of the latter, resulting in greatly improved meat yield.

Fig. 4. The sturgeon hybrid known as the "Bester" offers

The sturgeon hybrid known as the "Bester" Fig. 4) offers a classic textbook example of crossing two aquatic species in order to combine desirable traits. The Bester was first produced in Russia by Dr NI Nikolyukin in 1952 by crossing *Beluga sturgeon* (*Huso huso*) with sterlet (*Acipenser ruthenus*) in order to combine the Beluga's fast growth with the high-yielding caviar production of the sterlet.

The Bester is not only illustrative of the productive combination of desirable characters in a first generation hybrid, but also of how to develop a synthetic strain of fish. Beginning in Japan, first generation Bester were crossed among themselves to produce second generation fish (since parental lines were not available), and this strategy was subsequently adopted in a number of locations. As a result, not all Besters are created equal. Some Bester populations are now 7 generations removed from the original hybrids, and are routinely used for backcrossing with pure species. Some Russian and Ukrainian scientists now refer to the Bester as *Acipenser nikoljukini* in honor of Dr Nikolyukin. Similarly, most lines of red tilapia found throughout the world are the result of crossing first generation hybrids among themselves in relatively closed populations until certain desirable characters become fixed. Accordingly, a red tilapia from Colombia may be notably different from one in Thailand.

GYNOGENESIS, SELECTION AND HYBRIDIZATION

Gynogenesis

Gynogenesis is a method for producing fish in which all the genetic information originates from the female parent, i.e., from the egg. In salmonids, gynogenesis is accomplished by the fertilization of the egg with UV-irradiated sperm.

Gynogenetic diploid embryo contains a diploid chromosome complement, also derived from mother only and are obtained by treatments to prevent either the second division in an egg fertilized with UV-irradiated sperm.

In fish, the gynogenesis is achieved both naturally and artificially. In some fishes, gynogenesis is a natural mode of reproduction, these unisex fish may be often triploid. In a few fishes such as silver crucian carp (Prussian crucian carp of European goldfish) and Poeciliapsis, unisex form present in natural population.

Population, which are produced by natural gynogenesis. Artificial gynogenesis can also be produced under suitable experimental conditions. During artificial gynogenesis, manipulations in event of fertilization are made by destroying DNA of the spermatozoa and to restore diploidy to the ovum with either meiosis or cell cleavage.

In this process the male inheritance is excluded, the offspring obtained in this way possesses inheritance of female characters and they are of only one sex, i.e., female. The unisexual population obtained by artificial gynogenesis are used for fisheries management.

Gynogenesis can also be used to produce clonal lines of tilapia. Hybrid among such inbred lines are expected to exhibit significant heterosis and uniform performance with respect to growth rate and body conformation.

The most important breeding goal of tilapia is to improve growth rate and feed conversion efficiency, which are prime factor regulating profitability of the industry. Sex determination mechanisms, genetically male tilapia have higher growth rates than even hormonally reverse female.

Researchers observed that gible carp, Carassius gibelio exhibits gynogenetic reproduction by utilizing the sperm of other species which resulted in the increase population in the lake as gible carp females have little opportunity for mating with gibel carp male.

According to them Ylikiensis roach is the best available sperm donar option for gynogenetically gibel carp of the lake. These authors made artificial fertilization of gibel carp eggs with sperms of other species such as ylikiensis roach (CrXr), gold fish (Cr.XG) adb common carp (CrXC), hatching occurred after three days at 14 °C in all groups.

Selection

Aquaculture is predicted to play a major and ever increasing role in meeting human needs for animal source food. In terrestrial animal and plant species genetic improvement programmes have made a substantial contribution to agricultural productivity and viability. As a result of decades of selective breeding and domestication in these terrestrial species, thousands of genetically distinct breeds, strains and varieties now exist worldwide and are collectively termed genetic resources.

Theoretically, the principles of quantitative genetics and the studies of the variance components are the same which applied in the aquatic organisms, however the progress genetic obtained until now it is so superficial in comparison with terrestrial animals and plants. Although poorly exploited, fish have advantageous characteristics for breeding programs compared to terrestrial domestic animals, such as the high fertility and external fertilization, which guarantees flexibility in the definition of mating with the formation of full-sibs and halfsibling groups. As disadvantages, this animal group presents young forms much small, such as larvae and small fingerling, preventing early identification, with an increase in the cost of the program due to the need to grow whole-sibling families in separate structures, to a size sufficient for marking. This could provoke the emergence of the family effect among them (nongenetic, non-additive and maternal effects) and the rate of inbreeding would be easily increased by high prolificity of the animals.

According researchers, less than 20% (or even less than 10%, according to researchers of world aquaculture production originates from genetically improved stocks. The existence of few genetic improvement programs in the sector is due to three main reasons: (1) little information about the reproductive cycles of several cultivated species; (2) regular capture of wild specimens for breeding, which compromises the domestication of the species; (3) researchers, extensionists and fish farmers with little knowledge about quantitative genetics, theories and breeding programs.

Efficient breeding programs will be crucial to the development of fish farming, not only to achieve the necessary global demand, but also to reduce production costs, improve the resistance of farmed organisms to disease, improve food use and product quality [5]. Aquatic animals allow the implementation of several approaches to genetic improvement.

These include the hybridizations and cross-breeding, chromosome manipulation, sex control, transgenesis and selective breeding.

But of all existing genetic approaches, only selective breeding offers the opportunity of continued genetic gain, that the gain made can be permanent, that it is the only approach in which the gain can be transmitted from generation and that gains in a nucleus can be multiplied and expressed in thousands or millions of individuals in the production sector. The term selective breeding (or artificial selection) describes the various methods by which humans derive organisms with genetically based desirable traits through the use of chosen parents to make controlled crossed.

The aim of selection is to identify and select as parents for the next generation the individuals whose progeny, as a group, have the highest possible additive genetic merit for the trait or traits in question. This is almost equivalent to selecting as parents those whose own additive genetic merit is highest. The basic effect of selection is to change the gene frequencies, and the effects that can be observed are changes of the population mean.

According to researchers, one way to describe a selective breeding program is to document its breeding scheme. In the simplest and often least costly designs, such as individual selection, pedigree information is not required and the number of selected traits is therefore limited. In contrast, so-called family-based design require the family origin of the breeding candidates to be traced, which makes them more flexible regarding the number and the type of traits that can be selected. The advancement of molecular markers and their application in traditional breeding has led to the emergence of novel breeding approaches such as marker-assisted selection (MAS) and, more recently, genomic selection. The latter method in particular represents a great opportunity to improve selection routines and methodologies, although the technical sustainability of its implementation needs to be carefully assessed.

Breeding goals, the number and type of selected traits, are another essential feature of a breeding program. The traits of interest can be recorded directly on candidate breeders, as in the case of growth, morphology and some reproduction-related traits or on their sibs when the traits measurement requires a fish to be killed, assessment of disease resistance of processing yield. When a number of traits are targeted simultaneously, they are combined into an index in multi-generation and multi-trait selection, with each trait weighted according to its economic value.

In aquaculture, the traits of economic importance in aquaculture are generally quantitative, governed by a large number of genes. This is particularly evident with genes with additive effects. The effect of changing the frequencies of the favorable alleles at these genes can be observed as a change of population mean for the trait under selection. Researchers point out that the objectives of selection of breeding programs for aquaculture organisms can be changed with certain ease, including or withdrawing candidate characteristics, redirecting the program to new horizons quickly.

According to the researchers, this is possible due to the great additive genetic variability of the characteristics of interest in the aquaculture populations, the reproductive precocity of several species and their prolificacy. These aspects accelerate the annual genetic gain of a particular trait, since the high prolificacy is related to the increase of the degree of selection intensity, the precocity, the short generation interval and the additive genetic variability, the heritability.

In general terms, two traits are particularly important in aquaculture: the time required to reach commercial weight and the volume of food the animal will consume to reach that size. However, none of them are subject to direct evaluation, since the first former is replaced by weight or length at a fixed age and the second by weight gain per unit of time until that age. In this perspective, we can make the evaluation of these traits more easily, because the description quantification of this data, in both traits, having high correlation with each other.

Considering the complexity and the importance of the applied to these techniques on the production of fish, we review the main selection methods that have been developed for genetic improvement in aquaculture, and made a short discuss about their virtues and shortcomings. The efficiency of each method can be predicted by calculating the expected genetic response for a given set of parameters, which will partly dependent on how accurately the breeding values of individual animals will be evaluated.

According to researchers, the choice of selection method for a given situation (species, production environment, scale) will depend on a range of importance factors, which including:

- Target traits for which genetic improvement is desired
- Feasibility of recording the trait on live animals
- Magnitude of heritability for the traits in question
- Reproduction capacity of the species.

Individual or mass selection

The terms individual selection and mass selection are often used interchangeably, and they refer to selection solely based on the individual´s phenotype. This is a well-known and widely used method of selection in animal breeding, and for several aquaculture species. Can be used on characteristics that can be measured in the candidate animal itself, as weight gain, being most efficient in characteristics of high heritability.

Individual selection is usually the simplest method to operate and in many circumstances, it yields the most rapid response. It is does not require individual identification or the maintenance of pedigree records, hence, it may be considered the least costly method. In principle, it can produce rapid improvement if the heritability of the trait(s) under selection is high. For growth rate and morphological traits (easily assessed, expressed in both sexes), It can be suitable. By contrast, individual selection is not suitable for situations in which the estimation of breeding values requires slaughter of the animals (carcass and flesh quality traits) of challenge of some sort (selection for salinity tolerance for disease resistance).

When applying individual selection, it is necessary that the evaluated animals are under equal environmental conditions; therefore, it is very important that the environmental conditions have been controlled. In this case, the influence of external conditions should be minimal and kept the same for all individuals that are to be compared at any stage of the life cycle.

In this manner, the probability of correct ranking of the individuals based on thus genetic merit is maximized. Differences between individuals or groups of individuals for environmental factors like water temperature and salinity, different tanks, ponds or cages, density, light condition, and type of food and feeding regimes, may reduce the accuracy of the selection substantially, and in that way, can also reduce the possibility for genetic improvement.

In species with total spawning, it is not difficult to establish at the same time the same conditions for all candidate animals, since all hatching occur in a small-time interval. In species with split spawning, in which breeding herds reproduce asynchronously, it is important to form contemporary groups and select within them. To obtain as equal environmental conditions as possible all individuals to be compared should be hatched at the same day or within a few days' period and thereafter reared under identical environmental conditions. Therefore, it is important to form groups of animals with similar ages, grown in the same production systems, with similar control of water quality, storage density, handling and feeding.

However, mass selection may be unsuitable if there is large uncontrolled systematic environmental variation (for example age differences). Additionally, there is no control of inbreeding with mass selection, and this has caused problems in a number of fish breeding programs.

The lack of pedigree information often leads to a rapid increase in rates of inbreeding, especially in species where reproduction occurs naturally in a growing environment, such as tilapia. Normally in these situations, few dominant males reproduced with existing females. In tilapias, there is the possibility of contouring this problem with the practice of collecting eggs in the mouths of females, kept in hapas with males. In this way, it is easier to separate animals of both sexes that have already reproduced and contributed genetically to the progeny to be selected.

Family selection

Family selection refers to a selection method in which family groups are ranked according to the mean performance of each family and whole families are saved or discarded. The difficulty of the individual marking of the animals (which is still a fact in crustaceans) led to the recurrent use of these methods in fish farming, with the constant impediment of the emergence of the family effect.

The efficiency of family selection rests on the fact that the environmental deviations of the individuals tend to cancel each other out in the mean value of the family. Accordingly, the phenotypic mean of the family comes close to being a measure of its genotypic mean. The advantage of family selection over other types of selection is greater when environmental deviations constitute a large part of the phenotypic variance. This method has been widely used in fish farming for the selection of low heritability characteristics, in this situation; the use of family mean gives an increased accuracy when estimating breeding value.

Another great advantage in used family selection, it based on the programs where need improvement traits that cannot be currently measure on live individuals, like product quality traits, in this case, the use of family information is essential, recording these traits on sibs makes it possible to estimate breeding values with high accuracy. The improvement of traits like carcass quality and disease resistance only can be included in the breeding objective when applying family selection. Family selection is also far more effective than individual selection for threshold traits such as age at sexual maturity, particularly at low frequencies or incidence of the trait.

On the other hand, environmental variation common to members of a family impairs the efficiency of family selection. If this component is large it will tend to swamp the genetic differences between families and family selection will be correspondingly ineffective. Another important factor affecting the efficiency of family selection is the number of individuals in the families; the family size. The larger the family, the closer is the correspondence between mean phenotypic value and the mean genotypic value. The high reproductive capacity in aquatic animals makes family selection important for these species.

The other point is the difficult to keep the rate of inbreeding low and the intensity of selection high, the number of family groups bred and measured should not be smaller than 50-100. In the period prior to tagging, the family groups have to be kept separately. This makes family selection costly in terms of space. If breeding space is limited in this period, the intensity of selection that can be achieved under family selection may be quite small.

According to researchers, to apply family selection, it is necessary to know the parentage of each individual and hence important to maintain good pedigree records. This usually necessitates individual tagging of animals. Since it is not possible to physically tag the animals at hatching, each full-sib family must therefore be reared in separate units from egg stage through to the onset of feeding until they reach sufficient size to be physically tagged. During this period, each family will have a common environment that is different from other families.

This common environmental effect will to some extent reduce the accuracy of the use of the family average as a prediction of genetic merit. Therefore, this period should be made as short as possible. This stresses the importance of providing all families with as equal environmental conditions as possible during the testing period in order to minimize the common environmental effects. And the mating could be performed between individuals from the different group on a rotational basis to avoid inbreeding.

Within-family selection

This is the opposite of the previous method. The selection is based on the deviation of each animal from the average of your family. There is the choice of animals of all families, then the method requires identification of the families. This may be achieved by maintaining them in separate tank, cages, hapas or any other means of containment, without necessarily tagging the fish. The full-sib groups reared in unreplicated hapas or any other form of containment fall into this category.

Under such circumstances selection between families would be misleading from a genetic viewpoint because of the confounding between genetic merit and common environmental effects. If replacements are chosen so that every family contributes the same number of individual to the next generation (chose one female and one male from each family) the effective population size is twice the actual.

The criterion of selection is the deviation of each individual from the mean of the family to which it belongs. This type of selection is the reverse of family selection; the family means being given zero weight in the selection decision. The chief condition under which this method has an advantage over the others is when there is a large component of environmental variance common to members of a family. Selection within families would eliminate this large non-genetic component from the variation operated on by selection. So, the within-family selection is especially advantageous when there is a large component of environmental variance common to members of the same family.

However, not all the additive genetic variance is available for selection, but only a fraction equal to the coefficient of relationship, but only a fraction equal to the coefficient of relationship among the family relatives in question will be available (0.50 and 0.25 for full-sibs and half-sibs, respectively). The lower within-family heritability can be compensated for by the high within-family selection intensity that can be applied without increasing the rate of inbreeding.

The selection intensity within families will be limited only by the number of individuals tested per family. The number of families involved in the programme will determine the lower limit of inbreeding, which can easily be controlled by applying a rotational mating system such as that earlier suggested for selection within cohorts. An important practical advantage of selection within families, especially in laboratory experiments, is that it economizes breeding space, unlike family selection. The method reduces the need for tagging large numbers of individuals. They estimate that the implementation of a selection method that entailed the individual identification of large numbers of fish and a period of communal rearing would be more expensive and difficult to implement. Within family selection has low efficiency compared to most other selection methods.

Combined selection

The term, combined selection in a broad sense, meaning selection that is based on individual information as well as on information coming from relatives, combine in an optimal way all available sources of information that can add to our knowledge about the breeding value of an animal, as well information recorded on the animal itself, information about full-sibs and/or half-sibs and progenies as well as pedigree information. In this case, all of the additive genetic variance is available for selection and the use of information from relatives increases the accuracy of the estimation of breeding values.

It represents the general solution for obtaining the maximum rate of genetic gain, and the other simpler methods are special cases of this method. Combined selection is therefore in principle always the best method. [4]. Furthermore, relatives' records can be used to estimate breeding values for traits that require slaughter of the animals (carcass and flesh quality traits) or that entail a risky challenge (disease resistance, tolerance to some environmental component). This is not possible with the other methods (mass selection or within-family selection).

A selection index can be very useful in combining such information, but the approach has limitations that have been overcome with Best Linear Unbiased Prediction (BLUP). Those procedures rely on mixed model methodology for the estimation of individuals' genetic merit. In the case of aquaculture, all systematic effects (batch, sex, production environment, age variation) associated with traits of interest can be accounted for in the model fitted to the data.

One particular advantage of BLUP procedures is that genetic gain can be estimated from the mean of the estimated breeding values in each year or generation of selection provided there are genetic links. A drawback of BLUP selection is that if truncation selection on estimated breeding values is practiced, it also results in higher levels of inbreeding than individual selection, especially for lowly heritable traits. This is because BLUP uses family information leading to co-selection for relatives. Hence, a mating strategy should be used that results in genetic gain while controlling the rate of inbreeding.

However, the costs to apply this methodology are higher due to the need to mark the evaluated animals with the cultivation of families in separate structures up to a minimum weight for identification. Cultivation of the families separately up to part of the evaluation period may lead to the emergence of a family effect, causing confounding of existing genetic effects and reducing the accuracy of predictions of genetic values.

THE GENETIC IMPROVEMENT PROGRAMS IN AQUACULTURE

Most animal breeding programs have focused on cumulative short-term genetic changes of production traits, because breeding optimization has to a very large extent been based on market economic values. Sustainable genetic improvement by animal breeding is a long-term and complex process. Farmed fish is at an early stage of domestication and selection, but rapid selection responses for growth rate are already documented for several species. Precautions should be taken to avoid the same unwanted side effects, such as increased frequencies of diseases (e.g. mastitis in dairy cattle) and leg problems, which we have often seen in breeding programs for agricultural livestock. Careful monitoring of possible correlated responses is needed. More basic knowledge of, e.g. animal welfare and behavioural needs of fish may also be required, as we know very little about its perception of pain and satisfaction, and what it prefers.

Starting a selective breeding program should then focus on expected benefits of, e.g. faster growing fish with higher survival rates and the costs, by improving the aquaculture species in question. Because of the great fecundity of most aquaculture species, a single, well-organized breeding nucleus may supply a huge market with continuously upgraded seed by delivering selected brood stock to a number of multiplier farmers after each generation of selection. Consequently, the running costs of the breeding programs will be more or less constant and independent of the volume of the target industry.

The benefit/ cost ratio will then often depend more on the volume of the target industry than on the price of the end product, both from the consumers' point of view and from a private business point of view. The total impact of an aquaculture breeding program on food supply and resource efficiency for the society will usually be greater for a large volume species.

Genetically improvement strains are essential to aquaculture development. The application of proven quantitative genetic theory should continue for relevant species. There is an ample proof of the success such programmes can have. Greater emphasis is required in the areas of dissemination of the improved stock to farmers to ensure impact at the production level. However, consideration must be given to the potential impacts of fish escapees and the conservation of wild fish genetic resources. So, genetic management is necessary for both genetic improvement and conservation of genetic variation in cultured fish populations and should be emphasized in future capacity development initiatives

From an economic viewpoint, investment appraisal studies indicate very favorable benefit-to-cost ratios for genetic improvement programmes in *Nile tilápia* and common carp. This was show to be so even for situation in which there was genotype by environmental interaction and a single program had to service more than one environment. It is reasonable to think that these results can be generalized to other, similar, aquatic animal species.

Limitations and constraints during the implementation of genetic improvement programmes in fish often occur, particularly in developing countries. Issues are commonly related to financial resources, and to a paucity of human capacity in this field. Other constraints include technical issues related to individual identification of fish but these can be overcome with relatively small investment.

Before initiating a selective breeding programme, a farmer must ask the following question: Is a selective breeding programme appropriate or necessary? The adoption of a selective breeding programme should be restricted to farmers who are capable of keeping and maintaining records, who are good managers, who adopt and maintain new technologies and who are willing to incorporate some long-term planning into their management programmes.

In order to control fertilization, hatching and first feeding, it is essential that the entire reproduction cycle for a species is controlled in captivity. Furthermore, an inexpensive marking or tagging system is advantageous and a necessity for a breeding program based on testing and selection within and between families. In order to select the best individuals for breeding, we need records of the traits in the breeding objective or traits correlated to the breeding goal. The records should be obtained for many animals at a reasonable cost during a limited time interval. Knowledge and methodology to optimize selection by predicting and using breeding values is also essential.

The programmes starting with a population with ample genetic variation is a trademark of successful fish genetic improvement programmes. Although this in itself is not a sufficient condition for success, it is indeed a necessary condition. The failure of some attempts to achieve genetic improvement with aquatic animals may have been due more to weakness in the base population than to be selection method utilized. Irrespective of the method of choice, continued genetic improvement will hinge upon the adequate balance between high selection intensity and the maintenance of low inbreeding rate.

Before the selection is made, it is preponderant that the beginning of an improvement program originates from a base population with great additive genetic variability. In this way, it is important that the nucleus is formed by lots from several genetic sources, such as several sites or collection points of a particular river. Examples are the Norwegian Atlantic salmon program, which originated from 41 wild genetic sources and the GIFT program for Nile tilapia originated from the diallele crossing of 8 strains of the species.

The success of an aquaculture breeding program critically depends on the way which the base population of breeders is constructed. In particular, all the genetic variability that can be used for selection on the traits initially included in the breeding objective is that found in the original breeders. Also, the decisions taken when creating the base population will have consequences on the genetic progress for any other additional trait that may be part of future breeding goals, whatever production or fitness related traits.

Strains originating from the base population may be selected to serve crops in different environments, since there may be genotype-environment interaction, as evaluated by researchers for the characteristic weight at Nile tilapia farms cultivated in tanks-net and nurseries.

However, until then, little or no interaction was found in rainbow trout, Atlantic salmon and catfish, suggesting that it is not necessary to select several lineages for this species. However, there is always a need for previous studies to confirm the lack of genotype-environment interaction, mainly due to the great variation between and within existing production systems in fish farming.

The decision on the type of selection to be adopted will depend on the program's long-term objectives, available resources, the nature of the goal and the value of heritability. Ideal program designs maximize the likelihood of correct ordering of organisms as to the true genetic value for the target in question. However, perfect programs are rarely possible, restricted by the limitation of individual identification and the physical separation of families to communal storage Mair.

The major goal to most selective breeding programmes that are going to be used to improve quantitative phenotypes is to improve growth rate, and this will also increase yield. There are two basic approaches to improving yields. One is to use a breeding programme to improve the fish that are being cultured. The other is to improve the improvements in management are often inexpensive. For that, they must to conduct an assessment to determine its usefulness, needs to quantify his goals and know how the phenotype(s) will be measured, to determine the type of selective breeding programme he will use, needs to know how the results will be evaluated and must plan an annual evaluation and reassessment of the programme.

The overall aim of all selection schemes is to maximize the probability of correctly ranking all potential breeders with regard to their breeding value. An animal's breeding value can be defined as the average performance of an infinite number of its progeny or from a practical points of view its ability to produce good or bad progeny.

Selection that combines the individual information of the animal and its relatives. More efficient method, but more expensive. From this method, it is possible to predict the genetic values of the animals by BLUP (Best Linear Unbiased Predictions). Method adopted for the Norwegian National Program for Genetic Improvement of Atlantic salmon and the GIFT program for Nile tilapia. With the information of pedigree, it becomes possible the direct selection of characteristics of meat quality and survival to diseases.

One of the major difficulties of these programs in fish breeding is the mark used for the identification of animals. Most available and accessible such as numbered ribbons that span the musculature or nylon strands with small numbered labels that are also affixed in the same way, have high loss rates. Thus, more animals must be identified to compensate for information losses, increasing program costs. The emergence of identification tags improved pitfalls considerably (almost 0%), but it is the most expensive option on the market.

For each selection methods than can be chosen, always have the side of the advantages and disadvantages. That choice will be influence for a given particular breeding scenario. However, for most methods, there is a theoretical maximum value of the correlation between the true and estimated breeding values, given different heritability for the traits in question.

Final considerations

Genetically improved strains are essential to aquaculture development and the most important factors resulting in increased production, lower costs, increased consumption and, in some cases, improved overall nutritional status of certain sectors of the human population. The application of quantitative genetic theory to the improvement of farmed aquaculture species is still at an early stage, and its use should be expanded to develop important aquaculture species to further increase the productivity and quality of aquaculture products.

Selection was still poorly applied in fish, being restricted mainly in salmonids, carp and tilapia. This was mainly due to the lack of efficient methods of individual marking (until recently) and, consequently, controlled mating and pedigree information to improve the accuracy of the estimates. The maintenance of the additive genetic variability in the core population of the program should be guaranteed, but there is a reduction of the same that should be avoided, inbreeding. This is normally harmful promoting the reduction of performance of organisms to zootechnical characteristics, inbred depression.

In aquaculture, with the increase of endogamy, the worsening of several zootechnical characteristics, such as the reduction of reproductive performance and larval survival, as well as the increase in deformities of the same, have been confirmed. In the literature, values of 0.9% to >20% of performance reduction of inbred lines are found for each 10% increase in the coefficient of inbreeding of the original control population.

As we have pointed out, less than 20% of the fish production in the world comes from genetically improved stocks. This indicates that there is a great need for investment (public or private investment, training of personnel) to fill this gap. Considering the good results obtained in the few breeding programs implemented for some fish species (*Nilotica tilapia*, carp), it is expected that the dissemination and use of genetic improvement tools in fish farming will increase considerably in the coming years given the history (Even of traditional uses) and that gains in productivity improve the efficiency of the use of natural resources (water and land) needed for cultivation, which may contribute to the growing need for animal protein for human consumption.

The successful achievements of new genetic programmes depend not only on systematic steps involved in the design and conduct of experimental field work but also on the establishment of a base population with ample genetic variability to ensure a long-term response to selection. Choice of selection methods depends on species, objectives and the available resources of aquaculture enterprises. Ideally, a fully pedigreed population should be maintained to enable the applications and utilizations of advanced statistical methods, such as best linear unbiased prediction (BLUP), for genetic evaluation and selection. Within- and between-family selection, based on an individual's genetic merits and its relationship with other animals in the pedigree, shows several advantages over other methods.

Researchers compared the effect of individual, family and within-family selection with the efficiency of combined family and within-family selection. The correlation between phenotypic values of family members is known as the intraclass correlation (t) and is estimated as:

$$t = r.h^2 + c^2$$

where r is the correlation between family members, h^2 is heritability of the trait in question and c^2 is the proportion of the total variance accounted for by common environmental effects.

When the intraclass correlation (t) varies from around 0.25 to around 0.75, individual selection is more efficient than both family and within-family selection. This is because individual selection exploits the entire scope of genetic variation present, while family and within-family selection only utilizes part of the genetic variation.

When the intraclass correlation is low, family selection is more efficient than individual selection. Within-family selection compares favourably to both family and individual selection when the intraclass correlation is very high. Individual selection is approximately as efficient as combined selection when t=0.50 and for such and interclass correlation, family selection and within-family selection are equally efficient.

When the intraclass correlation is high due to members of a family being more alike phenotypically rather than genetically ($t>r=0.50$), within family selection is more efficient than both individual and family selection. However, this is a very rare scenario and the primary advantage of within-family selection is the reduction of common environmental effects between families.

Hybridization

Generally, hybridizations simply refer to generating a new form of plant or animal either naturally or by human intervention by combining the genes of two different species or subspecies. Similarly, fish hybridization is when two different species, genera or families can be crossed and the first filial generation then crossed, backcrossed or outcrossed to give the hybrid of desired qualities. Fish hybridization is one of the genetic techniques which help to remove undesirable characteristics while retaining only the desirable ones.

For instance, researchers investigated hybridization of the two African catfishes: *Clarias gariepinus* and *Heterobranchus longifilis*. They reported viability in reciprocal hybrids with their survival rates being similar to those found in the maternal species.

Moreover, Researchers investigated hybridization between two clariids: *Clarias batrachus* (Linn.)x *Clarias gariepinus* (Bur.) and performance of the offspring in rearing operations. Researchers reported that crosses between male *Heterobranchus longifilis* and female *Clarias anguillaris* were significantly heavier and longer than the reciprocal cross.

Types of hybridization in catfishes: Several studies have demonstrated that *Clarias gariepinus* x *Heterobranchus bidorsalis* hybrid exhibit superior growth, improved survival and general hardiness than true breed of either *Clarias gariepinus* or *Heterobranchus bidorsalis*. Most of these studies have focused on stock manipulations and growth performances at different dietary compositions in indoor and outdoor concrete tanks.

Intraspecific hybridization: Hybridization may involve combining different strains of a species (that is members of the same species with different characteristics). This is referred to as intraspecific hybridization. It is a hybridization exercise carried out between fish that belong to the same species; for example between strains within a species. A strain within a species is a population with common origin and history that possess a unique trait that distinguished it from other strains. Researchers reported that cross-breeding of *Clarias anguillaris* strains could be advantageous because of the better performance of the progeny. They also reported high hatching success and survival rates of *C. anguillaris* from their study.

Interspecific hybridization: Interspecific hybridization of different genetic type is an alternative to conventional selective breeding of fishes to produced qualitative or quantitative changes in commercial traits. Sometimes, an interspecific hybrid does not exhibit heterosis for any trait but is still quite important for aquaculture application as it expresses a good combination of beneficial traits from both parent species. Moreover, catfish hybrids were reported in *Clarias gariepinus*, *Heterobranchus longifilis*, *H. fossilis* and *C. batrachus* and *C. batrachus* × *C. gariepinus*. Researchers reported that the reciprocal intergeneric hybrid catfish between *C. gariepinus* and *H. longifilis* can be produced. Researchers reported a positive net gain and cost benefit Ratio recorded in all the diets fed to '*Heteroclarias*' which was an interspecific hybrid. They also showed that "*Heteroclarias*" can be economically reared on all diets.

Limit of hybridization: Many natural fish hybrids have been reported and numerous others have been produced in the laboratory. He recognized that hybridization success is correlated with close phylogenetic karyotypic affinities. That chromosome incompatibility is the primary block to any successful hybridization. Chromosome number in a species remains constant through successive generations and this result into constancy of characters. However, variations within the same organism have been reported (mosaicism-presence of two or more different cell lines in an organism). Variable chromosome number is a common phenomenon in some fish species. Researchers reported chromosome numbers of 52, 54 and 56 in *C. gariepinus*.

Producing catfish hybrid fry: The primary constraint to commercial production of most catfish hybrids has been the lack of reliable, cost-effective methods for producing large quantities of fry. However, refinements of techniques for producing hybrids and general superiority of hybrids catfish have spurred renewed interest in use of hybrids for commercial production. Traditional pond-spawning which is effectively used to produce some catfish fry, is ineffective for consistent, large-scale production of hybrid fry. Therefore, production of hybrid fry in most catfishes depends on the use of hormones to induce ovulation in females, manual 'stripping' of eggs and manual fertilization of the eggs with catfish sperm. Hormone often used includes pituitary extract: Common carp-CCP or catfish CP, lutenizing hormone releasing hormone analog-LHRHa, human chorionic gonadotropin and synthetic homorne-ovaprin, ovatide). Strategies for hormone-induced production of hybrids can be classified into two main categories: Pair-spawning and group-pawning. Important factors for successful production of hybrid fry include: Good broodstock quality, proper calculation and administration of hormone dosage, proper testes collection and sperm preparation, accurate determination of the time of ovulation in females, good stripping and fertilization techniques and aggressive egg treatment.

GLOSSARY

Abundance: Degree of plentifulness. The total number of fish in a population or on a fishing ground.

Alevin: The developmental life stage of young salmonids and trout that are between the egg and fry stage. The alevin has not absorbed its yolk sac and has not emerged from the spawning gravels.

Allele: Alternative form of a given gene or DNA sequence.

Amphidromous fish: Migratory fish that live in both fresh and salt water, independent of breeding.

Anadromous fish: Migratory fish that live in salt water but breed in fresh water.

Anchovies: Small silvery fishes that swim in big schools. Anchovies are eaten by tuna, salmon, and many other predatory fishes.

Antibiotic: A chemical compound which can inhibit reproduction or cause the destruction of bacteria.

Apex Predators: Species on the top of the food chain that depend on the availability of smaller fish for sustenance.

Aquaculture: The production and husbandry of aquatic animals and plants in a controlled environment.

Benthic: On or near the bottom of a lake, river or ocean.

Bivalve: A mollusk, such as an oyster or a clam, that has a shell consisting of two hinged valves.

Bluefin Tuna: A group of tuna species with iridescent blue skin. Bluefin are some of the ocean's biggest, fastest predators. They migrate across the world's oceans and can weigh up to 2,000 pounds (907 kilograms).

Bootstrapping Pseudorandom resampling of data with replacement to estimate the error of estimates of parameters of interest.

Brood stock: fish of any particular species which are raised for reproduction purposes.

Carnivorous: Term used to describe the diet of an organism who eats living animals

Carp: An herbivorous freshwater fish (Cyprinus carpio) originally from Asia that is frequently bred in ponds and lakes. The carp was also introduced into Europe, where it is extensively reared in artificial ponds.

Catadromous fish: migratory fish that live in fresh water but breed in salt water.

Catfish: Any of numerous scale-free, chiefly freshwater, bottom-living fishes of Eurasia and North America characteristically having whisker-like projections extending from the mouth. It is a food fish of the southern United States, farmed extensively in Mississippi. The omnivorous fish have been bred to require minimal amounts of fishmeal and fish oil.

Commodity: Product. Goods and services which are the result of production processes normally intended for sale on the market at a price that is designed to cover their costs of production.

Community: All of the plants and animals living in a specific area (habitat), often described by the most abundant or obvious organisms. The kelp forest community means all the animals and plants that are part of the kelp forest.

Compromise: An agreement between two parties in which each side makes concessions.

Conservation: The practice of protecting nature from loss or damage.

Contaminants: Both natural and artificial materials not typically found or found in unusually high concentrations which can be detrimental to the health of wild or farmed aquatic species

Continental Shelf: The submerged shelf of land that slopes gradually from the exposed edge of a continent to where the drop-off to the deep seafloor begins.

Coral: A group of invertebrate animals related to sea anemones. Individual coral animals have soft bodies topped by a ring of stinging tentacles for catching food. Some kinds of coral build hard limestone skeletons; when they die, other corals build on top until a great reef is formed.

Credibility Interval In Bayesian statistics, a credibility interval is a posterior probability interval. Credibility intervals differ from the confidence intervals in frequentist statistics in that they do not assume an underlying distribution (e.g., normal distribution) and are direct statements of probability: i.e., a %90 credibility interval has a %90 chance of containing the true answer.

Culture tanks: Artificial holding areas used to rear fish farm stocks.

Deficit: The quantity by which an amount is less than the needed amount.

Depletion: For renewable resources, the part of the harvest, logging, catch and so forth above the sustainable level of the resource stock; for non-renewable resources, the quantity of resources extracted.

Dioxins: A general term referring to a class of structurally and chemically related compounds, including the "dioxin-like" Biphenyls (PCBs). Dioxins are chemical contaminants that are known to cause cancer in humans.

Dissolved Oxygen: The volume of oxygen dissolved in water.

Dredge: A net attached to a frame that's dragged along the ocean floor to collect animals.

Eco-Label: A logo placed on product labels, usually indicating that a manufacturer has used "environment-friendly" practices when creating a product.

Ecosystem: The natural system in which energy and nutrients cycle between plants, animals and their environment.

Effective Population Size (Ne). The size of an ideal population that would be affected by genetic drift at the same rate as the actual population. This idealized population has discrete generations, an even sex ratio, constant size, random union of gametes and random survivorship of offspring (Kalinowski and Waples 2002).

Effluent: Discharge of liquid waste or pollution. Effluents can contain bacteria, viruses, excessive nutrient loads and chemicals that are harmful to wildlife and human health.

Escapes: When a farmed fish finds its way out of a netcage and into the open water. Escapes are often as the result of storm damage or human error.

Estuary: The area where a river meets the ocean.

Extensive aquaculture: managed aquaculture dependent on the local natural setting, such as a pond or coastal sea area.

Factory Trawler: A large ship equipped to catch, clean and freeze fish for market.

Filter Feeder: An aquatic animal, such as a clam, barnacle, or sponge, that feeds by filtering particulate organic material from water.

Fingerling: Development stage of fish following the fry stage and continuing into the first three to four months of life.

Fish farm: locations used to grow populations of aquatic organisms, primarily fish.

Fish Hatchery: one form of a fish farm, managed with the intent of resupplying native wild populations in natural environments.

Fish husbandry: the breeding and rearing of fish for a variety of reasons.

Fish kill: a description of the number of fish stock to die in a specific amount of time.

Fish meal: commercially processed food source used in fish farming as a source of protein for the fish stock.

Fisheries Management: The effort to regulate where, when and how people fish, and how many fish they catch. The intent is usually to protect fish populations so that people can continue to fish. Most fisheries management is done by government agencies such as the U.S. National Marine Fisheries Service.

Fishery: The occupation or industry of catching, processing, and selling fish and shellfish; an area where fish or shellfish are caught.

Fishing Pressure: The amount of fishing effort for a certain species of fish or shellfish. If there's heavy fishing pressure on sharks, it means that lots of sharks are being caught by fishers.

Food Safety: The consideration of human health risks associated with certain food products.

Fry: A stage of development in young salmon or trout. During this stage the fry is usually less than one year old, has absorbed its yolk sac, is rearing in the stream, and is between the alevin and parr stage of development.

F-statistics (FIS, FST, FIT). Measures used to hierarchically partition genetic diversity within and among populations. Common measures include: 1) FIS, which is the average departure of genotype frequencies from Hardy-Weinberg expectations within a population; 2) FST, which is the proportion of the variation due to allele-frequency differences among populations; and 3) FIT, which is the departure of genotype frequencies from Hardy-Weinberg expectations relative to the entire collection of populations. In this common hierarchy, the subscripts refer to comparisons between levels in the hierarchy: IS refers to individuals within populations, ST to subpopulations within the total population, and IT to individuals within the total population. Hierarchies and subscript notation can be extended to any level to accommodate different study designs.

Game fish: Fish, such as trout, raised to become replenishment stock for natural sport fisheries.

Gametic Disequilibrium (or Linkage Disequilibrium) A state that exists in a population when alleles at different loci are not distributed independently of each other in a population's gamete pool, often because the loci are physically linked on the same fragment of DNA.

Genetic Drift Chance changes in allele frequencies that result from the sampling of gametes from generation to generation in a finite population. The magnitude of these changes is inversely related to twice effective population size (2Ne), because each diploid fish carries two sets of genes.

Genetic Marker (Marker): A known DNA sequence that can be identified by a simple assay. For the purpose of this document, a marker can be a single SNP or a combination of SNPs.

Genotype The set of alleles at one or more loci for an individual.

Genotyping: determining the genetic makeup of a fish or stock or the purpose of managing the efficiency of aquaculture production.

Gill Net: A type of fishing net that catches fish by their gills or gill covers.

Habitat: An organism's home; for example: in the midwater, on the seafloor, near the surface or in a tide pool.

Hardy-Weinberg Expectations (HWE). Genotypic frequencies expected from given allele frequencies, assuming random mating, no mutation (the alleles do not change), no migration or emigration (no exchange of alleles between populations), infinitely large population size, and no selective pressure for or against any of the genotypes.

Harvesting: The catching of seafood.

Hatchery: A place where the eggs of fish are hatched

Herbivore: An animal that feeds chiefly on plants.

Heterozygosity The proportion of individuals in a population that carry two different alleles at a particular marker. This statistics can be used as a measure of genetic variability in a population.

***In vitro* fertilization**: artificially fertilizing fish eggs in a laboratory setting.

Infectious Salmon Anemia: A salmon disease that originates in the wild, but tends to be greatly amplified within the confines of a crowded net cage or pond with high stocking densities.

Intensive aquaculture: Managed aquaculture controlled through human engineered means, such as managing water quality and sources of food.

Larva: The young and immature form of an animal, which must change to become an adult.

Locus (plural Loci). A fixed position or region on a chromosome that may contain more than one genetic marker.

Mangrove: A tree that grows along tropical coasts in salty ocean water, Their extensive root systems provide a breeding ground for plant and marinelife biodiversity, and also aid in building up coastlines.

Mendelian (Mendelian inheritance): Pattern of inheritance that follows the principles proposed by botanist Gregor Mendel. Mechanistically, Mendelian inheritance follows the law of segregation and the law of independent assortment.

Microsatellite: Regions of repeated DNA found abundantly throughout the genome, often in non-coding areas, and which are highly variable and prone to mutation (see short tandem repeats). The various forms of a given microsatellite (identified by differences in the number of repeats) are referred to as alleles, and these are inherited in a Mendelian fashion. Genetic variation at the microsatellite level can be used to identify individuals. DNA sequences containing short (2–5 base pairs) tandem repeats of nucleotides (e.g., GTGTGTGT).

Mitochondria: Mitochondria (singular: mitochondrion) are membrane bound organelles, or specialized parts of cells. They play a key role in generating cellular energy known as adenosine triphosphate (ATP). Mitochondria possess their own independent genome separate from the DNA found in the nucleus (see: mitochondrial DNA).

Mitochondrial DNA (mtDNA). Circular, maternally inherited DNA located outside the nucleus in organelles in a cell called mitochondria.

Mixed Stock Analysis (MSA). Method using allele frequencies from populations and genotypes from mixture samples to estimate stock compositions.

Mollusk: Any of numerous chiefly marine invertebrates of the phylum Mollusca, typically having a soft unsegmented body, a mantle, and a hard, protective shell and including the edible shellfish and the snails.

Mutation: A change in the nucleotide sequence of DNA. A mutation can be small (e.g. single base pair) or large (e.g. segment of a chromosome encompassing multiple genes). Mutations create alleles, which are alternate versions of a gene.

Nucleotide: An organic molecule that is the basic structural unit of DNA (and RNA). Nucleotides consist of a nitrogenous base, a five-carbon sugar molecule, and a phosphate group. In DNA, the nitrogenous bases are Adenine (A), Guanine (G), Thymine (T), and Cytosine (C). In RNA, Uracil (U) replaces Thymine (T).

Oceanodromous fish: migratory fish that move within salt water only.

Offshore Fish Farms: Aquaculture operations located beyond three miles of a coastline, often involving submerged netcages anchored to the seafloor.

Organic: From a living thing or organism; originating in nature rather than being made artificially.

Ornamental fish: are small fish, such as koi, which are typically grown to be kept in aquariums or small landscape ponds.

Overfishing: Catching too many fish; fishing so much that the fish cannot sustain their population. The fish get fewer and fewer, until finally there are none to catch.

Pisciculture: aquaculture practice involving finned fish.

Plankton: Plants and animals (mostly tiny) that swim weakly, or not at all, and drift with ocean currents. Plankton are an important food source for many fish and other organisms that live in the sea.

Pollock: A fish related to cod, heavily fished by people for food. Pollock are made into fish sticks and imitation crab meat.

Pollution: Degradation of the natural environment by chemicals, oil, trash or other substances.

Polyculture: A system of pond aquaculture in which more than one species is raised. In China a type of polyculture has been developed over 4000 years in which different species of carp inhabit separate niches within the pond ecosystem, with beneficial results.

Polymerase Chain Reaction (PCR). The enzymatic amplification a single or few copies of a DNA fragment several orders of magnitude, generating millions of copies.

Pond acreage: area committed to extensive aquaculture holdings at a fish farm or company.

Population A randomly mating group of fish that are largely reproductively isolated from other populations.

Potamodromous fish: migratory fish that move within fresh water only.

Reef: An underwater structure; something that extends up from the seafloor but does not rise above the surface of the water like an island. Coral reefs are formed from the hard skeletons of coral. There are also rocky reefs, which are piles of rock under water.

Reporting Group A group of populations in a genetic baseline to which portions of a mixture are allocated during mixed stock analyses. In this study, reporting groups were constructed based on stakeholder needs, on the extent of genetic distinction among populations, and on advice from the WASSIP Technical Committee and Advisory Panel.

Salmon: A family of anadromous fishes that breed in rivers but live most of their adult lives at sea. Salmon have orange or pink flesh. For centuries, salmon have been important food fish to people of many nations. When they're ready to breed, most salmon find their way from the ocean back to the same stream where they were born.

Shrimp: Any of various small, chiefly marine decapod crustaceans, many of which are edible. The larger kinds are called also prawns. Shrimp are carnivorous. In Thailand the most commonly farmed species are black tiger shrimp (Penaeus monodon). Some other species of shrimp, including Pacific white shrimp (Litopenaeus vannamei) and some freshwater species are being bred to be more disease resistant.

Single nucleotide polymorphism (SNP). DNA sequence variation occurring when a single nucleotide (A, T, C, or G) differs among individuals or within an individual between paired chromosomes.

Spawn: To breed; especially, to breed by releasing eggs and sperm into the water.

Sustainability: Meeting the needs of the present without compromising the ability of future generations to meet their own needs, or the health of the planet.

Swordfish: A large predatory fish with a long, sword-like bill at the tip of its snout. Swordfish are famous for their speed, strength and trans-ocean migrations. They are heavily fished and many populations are vulnerable to over-fishing.

Tilapia: Any of various cichlid fishes of the genus Tilapia, native to Africa's Nile River but introduced elsewhere as a valuable food fish. Some historians believe that tilapia were the fish that Jesus fed to the multitudes in the Biblical account. These omnivorous fish have been bred to require minimal amounts of fishmeal and fish oil.

Transgenic Fish: Fish that have been genetically modified.

Water Quality: the characteristics of water, such as water temperature and contamination, which define its ability to sustain life and its purity from chemicals.

Wild fishing harvest: the commercial and personal fishing consumption in a specific area and over a specific time frame.

Yearling: Development stage of fish following the fingerling stage and lasting until approximately one year of age.

Yellowfin Tuna: One of the larger species of tuna, which has yellow markings on its fins and tail. Yellowfin live in warm, tropical waters. Much of the world's canned tuna is yellowfin.

Zooplankton: Animal plankton.

BIBLIOGRAPHY

Adams, K. L. and Wendel, J. F. Polyploidy and genome evolution in plants. *Current Opinion in Plant Biology* **8**, 135-141 (2005)

Akashige, S., and T. Fushimi. 1992. "Growth, survival, and glycogen content of triploid Pacific oyster *Crassostrea gigas* in the waters of Hiroshima, Japan." *Nippon Suisan Gakkaishi* 58: 1063–71.

Allen, S. K. 1983. "Flow cytometry—Assaying experimental polyploid fish and shellfish." *Aquaculture* 33: 317–28.

Allen, S. K., A. J. Erskine, E. J. Walker and G. A. DeBrosse. 2005. "Production of tetraploid suminoe oysters, *Crassostrea ariakensis*." Aquaculture 247: 3.

Allen, S. K., Jr., S. L. Downing and K. K. Chew. 1989. *Hatchery Manual for Producing Triploid Oysters*. University of Washing Press, Seattle, Washington.

Allen, S. K., S. L. Downing, J. Chaiton, and J. H. Beattie. 1986. "Chemically and pressure-induced triploidy in the Pacific oyster *Crassostrea gigas*." *Aquaculture* 57: 359–60.

Amberger J, Bocchini CA, Scott AF, Hamosh A (2009) McKusick's Online Mendelian Inheritance in Man (OMIM). Nucleic Acids Res 37: D793–6

April, J. , Hanner, R. H. , Dion Côté, A. M. and Bernatchez, L. (2012). Glacial cycles as an allopatric speciation pump in north eastern American freshwater fishes. *Molecular Ecology*, 22, 409–422. 10.1111/mec.12116.

Barber, B. J., and R. Mann. 1991. "Sterile triploid *Crassostrea virginica* (Gmelin, 1791) grow faster than diploids but are equally susceptible to *Perkinsus marinus*." *Journal of Shellfish Research* 10: 445–50.

Barrett, R. D. H., *et al.* Experimental evolution of *Pseudomonas fluorescens* in simple and complex environments. *American Naturalist* **166**, 470–480 (2005)

Berg, J. M., Tymoczko, J. L. and Stryer, L. (2002). DNA, RNA, and the Flow of Genetic Information. In Biochemistry (5th ed.). chapter, W H Freeman & Co.

Bernatchez, L. and Duchesne, P. (2000). Individual based genotype analysis in studies of parentage and population assignment: How many loci, how many alleles? *Canadian Journal of Fisheries and Aquatic Science*, 57, 1–12. 10.1139/f99-271.

Bossi A, Lehner B (2009) Tissue specificity and the human protein interaction network. Mol Syst Biol 5: 260

Brake, J., J. Davidson, and J. Davis. 2004. "Field observations on growth, gametogenesis, and sex ratio of triploid and diploid *Mytilus edulis*." *Aquaculture* 236: 179–91.

Burkard TR, Planyavsky M, Kaupe I, Breitwieser FP, Bürckstümmer T, et al. (2011) Initial characterization of the human central proteome. BMC Syst Biol 5: 17

Chapman, J. R. , Nakagawa, S. , Coltman, D. W. , Slate, J. and Sheldon, B. C. (2009). A quantitative review of heterozygosity–fitness correlations in animal populations. *Molecular Ecology*, 18, 2746–2765.

Christie, M. R. (2010). Parentage in natural populations: Novel methods to detect parent offspring pairs in large data sets. *Molecular Ecology Resources*, 10, 115–128. 10.1111/j.1755-0998.2009.02687.x.

Christie, M. R. , Johnson, D. W. , Stallings, C. D. and Hixon, M. A. (2010). Self recruitment and sweepstakes reproduction amid extensive gene flow in a coral reef fish. *Molecular Ecology*, 19, 1042–1057. 1

Ciccarelli FD, Doerks T, von Mering C, Creevey CJ, Snel B, et al. (2006) Toward automatic reconstruction of a highly resolved tree of life. Science 311: 1283–1287

Cohen-Gihon I, Lancet D, Yanai I (2005) Modular genes with metazoan-specific domains have increased tissue specificity. Trends Genet 21: 210–213.

Coltman, D. W. and Slate, J. (2007). Microsatellite measures of inbreeding: A meta analysis. *Evolution*, 57, 971–983.

Comai, L. *et al.* Phenotypic instability and rapid gene silencing in newly formed *Arabidopsis* allotetraploids. *Plant Cell* 12, 1551-1568 (2000)

Croft D, O'Kelly G, Wu G, Haw R, Gillespie M, et al. (2011) Reactome: a database of reactions, pathways and biological processes. *Nucleic Acids R*es 39: D691–7

Crozier, L. G. and Hutchings, J. A. (2014). Plastic and evolutionary responses to climate change in fish. *Evolutionary Applications*, 7, 68–87. 10.1111/eva.12135.

Csete M, Doyle J (2004) Bow ties, metabolism and disease. Trends Biotechnol 22: 446–450

Dahm, R. (2005). Friedrich Miescher and the discovery of DNA. Developmental Biology, 278(2), 274–288. https://doi.org/10.1016/j.ydbio.2004.11.028(opens in new tab)

De Decker, S., J. Normand, D. Saulnier, F. Pernet, S. Castagnet, and P. Boudry. 2011. "Responses of diploid and triploid Pacific oysters *Crassostrea gigas* to *Vibrio* infection in relation to their reproductive status." *Journal of Invertebrate Pathology* 106: 179–91.

Degremont L., C. Ledu, E. Maurouard, M. Nourry, and A. Benabdelmouna. 2016. "Effect of ploidy on the mortality of *Crassostrea gigas* spat caused by OsHV-1 in France using unselected and selected OsHV-1 resistant oysters." *Aquaculture Research* 47: 777–86.

Degremont, L., C. Garcia, A. Frank-Lawale, and S. K. Allen. 2012. "Triploid oysters in the Chesapeake Bay: Comparison of diploid and triploid *Crassostrea virginica*." *Journal of Shellfish Research* 31: 21–31.

DeWoody, J. A. and Avise, J. C. (2000). Microsatellite variation in marine, freshwater and anadromous fishes compared with other animals. *Journal of Fish Biology*, 56, 461–473. 10.1111/j.1095-8649.2000.tb00748.x.

Downing, S. L., and S. K. Allen. 1987. "Induced triploidy in the Pacific oyster, *Crassostrea gigas*—Optimal treatments with cytochalasin B depend on temperature." *Aquaculture* 61: 1–15.

Eisenberg E, Levanon EY (2013) Human housekeeping genes, revisited. Trends Genet 29: 569–574

Ellegren, H. & Galtier, N. (2016). Determinants of genetic diversity. *Nature Reviews Genetics*, 17, 422–433. 10.1038/nrg.2016.58.

Eriksson, N., Wu, S., Do, C. B., Kiefer, A. K., Tung, J. Y., Mountain, J. L., Hinds, D. A. and Francke, U. (2012). A genetic variant near olfactory receptor genes influences cilantro preference. Flavour, 1(1). https://doi.org/10.1186/2044-7248-1-22(opens in new tab)

Etterson, J. R. Evolutionary potential of *Chamaecrista fasciculata* in relation to climate change. I. Clinal patterns of selection along an environmental gradient in the Great Plains. *Evolution* 58, 1446–1458 (2004)

Eudeline, B., S. K. Allen, and X. M. Guo. 2000/ "Optimization of tetraploid induction in Pacific oysters, *Crassostrea gigas*, using first polar body as a natural indicator." *Aquaculture* 187: 73–84.

Falconer, D. S. and Mackay, T. C. F. *Introduction to Quantitative Genetics* (London, Longman, 1996)

FAO. 2016. The State of World Fisheries and Aquaculture 2016, contributing to food security and nutrition for all. p. 200, Rome.

Feldman I, Rzhetsky A, Vitkup D (2008) Network properties of genes harboring inherited disease mutations. Proc Natl Acad Sci U S A 105: 4323–4328

Fisher, K., *et al.* Genetic and environmental sources of egg size variation in the butterfly *Bicyclus anynana. Heredity* 92, 163–169 (2004)

Forbes SA, Bindal N, Bamford S, Cole C, Kok CY, et al. (2011) COSMIC: mining complete cancer genomes in the Catalogue of Somatic Mutations in Cancer. Nucleic Acids Res 39: D945–50

Forstmeier, W. , Schielzeth, H. , Mueller, J. C. , Ellegren, H. and Kempenaers, B. (2012). Heterozygosity fitness correlations in zebra finches: Microsatellite markers can be better than their reputation. *Molecular Ecology*, 21, 3237–3249. 10.1111/j.1365-294X.2012.05593.x.

Frankham, R. , Bradshaw, C. J. A. and Brook, B. W. (2014). Genetics in conservation management: Revised recommendations for the 50/500 rules, Red List criteria and population viability analyses. *Biological Conservation*, 170, 56–63. 10.1016/j.biocon.2013.12.036.

Fransz, P., *et al.* Interphase chromosomes in *Arabidopsis* arc organized as well defined chromocenters from which euchromatin loops emanate. *Proceedings of the National Academy of Sciences* 99, 14584-14589 (2002)

Freilich S, Massingham T, Bhattacharyya S, Ponsting H, Lyons PA, et al. (2005) Relationship between the tissue-specificity of mouse gene expression and the evolutionary origin and function of the proteins. Genome Biol 6: R56

Froese, R. and Pauly, D. (2017). FishBase. Retrieved from: www.fishbase.org (accessed 27 January 2018).

Galtsoff, P. S. 1964. *The American oyster, Crassostrea virginica Gmelin, Fishery Bulletin Volume 64.* United States Government Printing Office, Washington, D. C.

Garnier-Gere, P. H., Y. Naciri-Graven, S. Bougrier, A. Magoulas, M. Heral M, G. Kotoulas, A. Hawkins, and A. Gerard. 2002. "Influences of triploidy, parentage and genetic diversity on growth of the Pacific oyster *Crassostrea gigas* reared in contrasting natural environments." *Molecular Ecology* 11: 1499–514.

Geiger T, Velic A, Macek B, Lundberg E, Kampf C, et al. (2013) Initial quantitative proteomic map of 28 mouse tissues using the SILAC mouse. Mol Cell Proteomics 12: 1709–1722

Geiger T, Wehner A, Schaab C, Cox J, Mann M (2012) Comparative proteomic analysis of eleven common cell lines reveals ubiquitous but varying expression of most proteins. Mol Cell Proteomics 11: M111.014050

Gienapp, P., *et al.* Climate change and evolution: Disentangling environmental and genetic responses. *Molecular Ecology* 17, 167–178 (2008)

Goh KI, Cusick ME, Valle D, Childs B, Vidal M, et al. (2007) The human disease network. Proc Natl Acad Sci U S A 104: 8685.

Gosling, E. 2003. *Bivalve Mulluscs Biology, Ecology and Culture*. Fishing News Books, Malden, MA.

Grant, W. S. and Bowen, B. W. (1998). Shallow population histories in deep evolutionary lineages ofmarine fishes: Insights from sardines and anchovies and lessons for conservation. *Journal of Heredity*, 89, 415–426. 10.1093/jhered/89.5.415

Gregory, T. R. and Witt, J. D. S. (2008). Population size and genome size in fishes: A closer look. *Genome*, 51, 309–313. 10.1139/G08-003.

Guo, M., Davis, D. and Birchler, J. A. Dosage effects on gene expression in a maize ploidy series. *Genetics* 142, 1349-1355 (1996)

Guo, X. 2004. "Oyster breeding and the use of biotechnology." *Bulletin of the Aquaculture Association of Canada* 104: 26–33.

Guo, X. 2012. "Production and breeding of tetraploid eastern oyster *Crassostrea virginica*." *Journal of Shellfish Research* 31: 292.

Guo, X. M., G. A. DeBrosse, and S. K. Allen. 1996. "All-triploid Pacific oysters *Crassostrea gigas* Thunberg produced by mating tetraploids and diploids." *Aquaculture* 142: 149–61.

Guo, X., and S. K. Allen Jr. 1994. "Viable tetraploids in the Pacific oyster (*Crassostrea gigas* Thuneberg) produced by inhibiting polar body 1 in eggs from triploids." *Molecular Marine Biology abd Biotechnology* 3: 42–50.

Guo, X., W. K. Hershberger, K. Cooper, and K. K. Chew. 1992. "Genetic consequences of blocking polar body I with cytochalasin B in fertilized eggs of the Pacific oyster, *Crassostrea gigas*. 2. Segregation of chromosomes." *Biological Bulletin* 183: 387–93.

Guo, X., Y. Wang, G. A. Debrosse, D. Bushek, and S. E. Ford. 2008. "Building a superior oyster for aquaculture." *Jersey Shoreline* 25: 7–9.

Guo, X., Y. Wang, Z. Xu, and H. Yang. 2009. "Chromosome Set Manipulation in Shellfish." In: *New Technologies in Aquaculture: Improving Production Efficiency, Quality and Environmental Management* (eds. by Burnell G & Allen G), pp. 165–94. Woodhead Publishing Limited, Abington, Cambridge.

Hand, R. E., and J. A. Nell. 1999. "Studies on triploid oysters in Australia - XII. Gonad discolouration and meat condition of diploid and triploid Sydney rock oysters *Saccostrea commercialis* in five estuaries in New South Wales, Australia." *Aquaculture* 171: 181–94.

Hand, R. E., J. A. Nell, and P. A. Thompson. 2004. "Studies on triploid oysters in Australia XIII: Performance of diploid and triploid Sydney rock oyster, *Saccostrea glomerata* (Gould, 1850), progeny from a third generation breeding line." *Aquaculture* 233: 93–107.

Harding, J. M. 2007. "Comparison of growth rates between diploid deby eastern oysters *Crassostrea virginica*, (Gmelin 1791), triploid eastern oysters, and triploid Suminoe oysters *C. ariakensis* (Fugita 1913)." *Journal of Shellfish Research* 26: 961–72.

Hauser, L. and Carvalho, G. R. (2008). Paradigm shifts in marine fisheries genetics: Ugly hypotheses slain by beautiful facts. *Fish and Fisheries.*, 9, 333–362. 10.1111/j.1467-2979.2008.00299.x

Hauser, L. , Adcock, G. J. , Smith, P. J. , Bernal Ramírez, J. H. and Carvalho, G. R. (2002). Loss of microsatellite diversity and low effective population size in an overexploited population of New Zealand snapper (*Pagrus auratus*). *Proceedings of the National Academy of Sciences*, 99, 11742–11747. 10.1073/pnas.172242899.

Hawkins, A. J. S., A. J. Day, A. Gerard A, Y. Naciri, C. Ledu, B. L. Bayne, and M. Heral. 1994. "A genetic and metabolic basis for faster growth among triploids induced by blocking meiosis I but not meiosis II in the larviparous European flat oyster, *Ostrea edulis* L." *Journal of Experimental Marine Biology and Ecology* 184: 21–40.

Hedgecock, D. (1994). Does variance in reproductive success limit effective population sizes of marine organisms? In A. *Genetics and Evolution of Aquatic Organisms*, 122, 122–134.

Hedgecock, D. and Pudovkin, A. I. (2011). Sweepstakes reproductive success in highly fecundmarine fish and shellfish: A review and commentary. *Bulletin of Marine Science*, 87, 971–1002. https://doi.org/info:doi/10.5343/bms.2010.1051 .

Helfman, G. , Collette, B. B. , Facey, D. E. and Bowen, B. W. (2009). *The diversity of fishes: Biology, evolution, and ecology*. Hoboken, NJ: John Wiley & Sons.

Hoagstrom Christopher, W. and Turner Thomas, F. (2015). Recruitment ecology of pelagic broadcast spawning minnows: Paradigms from the ocean advance science and conservation of an imperilled freshwater fauna. *Fish and Fisheries*, 16, 282–299. 10.1111/faf.12054.

Holderegger, R. , Kamm, U. and Gugerli, F. (2006). Adaptive vs. neutral genetic diversity: Implications for landscape genetics. *Landscape Ecology*, 21, 797–807. 10.1007/s10980-005-5245-9.

Huang DW, Sherman BT and Lempicki RA (2009) Systematic and integrative analysis of large gene lists using DAVID bioinformatics resources. Nat Protoc 4: 44–57

Hughes, A. R. , Inouye, B. D. , Johnson, M. T. J. , Underwood, N. and Vellend, M. (2008). Ecological consequences of genetic diversity. *Ecology Letters*, 11, 609–623. 10.1111/j.1461-0248.2008.01179.x.

Incorvaia, D. (2021, September 28). *Sex is more complex than a simple binary suggests*. Discover Magazine. Retrieved January 14, 2022, from https://www.

discovermagazine.com/the-sciences/sex-is-more-complex-than-a-simple-binary-suggests(opens in new tab)

IUCN (2018). The IUCN Red List of Threatened Species. Retrieved from http://www.iucnredlist.org/ (accessed 22 January 2018).

Jayaprasad, P. P., T. C. Srijaya, D. Jose, A. Papini, A. Hassan, and A. K. Chatterji. 2011. "Identification of diploid and triploid red tilapia by using erythrocyte indices." *Caryologia* 64, 485–92.

Jukam D, Xie B, Rister J, Terrell D, Charlton-Perkins M, et al. (2013) Opposite feedbacks in the Hippo pathway for growth control and neural fate. Science 342: 1238016

Kalinowski, S. T. (2004). Counting alleles with rarefaction: private alleles and hierarchical sampling designs. *Conservation Genetics*, 5, 539–543.

Kandoth C, McLellan MD, Vandin F, Ye K, Niu B, et al. (2013) Mutational landscape and significance across 12 major cancer types. Nature 502: 333–339

Kiel C, Serrano L (2009) Cell type-specific importance of ras-c-raf complex association rate constants for MAPK signaling. Sci Signal 2: ra38

Kitano H (2004) Biological robustness. Nat Rev Genet 5: 826–837

Komurov K, Ram PT (2010) Patterns of human gene expression variance show strong associations with signaling network hierarchy. BMC Syst Biol 4: 154

Korcsmáros T, Farkas IJ, Szalay MS, Rovó P, Fazekas D, et al. (2010) Uniformly curated signaling pathways reveal tissue-specific cross-talks and support drug target discovery. Bioinformatics 26: 2042–2050

Kritzer, J. P. and Sale, P. F. (2004). Metapopulation ecology in the sea: From Levins' model to marine ecology and fisheries science. *Fish and Fisheries*, 5, 131–140. 10.1111/j.1467-2979.2004.00131.x.

Leidy, R. A. and Moyle, P. B. (1998). Conservation status of the World's Fish Fauna: An overview In Fiedler P. L. and Kareiva P. M. (Eds.), *Conservation biology* (pp. 187–227). Boston, MA: Springer.

Liang, Y., R. Wang, C. Tian, R. Yu, and X. Qu. 1994. "A preliminary study on the triploid oyster *Crassostrea talienwanese.*" *Journal of Fisheries of China* 18: 237–40.

Luikart, G. , Sherwin, W. B. , Steele, B. M. and Allendorf, F. W. (1998). Usefulness of molecular markers for detecting population bottlenecks via monitoring genetic change. *Molecular Ecology*, 7, 963–974. 10.1046/j.1365-294x.1998.00414.x.

Lynch, M. and Walsh, B. *Genetics and Analysis of Quantitative Traits* (Sunderland, MA, Sinauer Associates, 1998)

Maguire, G. B., N. C. Gardner, J. A. Nell, G. N. Kent, and A. S. Kent. 1995. "Studies on triploid oysters in Australia. 2. Growth, condition index gonad area, and glycogen content of triploid and diploid Pacific oysters, *Crassostrea gigas*, from oyster leases in Tasmania, Australia." *Aquaculture* 137: 357.

Mallia, J. V., P. Muthiah, and P. C. Thomas. 2006. Growth of triploid oyster, *Crassostrea madrasensis* (Preston)." *Aquaculture Research* 37: 718–24.

Markert, J. A. , Champlin, D. M. , Gutjahr Gobell, R. , Grear, J. S. , Kuhn, A. , McGreevy, T. J. , ... Nacci, D. E. (2010). Population genetic diversity and fitness in multiple environments. *BMC Evolutionary Biology*, 10, 205 10.1186/1471-2148-10-205.

Martin GS (2003) Cell signaling and cancer. Cancer Cell 4: 167–174

Matthiessen, G. C., and J. P. Davis. 1992. "Observations on growth rate and resistance to MSX *Haplosporidium nelsoni* among diploid and triploid Eastern oysters *Crassostrea virginica* (Gmelin, 1797) in New England." *Journal of Shellfish Research* 449–54.

Melaragno, J. E., Mehrotra, B. and Coleman, A. W. Relationship between endopolyploidy and cell size in epidermal tissue of *Arabidopsis*. *Plant Cell* 5, 1661-1668 (1993)

Merilä, J., *et al.* Variation in the degree and costs of adaptive phenotypic plasticity among *Rana temporaria* populations. *Journal of Evolutionary Biology* 17, 1132–1140 (2004)

Miller, P. A., N. G. Elliott, R. E. Vaillancourt, A. Koutoulis, and J. M. Henshall. 2016. "Assignment of parentage in triploid species using microsatellite markers with null alleles, an example from Pacific oysters *Crassostrea gigas*." *Aquaculture Research* 47: 1288–98.

Miller-Jensen K, Janes KA, Brugge JS, Lauffenburger DA (2007) Common effector processing mediates cell-specific responses to stimuli. Nature 448: 604–608

Mittelsten Scheid, O., *et al.* A change of ploidy can modify epigenetic silencing. *Proceedings of the National Academy of Sciences* 93, 7114-7119 (1996)

Mitton, J. B. and Lewis, W. M. (1989). Relationships between genetic variability and life history features of bony fishes. *Evolution*, 43, 1712–1723.

Mousadik, A. E. and Petit, R. J. (1996). High level of genetic differentiation for allelic richness among populations of the argan tree [*Argania spinosa* (L.) Skeels] endemic to Morocco. *Theoretical and Applied Genetics.*, 92, 832–839.

Mousseau, T. A. and Fox, C. W., eds. *Maternal Effects as Adaptations* (New York, Oxford University Press, 1998)

Mullen AC, Orlando DA, Newman JJ, Lovén J, Kumar RM, et al. (2011) Master transcription factors determine cell-type-specific responses to TGF-□ signaling. Cell 147: 565–576

NCBI Resource Coordinators (2013) Database resources of the National Center for Biotechnology Information. Nucleic Acids Res 41: D8–D20

Nei, M. , Maruyama, T. and Chakraborty, R. (1975). The bottleneck effect and genetic variability in populations. *Evolution*, 29, 1–10. 10.2307/2407137.

Nell, J. A., and B. Perkins. 2005. "Studies on triploid oysters in Australia: Farming potential of all-triploid Pacific oysters, *Crassostrea gigas* Thunberg, in Port Stephens, New South Wales, Australia." *Aquaculture Research* 36: 530–6.

Nell, J. A., E. Cox, I. R. Smith, and G. B. Maguire. 1994. "Studies on triploid oysters in Australia. 1. The farming potential of triploid Sydney rock oysters *Saccostrea commercialis* (Iredale and Roughley)." *Aquaculture* 126: 243–55.

Nelson, J. S. , Grande, T. C. and Wilson, M. V. H. (2016). *Fishes of the world*, 5th edn. Hoboken, NJ: Wiley; 10.1002/9781119174844.

Nevo, E. (1978). Genetic variation in natural populations: Patterns and theory. *Theoretical Population Biology*, 13, 121–177. 10.1016/0040-5809(78)90039-4.

NIH National Human Genome Research Institute. (2021, February 12). Human genome project timeline of events. Genome.gov. Retrieved January 14, 2022, from https://www.genome.gov/human-genome-project/Timeline-of-Events(opens in new tab)

NIH National Human Genome Research Institute. (n.d.). *Sex chromosome*. Genome.gov. Retrieved January 14, 2022, from https://www.genome.gov/genetics-glossary/Sex-Chromosome(opens in new tab)

Oda K, Kitano H (2006) A comprehensive map of the toll-like receptor signaling network. Mol Syst Biol 2: 2006.0015

Oda K, Matsuoka Y, Funahashi A, Kitano H (2005) A comprehensive pathway map of epidermal growth factor receptor signaling. Mol Syst Biol 1: 2005.0010

Palumbi, S. R. and Baker, C. S. (1994). Contrasting population structure from nuclear intron sequences and mtDNA of humpback whales. *Molecular Biology and Evolution*, 11, 426–435. 10.1093/oxfordjournals.molbev.a040115.

Pauly, D. and Zeller, D. (2016). Catch reconstructions reveal that global marine fisheries catches arehigher than reported and declining. *Nature Communications*, 7, 10244 10.1038/ncomms10244.

Peachey, B. L., and S. K. Allen. 2016. "Evaluation of cytochalasin B and 6-dimethylaminopurine for tetraploidy induction in the Eastern oyster, *Crassostrea virginica*." *Aquaculture* 450: 199–205.

Pennell, M. W. (2014). An introduction to the phylogenetic comparative method In Garamszegi L. Z. (Ed.), *Modern phylogenetic comparative methods and their application in evolutionary biology: Concepts and practice*. New York, NY: Springer.

Pinheiro, J. , Bates, D. , DebRoy, S. , Sarkar, D. and R Core Team (2017). nlme: Linear and Nonlinear Mixed Effects Models.

Pires-daSilva A, Sommer RJ (2003) The evolution of signalling pathways in animal development. Nat Rev Genet 4: 39–49

Pusack, T. J. , Benkwitt, C. E. , Cure, K. and Kindinger, T. L. (2016). Invasive Red Lionfish (*Pterois volitans*) grow faster in the Atlantic Ocean than in their native Pacific range. *Environmental Biology of Fishes*, 99, 571–579. 10.1007/s10641-016-0499-4.

Reed, D. H. and Frankham, R. (2003). Correlation between fitness and genetic diversity. *Conservation Biology*, 17, 230–237. 10.1046/j.1523-1739.2003.01236.x.

Revell, L. J. (2010). Phylogenetic signal and linear regression on species data. *Methods in Ecology and Evolution*, 1, 319–329. 10.1111/j.2041-210X.2010.00044.x.

Saastamoinen, M. Heritability of dispersal rate and other life history traits in the Glanville fritillary butterfly. *Heredity* 100, 39–46 (2008)

Schaefer MH, Fontaine J-F, Vinayagam A, Porras P, Wanker EE, et al. (2012) HIPPIE: Integrating protein interaction networks with experiment based quality scores. PLoS One 7: e31826

Schaefer MH, Lopes TJS, Mah N, Shoemaker JE, Matsuoka Y, et al. (2013) Adding protein context to the human protein-protein interaction network to reveal meaningful interactions. PLoS Comput Biol 9: e1002860

Schwanbeck R, Martini S, Bernoth K, Just U (2011) The Notch signaling pathway: molecular basis of cell context dependency. Eur J Cell Biol 90: 572–581

Shaked, H., *et al.* Sequence elimination and cytosine methylation are rapid and reproducible responses of the genome to wide hybridization and allopolyploidy in wheat. *Plant Cell* 13, 1749-1759 (2001)

Spielman, D. , Brook, B. W. and Frankham, R. (2004). Most species are not driven to extinction before genetic factors impact them. *Proceedings of the National Academy of Sciences*, 101, 15261–15264. 10.1073/pnas.0403809101.

Stanley, J. G., H. Hidu, and S. K. Allen. 1984. "Growth of American oysters increased by polyploidy induced by blocking meiosis I but not meiosis II." *Aquaculture* 37: 147–55.

Stanley, J. G., S. K. Allen, and H. Hidu. 1981. "Polyploidy induced in the American oyster *Crassostrea virginica* with cytochalasin B." *Aquaculture* 23: 1–10.

Stockwell, C. A. , Hendry, A. P. and Kinnison, M. T. (2003). Contemporary evolution meets conservation biology. *Trends in Ecology & Evolution*, 18, 94–101. 10.1016/S0169-5347(02)00044-7.

Stone, B. W., N. H. Hadley, and P. R. Kingsley-Smith. 2013. "Evaluating the potential growth advantage of triploid eastern oysters *Crassostrea virginica* in South Carolina relative to commercially cultured diploid native stocks." *Journal of Shellfish Research* 32: 647–55.

Sturmer, L. N., C. Cyr, R. Markham, N. Brandimarte, and S. Laramore. 2018. "Application of triploidy to an emergent oyster culture industry on the Florida west coast: Results of grower trials." Program and Abstracts of the 110th annual meeting of the National Shellfisheries Association: 169.

Thomaz, A. T. , Christie, M. R. and Knowles, L. L. (2016). The architecture of river networks can drive the evolutionary dynamics of aquatic populations. *Evolution*, 70, 731–739. 10.1111/evo.12883.

Trompouki E, Bowman TV, Lawton LN, Fan ZP, Wu D-C, et al. (2011) Lineage regulators direct BMP and Wnt pathways to cell-specific programs during differentiation and regeneration. Cell 147: 577–589

Troup, A. J., S. C. Cairns, and R. D. Simpson. 2005. "Growth and mortality of sibling triploid and diploid Sydney rock oysters, *Saccostrea glomerata* (Gould), in the Camden Haven River." *Aquaculture Research* 36: 1093–103.

U.S. National Library of Medicine. (2020, September 28). HBA1 gene. MedlinePlus. Retrieved January 14, 2022, from https://medlineplus.gov/genetics/gene/hba1/ (opens in new tab)

U.S. National Library of Medicine. (2021, January 19). What is DNA?: Medlineplus Genetics. MedlinePlus. Retrieved January 14, 2022, from https://medlineplus.gov/genetics/understanding/basics/dna/(opens in new tab)

U.S. National Library of Medicine. (2022, January 5). OR6A2 olfactory receptor family 6 subfamily a member 2 [homo sapiens (human)] - gene - NCBI. National Center for Biotechnology Information. Retrieved January 14, 2022, from https://www.ncbi.nlm.nih.gov/gene/8590(opens in new tab)

USDA. 2014. 2012 Census of Agriculture: Census of Aquaculture (2013), released on September, 2014. p. 98.

Van de Peer, Y. and Meyer, A. Chapter 6: Large-scale gene and ancient genome duplications. In *The Evolution of the Genome*, ed. T. R. Gregory, 330-363 (San Diego, Elsevier, 2005)

Vandewoestijne, S. , Schtickzelle, N. and Baguette, M. (2008). Positive correlation between genetic diversity and fitness in a large, well connected metapopulation. *BMC Biology*, 6, 46 10.1186/1741-7007-6-46.

Via, S. and Hawthorne, D. J. Back to the future: Genetic correlations, adaptation and speciation. *Genetica* 123, 147–156 (2005)

Villanueva-Fonseca, B. P., A. M. Gongora-Gomez, N. P. Munoz-Sevilla, A. L. Dominguez-Orozco, J. A. Hernandez-Sepulveda, M. Garcia-Ulloa, and J. T. Ponce-Palafox. 2017. "Growth and economic performance of diploid and triploid Pacific oysters

Crassostrea gigas cultivated in three lagoons of the Gulf of California." *Latin American Journal of Aquatic Research* 45: 466–80.

Wadsworth, P. 2018. Comparing triploid and diploid growth and mortality in farmed oysters, *Crassostrea virginica*, in the northern Gulf of Mexico.

Wagner JP, Wolf-Yadlin A, Sevecka M, Grenier JK, Root DE, et al. (2013) Receptor tyrosine kinases fall into distinct classes based on their inferred signaling networks. Sci Signal 6: ra58

Wang, J., *et al.* Genome-wide nonadditive gene regulation in *Arabidopsis* allotetraploids. *Genetics* 172, 507-517 (2006)

Wang, Y., X. Guo, G. A. Debrosse, and S. E. Ford. 2005. "Growth in natural triploid eastern oyster produced by diploid × tetraploid crosses." *Journal of Shellfish Research* 24: 1274.

Wang, Z., X. Guo, S. Allen, and R. Wang. 2002. "Heterozygosity and body size in triploid Pacific oysters, *Crassostrea gigas* Thunberg, produced from meiosis II inhibition and tetraploids." *Aquaculture* 204: 337–48.

Waples, R. S. , Luikart, G. , Faulkner, J. R. and Tallmon, D. A. (2013). Simple life history traits explain key effective population size ratios across diverse taxa. *Proceedings of the Royal Society of London. Series B: Biological Sciences*, 280, 20131339 10.1098/rspb.2013.1339.

Ward, R. D. , Woodwark, M. and Skibinski, D. O. F. (1994). A comparison of genetic diversity levels in marine, freshwater, and anadromous fishes. *Journal of Fish Biology*, 44, 213–232. 10.1111/j.1095-8649.1994.tb01200.x.

Weiss M, Schrimpf S, Hengartner MO, Lercher MJ, von Mering C (2010) Shotgun proteomics data from multiple organisms reveals remarkable quantitative conservation of the eukaryotic core proteome. Proteomics 10: 1297–1306

Willoughby, J. R. , Sundaram, M. , Wijayawardena, B. K. , Kimble, S. J. A. , Ji, Y. , Fernandez, N. B. , … DeWoody, J. A. (2015). The reduction of genetic diversity in threatened vertebrates and new recommendations regarding IUCN conservation rankings. *Biological Conservation*, 191, 495–503. 10.1016/j.biocon.2015.07.025.

Winter EE, Goodstadt L, Ponting CP (2004) Elevated rates of protein secretion, evolution, and disease among tissue-specific genes. Genome Res 14: 54–61

Yang, H., and X. Guo. 2017. "Triploid hard clams *Mercenaria mercenaria* produced by inhibiting polar body I or polar body II." *Aquaculture Research*. (in press).

Yang, H., F. Zhang, and X. Guo. 2000b. "Triploid and tetraploid Zhikong scallop, *Chlamys farreri* Jones et Preston, produced by inhibiting polar body I." *Marine Biotechnology* 2: 466–75.

Yang, H., H. Que, Y. He, and F. Zhang. 2000a. "Chromosome segregation in fertilized eggs from zhikong scallop *Chlamys farreri* (Jones & Preston) following polar body 1 inhibition with cytochalasin B." *Journal of Shellfish Research* 19: 101–5.

Yang, H., L. N. Sturmer, and S. Baker. 2016. *Molluscan shellfish aquaculture and production.* FA191. Gainesville: University of Florida Institute of Food and Agricultural Sciences. p. 8. https://edis.ifas.ufl.edu/fa191

Zhang, D. X. and Hewitt, G. M. (2003). Nuclear DNA analyses in genetic studies of populations: Practice, problems and prospects. *Molecular Ecology*, 12, 563–584. 10.1046/j.1365-294X.2003.01773.x.

Zhang, Y., J. Li, Y. Qin, Y. Zhou, Y. Zhang, and Z. N. Yu. 2017. "A comparative study of the survival, growth and gonad development of the diploid and triploid Hong Kong oyster, *Crassostrea hongkongensis* (Lam & Morton 2003)." *Aquaculture Research* 48: 2453–62.

INDEX

A

Absorb 222

Abundance 85-87

Additive 65

Affects 99

Allele 65, 79, 137

Alleles 71, 91, 120, 124, 204, 206, 295

Allozyme 77

Allozymes 88

Androgen 159

Androgens 22, 128, 147

Animals 54, 253, 268, 276, 279-280, 292

Antibodies 89

Antibody 90

Antifreeze 251

Antigens 89

Approach 114, 232

Aromatase 159-160

Asexual 51-52

B

Bacteria 49

Binary 84

Bivalent 105

Bottom 179

Breeder 98

Budding 52

C

Catfish 189, 192, 218, 266, 289

Cellular 85

Chromatids 50

Chromosome 47-48, 105, 107, 162, 165-166, 195, 200

Clarias 288

Cleavage 125

D

Diversity 66-67, 260, 304

E

Ecology 304

Ecosystems 183

Embryos 120-122

Energy 102, 222

Enzyme 82, 88

Estrogens 147

Estuarine 183

Evolution 308

F

Family 279-281, 287

Fisheries 59, 177-178, 180-181, 294

Fission 52

Flower 92

Fragment 83, 165

Fragments 81

G

Gametes 200

Gene-one 62

Genetics 6

Genomes 111, 226

Genomics 17-19, 213

Genotype 63, 92-93

Genotypes 64

H

Haploid 124

Hatchery 79, 193

Hereditary 176

Heterosis 274

Histone 224

Homologous 50

Hormones 147, 171

Hybrids 104, 187-191, 193-194, 273-274, 289

I

Immunity 31

Inbreeding 177, 261-263, 269, 271

Inhibition 114

Insulin 8

L

Linkage 199, 201

M

Markers 68-77, 88, 118, 138, 196, 199-204, 208, 216

Matings 262

Maturation 264

Meiosis 200

Microarray 221

Mitotic 22

Moisture 93

Mutations 69

O

Oocytes 149

P

Pathways 84-86

Phenotypes 63, 206, 231

Pigment 92

Pituitary 97

Polyploids 96

Polyploidy 95, 104, 107

Progeny 132, 136-137, 199

R

Radiation 74

Ribosomes 231

S

Salinity 183

Salmon 101, 173, 187, 190, 244, 254, 266, 297

Sequencing 16, 18-19, 217

Sexual 51, 141, 146, 161, 264, 267

Shellfish 98, 112

Shrimp 130-131, 297

Software 84

Steroids 102

T

Tetraploid 21, 99, 265-266

Tilapia 144-145, 157-158, 161, 163, 166-168, 188-190, 193-194, 244, 254, 266, 275

Traits 20, 26, 63-64, 70, 187, 208-210, 277-278, 280, 282, 284

Transgenes 249

Triploids 99, 101-103, 114-117, 265-266

Tyrosinase 227-228

V

Vaccines 30-31

Variations 25

Z

Zebrafish 133, 138, 149, 154, 225

www.ingramcontent.com/pod-product-compliance
Lightning Source LLC
LaVergne TN
LVHW062116200926
848484LV00007B/1685
9781787150621